MW01518243

Bushmen, Botany and Baking Bread

First edition 2018

Text and the work copyright © the editors 2018

All rights reserved. No part of this publication may be reproduced or transmitted in any form or by electronic or mechanical means, including any information storage or retrieval system, without the prior permission in writing of the copyright holder.

The editors and the publisher have made every effort to obtain permission for and acknowledge use of copyright material. Should an inadvertent infringement of copyright have occurred, please contact the publisher and we will rectify omissions or errors in any subsequent reprint or edition.

Published in South Africa on behalf of the editors by
NISC (Pty) Ltd, PO Box 377, Grahamstown, 6140, South Africa

ISBN 978-1-920033-30-9 (print)
ISBN 978-1-920033-75-0 (pdf)

Design, typesetting and layout: NISC (Pty) Ltd
Cover design: Advanced Design Group

Printed by Digital Action (Pty) Ltd

Cover photograph: Kuyella playing the bow, Kaiongo, 2nd September 1925. 'This is not a special musical instrument, but the singer's own hunting bow with the string pulled in and tied back at about a third of its length. By slightly altering the position of the left hand and calabash, he can vary the note a little. The performer sings as he plays; the time kept is good, the melody very slight, merely anaccompaniment to the voice'. Bleek, D.F. 1928. Bushmen of Central Angola. *Bantu Studies* 3, 2: 105–125.
Cover map: Gossweiler, J. & Mendonça, F.A. 1939. *Carta Fitogeográfica de Angola*. República Portuguesa Ministério das Colónias.

Bushmen, Botany and Baking Bread

Mary Pocock's record of a journey
with Dorothea Bleek across Angola in 1925

compiled and edited

by

Tony Dold and **Jean Kelly**

Acknowledgements

We thank David Goyder, Duncan Greaves, Eric Kelly, Estelle Brink, Margaret and Edward Cock, Peter Loudon and Sarah Gess for their kind help in the preparation of this book.

Thanks to Clive Kirkwood (Special Collections and Archives, University of Cape Town Library) for permission to use the portrait of Dorothea Bleek. Thanks also to Erica Henderson, Allen Press, for permission to reproduce Mary Pocock's obituary from *Phycologia* 17,4 (1978).

Special thanks to Mike Schramm at NISC for his patience and kindness in designing and producing this book.

We are grateful to Rhodes University and the Albany Museum for institutional support.

Contents

INTRODUCTION

The Selmar Schonland Herbarium in Grahamstown is the repository of a unique treasure – a collection of glass plate lantern slides, photographs and negatives, water colour paintings and a number of herbarium specimens, as well as five precious, dilapidated exercise books – the record of a remarkable journey across Angola in 1925 documented by Mary Agard Pocock, the renowned botanist and world-famous algologist.

The material documenting this journey has been in the safe-keeping of Rhodes University Botany Department for decades, during which time there have been several suggestions that the written account along with the illustrations and photographs should be shared with a wider readership. Perhaps now, after nearly a hundred years, is the time to share this treasure trove as a fitting memorial to a remarkable woman and as a reminder of a virtually pristine part of Africa which has gone forever.

Mary labelled her account in the five exercise books a 'diary' but what she wrote is far more than a prosaic record of humdrum activities. She possessed remarkable powers of observation and a wonderful facility in the use of words to bring what she saw to vivid life. She had an artistic flair which enables the reader to visualise the splendours of the scenery, the sunsets and the vegetation which she describes in precise and colourful detail. Her enjoyment of life and acuity of observation permeate her writing, so that her personality shines through on every page. She kept this record not only for her own purposes but to inform and entertain her family and friends to whom it was sent. Each volume, as it was completed, was entrusted to the vagaries of whatever postal service

Mary's diaries relating to her expedition through Angola with Dorothea Bleek in 1925

existed in Angola at that time. Amazingly, they all reached their destination! She put her drawing skills to good effect and added quirky field-sketches to the pages to illustrate what she was writing about.

Mary was a highly intelligent and very well educated woman. She was most articulate and presumably enjoyed writing, as her effortless daily chronicles flow like a masterly narrative. During the first part of the journey by train (from Cape Town to Livingstone) and then by river boat from Kazungula to Kama, she whiled away hours of travel by describing the passing scene. So, from nearly a century ago, we have a vivid example of what today would be described as 'live-streaming.' Once the cross-country journey started, she spent much of the evening writing up the events of the day. Even on days which she records as 'uneventful' she manages to produce several pages of descriptions of the surroundings, or of 'goings-on' in the camp or in nearby villages, often enlivened by her subtly humorous comments.

Mary wrote in a 'diaristic' style, often omitting the subject of her sentences and showing little concern for punctuation, frequently employing a dash, rather than more conventional marks. The reader is left with an impression of an almost breathless flurry of words as her fountain pen races over the pages. It has been necessary to prune some of the rambling sentences in the interests of clarity and to insert a few punctuation marks where necessary, but the use of dashes has been retained where it adds to the character of her writing. Mary grew up at the height of the era of British imperialism and in a climate of colonial expansionism. She was a woman of her time, as revealed occasionally in the opinions she expresses and the terminology she uses. How could she be otherwise?

The five exercise books in which this journey is recorded were not bought new for the occasion. One visualises Mary, as she prepared for this expedition, ransacking a cupboard in the family home and finding relics of the children's school days. One book has the name of her sister Florence Edith Pocock laboriously inscribed in childish script. Another belonged to her brother John Pocock and has some pages of technical drawings done by him. Another bears evidence of pages having been torn out so that the remainder of the book could be utilised for her present purpose.

What makes this epic journey particularly unusual is that it was undertaken by two women on their own, assisted by relays of porters they engaged to carry their equipment and luggage and set up camp for them. Mary was the junior member of this formidable party, the leader being Miss Dorothea Bleek, daughter of the world-famous authority on the Bushmen, Wilhelm Bleek. Miss Bleek's objective in crossing Angola was to make contact with the elusive remnants of the Bushmen and document their language and customs. Some years later she published the first dictionary of Bushman Languages. Mary's role was to oversee the commissariat and ensure a supply of freshly baked bread each day. Her botanical collecting was fitted in alongside her domestic duties.

It is astonishing to discover that they had only the most rudimentary first aid supplies with them, treating minor ailments with permanganate of potash or hydrogen peroxide, quinine and occasionally, very daringly, with a tablespoonful of brandy (carried purely for medicinal purposes!). When they left the mission stations at which they stayed in several places and set up their camp in the bush, they were virtually cut off from communication with the outside world. They were dependent on messages being sent to, and received from, the nearest mission station, and it was through the missionaries that they received their longed-for post from home which, surprisingly, reached them on a fairly regular basis.

We have no record of how these two focussed and energetic scientists came to join forces on this expedition. Miss Bleek at that time was 52 and the veteran of a number of arduous expeditions across southern and central Africa in pursuit of her studies of the Bushmen and their language. Mary was 39. It would appear from her record of the expedition that the two women were not particularly well acquainted and each was clearly pursuing her own interests single-mindedly. Mary refers to her companion throughout as 'Miss Bleek' or, occasionally, 'Miss B'.

The diarist: Mary Agard Pocock (1886–1977)

Mary, known to family and friends as Mamie, was born in Cape Town but received her higher education abroad. She went to England in 1899 to Bedford High School for Girls and from there to Cheltenham Ladies' College. Working from there, she gained a BSc degree from London University in 1908. She also obtained a Teachers' Diploma from London University in 1911.

Interviewed in America in 1969 she remarked, 'I guess my scientific interest came from my father. He was a chemist (pharmacist)...'[1]. In the same interview she recalled growing up in a home where there were seven children as well as numerous pets. She was the second eldest of that brood, the children of William Frederick Henry Pocock and Elizabeth Lydia Dacomb. Mary returned to South Africa in 1913 and spent four

years teaching science and mathematics at Wynberg Girls' High School in Cape Town. After the First World War she returned to England and spent two years working in the Botany School at Cambridge as a student of Newnham College. She obtained her BSc (Hons) through London University, as Cambridge at that time did not award degrees to women.

Mary's passion for botanical research and the related field-work resulted in a lifestyle in which periods of teaching at secondary or tertiary level alternated with extended field trips in pursuit of her scientific interests. It was in this way that her long association with Rhodes University started, with stints as a temporary lecturer from 1924 onwards.

The diarist: Mary Agard Pocock

The topic of her doctoral research, which she undertook in 1928, was the genus *Volvox* and thereafter Volvocales and other fresh-water algae, and later still, marine algae, became the dominant scientific focus of her career. Mary obtained her PhD from the University of Cape Town in 1932. In later life she settled in Grahamstown and worked in the Rhodes Botany Department as a researcher and lecturer. In 1967 she was awarded a DSc *honoris causa* by Rhodes University.

Mary took every opportunity of travelling that was offered and in the course of her long life journeyed far and wide, often to remote and inaccessible places in her zeal for collecting new specimens. No doubt it was this adventurous spirit and her hardiness when faced with daunting conditions which led her to ask to join Dorothea Bleek in 1925 on the expedition to Angola which she recorded so vividly in her diaries.

The leader of the expedition: Dorothea Frances Bleek (1873–1948)

Dorothea Bleek, the sixth of seven children of Dr Wilhelm Bleek and Jemima Lloyd was born in Cape Town. Her father was a noted linguist with an extensive knowledge of African languages. He worked as an interpreter and linguist in various parts of Africa before moving to Cape Town where he married Jemima Lloyd in 1862.

Wilhelm Bleek (1827-1875) and his wife's sister, Lucy Lloyd (1834-1914), worked together to learn Bushman languages. They were pioneers in recording Bushman narratives and folklore for posterity. Lucy Lloyd, with support from her sister Jemima, continued this valuable work after Wilhelm Bleek's death.

In 1884 Jemima Bleek moved to Germany with her children, followed by her sister Lucy about 1887. It seems that it was at that time that Lucy trained Dorothea, her niece, in translation and methods of research and taught her to speak /Xam and !Kung. Quite apart from the legacy of her father and aunt, Dorothea Bleek became an internationally recognised authority in linguistic and anthropological fields in her own right. She attended international conferences and published numerous papers and is best known for her book *The Mantis and his Friends: Bushman Folklore*[2], co-authored by Lucy Lloyd, and her *Bushman Dictionary*, published posthumously in 1956[3].

The leader of the expedition: Dorothea Frances Bleek

Dorothea was educated in Switzerland and Germany and trained as a teacher at Berlin University. Her interest in African languages was fostered during time she spent at the School of Oriental and African

Languages in London. On her return to South Africa in 1904 she was appointed to a teaching post at Rocklands Girls' High School in Cradock where she worked until 1907. She became intensely interested in Bushman rock art and went on field trips to many parts of the country recording these. She devoted the rest of her life to studying the Bushmen. She went on several expeditions, not only to parts of South Africa, but further afield to Tanganyika, Angola and the Kalahari. She studied and recorded Bushman art and language, genealogies, stories, their culture, customs and way of life. In 1936 Dorothea was offered an Honorary PhD by the University of the Witwatersrand, but declined the honour, saying her father should be the only Dr Bleek.

The work of Bleek and Lloyd on the Southern African Bushmen has been popularized in a number of handsomely presented books and their entire archival collection has been digitised and made available electronically by the University of Cape Town. On the other hand, Mary Pocock's descriptions and photos of the Bushmen in Angola have not been published before and are presented here for the first time.

The Journey

In a paper published in the journal *Bantu Studies*, July 1928[4], Dorothea gives a succinct account of the journey: *In 1925 I spent six months travelling across Angola from the Rhodesian border to Lobito Bay. Miss MA Pocock was with me, pursuing botanical studies. From Livingstone we went up the Zambesi to about the 15 deg S latitude (opposite Lialui), then by carriers into and across Angola keeping between the 14 deg and 15 deg S latitude till we reached the Cwelei river near the 16 [17] deg E longitude. Thence we struck north to the railway, which we reached at Vila Silva Porto, better known as Bie. Thence we took train for Lobito Bay.* Thus in a few business-like sentences she disposes of a marathon trek which took over six months. Mary's sketch map of their route is at the beginning of Volume 2.

The two women travelled from Cape Town by train as far as the Victoria Falls. From there they and their luggage were transferred by lorry to Kazungula where they embarked on the river boat in which they were rowed and poled up the Zambesi to Kama near Lialui. There their overland adventure began, assisted by relays of porters, including the carriers for the *machilas* in which they rode for part of each day, alternating with periods of walking. (Once in Angola the terminology changed and their conveyances were then referred to as '*tipoias*'.) As Mary explained in an article she wrote for *The Cape Times*, *The machila is a canvas hammock slung on a long pole which is carried by a couple of sturdy boys, either on the shoulder or the head* (Pocock *Cape Times*, 5th December 1925).

Their porters were only engaged for specified sections of the trek. The Zambian (then Northern Rhodesian) porters had to be dismissed when they crossed into Angola, and once in Angola they could only engage porters within a particular administrative district. The employment of these men was regulated by the administrative officials of each district through which they travelled. When they set up camp for several weeks at a time, they dismissed the porters and retained the services of a few chosen men to assist with the domestic chores and help them interact with the local community. Then, when they decided to move on, they had to engage a new team of bearers.

'This is our boat'. The *wata* (boat) in which Mary and Miss Bleek travelled for 18 days from
Kazungula to Kama

Porters with loads outside Mr Pontier's house, Cwelei Mission, 30 September

They had set out intending to travel as far as the Kutsi river close to Muié and then to return by the same route. In the event, as they heard news of Bushmen further to the west, they continued their trek westward as far as the mission station at Cwelei and then northward from there to Bihe (Bié, Bihé), from where they were able to travel by train to Lobito Bay and take passage on a ship to the Cape. It was indeed an epic journey by two intrepid women.

Looking back on their journey, Dorothea Bleek wrote: *My object in undertaking this journey was to look up the Bushmen of whose presence here I had been told by members of the South African General Mission working among the Mbunda, Luchazi, Kangali and Nyemba, Bantu tribes who inhabit this part of Central Angola. With the help of the missionaries I succeeded in finding them. ... I owe a deep debt of gratitude to the members of the South African General Mission for much hospitality, help and advice, without which our expedition could never have been carried out*[5].

Myths and Inaccuracies

In the few published accounts by other writers of this journey, various inaccuracies have been perpetuated. The most glaring of these is the oft-reiterated information that the travellers lived on a diet of burnt porridge[6]. A cursory reading of the diaries immediately dispels this fallacy. They had a varied menu and plenty of produce available for their food. Mary baked bread, biscuits and scones regularly, and supervised the cook who roasted fowls or venison accompanied by whatever local vegetables they could obtain

Bushman hut of the winter (dry season) type at Kaiongo, 27 August

for their evening meal. They had a plentiful supply of eggs available and acquired a flock of several hens. Mary often mentions having made a delicious omelette for their lunch. They were given lemons and sometimes oranges by the missionaries and, ever resourceful, Mary squeezed the fruit to make lemon syrup and then used the peel to make marmalade. On June 15th she describes their evening meal: *...we have just had a most sumptuous repast – excellent bean soup (beans, onions and a little of the pork), fried pork cutlets and sweet potatoes, followed by strawberries and cream, and a piece of Mrs Wilson's excellent cake and coffee. How's that for the wilds of central Africa in June?*

There is also the misapprehension that the two women crossed Angola 'on foot'. While they both walked for much of the time, the record shows that their walking was alternated by periods when they were carried in the *machilas/tipoias*: *To begin with, we usually walk a mile or so (the 'so' sometimes is zero or even a minus quantity). This morning we walked for over half an hour through the woodland, and then we got into the machilas* (May 9th). Furthermore we are led to believe that the travellers spent the full six months 'on the move' through the bush. In fact more than half of their time was spent based at a camp or a mission station.

Five years later Mary went on another expedition, this time to the Linyanti River in north-west Botswana. This was also well-documented photographically but we have not found any written record of this trip. Unfortunately, it appears that some confusion later arose, resulting in some of her photographs and their captions from the later expedition being attributed by several authors to the journey across Angola.

'The *machila* is a canvas hammock slung on a long pole which is carried by a couple of sturdy boys, either on the shoulder or the head'. Demonstrated here by Mary and two porters

Botanical specimens

In 1925 the botany of Angola was poorly known. The most notable collector before this time was the German botanist Hugo Baum who visited the same area in 1900 and collected about 1000 specimens[7] but the first comprehensive study of the region was only published in 1939[8]. Mary's main objective in joining Miss Bleek's expedition was to collect and study plants, which she did whenever the opportunity arose, even pressing specimens while being carried in her *machila: However, with the exception of a brief spell in the machila to put away my specimens, we continued walking...* (May 23rd). She even enlisted the help of two Bushmen: *Then I provided Golli and Baita-de with needle and silk and set them to threading labels for my specimens* (June 2nd).

Despite her efforts to 'change papers' regularly, that is changing wet blotter sheets for dry ones and drying wet ones in the sun, Mary had great difficulty getting her specimens dry: *Such a lot have got spoilt, most of the water ones, and a good many of the others* (June 3rd); *I could only pack away very few – they are so wet and are going mouldy* (October 10th). Compounding this were lesser difficulties with ants and chickens: *It was as well that I moved my specimens this morning, as the ants were getting to them. They have eaten holes in the bottom sheet of drying paper...* (July 23); *After lunch set to work putting my specimens away, a tedious job and much interrupted by the fowls – who, missing their recent filling meals of wunga, gave me no peace but persisted in walking over and sampling my specimens* (July 31).

At Benguela Mary assessed her specimens and was disappointed to find that many of them had to be discarded: *Alas! – nearly all have gone mouldy and I've had to throw away the bulk of specimens and paper – all my work at Kamundongo worse than useless* (October 16). It has been estimated that Mary collected nearly 1000 plant specimens on the expedition[9]. This estimate was based on her collecting register (three notebooks that have inexplicably disappeared) written in the field and does not take into account the many specimens that were later discarded because of mould. We have five notebooks that Mary wrote two years later while she was identifying her specimens at Kew and at the British Museum.

In June 1961, Mary added a postscript to her diary indicating where the various specimens had finally been deposited. As is usual for botanists, the material was separated into duplicate sets and allocated to various herbaria. The first and most complete set was intended for the Bolus Herbarium at the University of Cape Town. In 1936, however, Mary had parcelled up the specimens ready to take to the herbarium when they were stolen

Mary's specimen labels, some of which were threaded by the Bushmen Golli and Baita-de

from her car! Apparently the parcel was recovered and today the Bolus Herbarium has at least 160 Pocock specimens from Angola on their database and probably more yet to be databased (pers. comm. Terry Trinder-Smith, 2015).

Four hundred and twenty-one duplicate specimens were purchased by the botanist Rudolph Marloth for the National Herbarium in Pretoria for the sum of £12. A small number of specimens was given to the Rhodes University Herbarium (Mary Pocock was instrumental in founding this herbarium in 1942. It was amalgamated with the Albany Museum Herbarium in 1993 to constitute the Selmar Schonland Herbarium). Some specimens went to Stellenbosch University Herbarium (now at the Compton Herbarium, Kirstenbosch).

There is also a number of specimens at Kew Herbarium where Mary worked on identifying her collections in 1927. Here she consulted Hugo Baum's important collection and cites his numbers extensively. She was assisted by renowned South African botanists Dr Inez Verdoorn and Neville Pillans who must have been there at the same time. The British Museum Herbarium also has a small number of Mary's specimens that she donated while consulting their African collections, notably those of Baum, Welwitsch and Gossweiler, in the same year.

Mary's notes indicate that a small number of bulbs and corms collected on the journey were grown in her garden in Cape Town and specimens prepared when they produced flowers.

Sketching and painting

Mary also found time to paint and draw flowering plants and Bushmen. All her paintings were given to the Rhodes University Herbarium and are reproduced here for the first time. Although she was most critical of her own work, the Bushmen were fascinated by both sketches and paintings, visiting from afar just to see them. Most of the paintings are captioned with a locality and a name. Under the circumstances, painting was challenging with both paper and pigment at a premium: *I treated myself to a piece of the thick drawing paper, really a treat, but oh for some yellow ochre! I want it for the sky, the trees, the ground, above all the grass, the people, specially the Bushmen, tree trunks, everything! Fortunately, I have plenty of blues, but my browns and reds, too, are not too good* (August 11).

Photography

Mary was an accomplished photographer and took two cameras on the journey: her 'VPK', a compact Vest Pocket Kodak, and her bulkier glass plate camera which apparently could also operate using film. As her diary shows, taking photos was exacting and time consuming, involving the changing of fragile glass plates in the dark under a blanket. She also developed the negatives: *I spent the morning developing films, chiefly Bushmen, the negatives are excellent but the differences in temperature of the various washes nearly ruined them. Fortunately Mrs Proctor had some alum (of which I had none) and I hope I've saved them – must wait till morning to see, when the films are dry* (September 11).

Towards the end of May, Mary's smaller film camera, the Vest Pocket Kodak, stopped

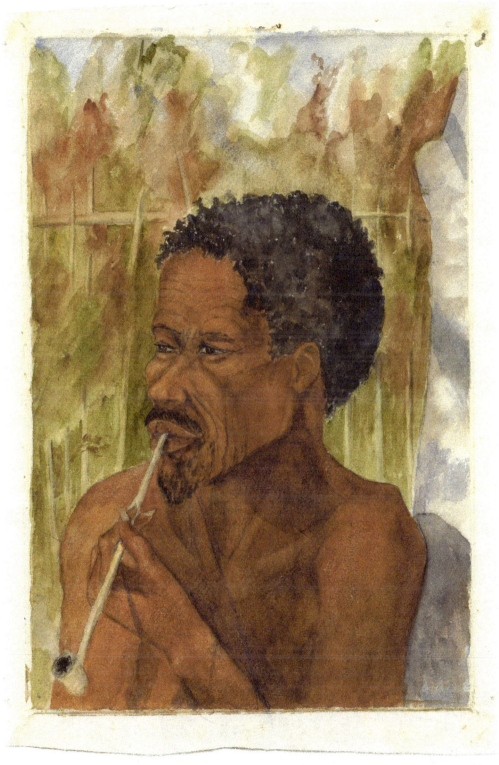

Mary's painting of an old [unnamed] Bushman from Kunzumbia

Mary's painting of *Barleria buddleioides* collected on the Luahuka River near Menongue on the 20th of August. The herbarium specimen and hand-coloured glass plate slide are shown on the page alongside

Barleria buddleioides Sp Moore.

Acanthaceae

Angola

Herb. Rhodes University

No. 6771

BARLERIA BUDDLEIOIDES Sp. Moore

Loc. Angola

Legit M.A.Pocock (leg) Date 1938

The herbarium specimen and hand-coloured glass plate slide of *Barleria buddleoides*

working and she was unable to correct the fault. The camera was sent by post to the Kodak agent in Cape Town and, remarkably, returned by the end of August: *By the way, I set my camera up this morning and found, to my great annoyance, that Kodak's have sent it back still defective – sometimes the shutter works all right, sometimes not. I think I can manage to make it go, by dint of poking up the spring each time, but it is disappointing in the extreme* (August 30).

The photographs reproduced here are copies of Mary's collection of negatives and lantern slides, some of which are hand-coloured. The pigments have unfortunately faded and changed over time resulting in an unnatural appearance in some slides. These slides would have been viewed using a lantern projector, the fore-runner of the modern-day slide projector, now also becoming obsolete in the digital age.

Miss Bleek also took photographs of the Bushmen and these have been digitised by the University of Cape Town. Three of Mary's photos ('Bushmen slaves') in our possession are identical to those of Miss Bleek's. We assume that these were taken by Mary as the originals are in her archive.

Place names

It has proved impossible to verify the correct form of names recorded during the journey across Angola. Many of these have changed completely since the end of the colonial era, and there are many variations of spelling of the names which can be identified. As far as possible a standardised version has been adopted.

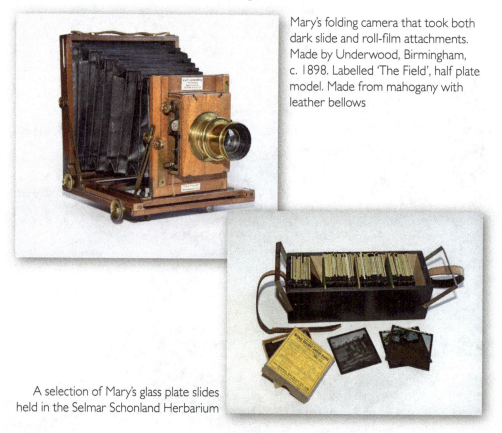

Mary's folding camera that took both dark slide and roll-film attachments. Made by Underwood, Birmingham, c. 1898. Labelled 'The Field', half plate model. Made from mahogany with leather bellows

A selection of Mary's glass plate slides held in the Selmar Schonland Herbarium

Newspaper reports

In conjunction with her daily diary entries, Mary also wrote a series of articles describing the journey which she sent by post to her 'Uncle John' in Oudtshoorn. John Pocock was the owner and editor of the local newspaper, *The Courant* (still in circulation today), in which he published a series of seven articles under the title *A Trip into the Interior*. After her return, Mary submitted three lengthy illustrated articles entitled *Overland to Lobito Bay* to the *Cape Times*. Edited transcriptions of both series of newspaper articles are appended here.

Overland to Lobito Bay in the Cape Times, December 1925–January 1926

Diagram of the Journey

SUNDAY	MONDAY	TUESDAY	WEDNESDAY	THURSDAY	FRIDAY	SATURDAY
	6/04 Cape Town	7/04	8/04	9/04 Bulawayo	10/04 Victoria Falls	11/04
12/04	13/04 Livingstone	14/04	15/04 Katombora	16/04	17/04 Kazungula	18/04
19/04 Sesheke †	20/04	21/04 Katonja	22/04	23/04	24/04 Monomé	25/05
26/04	27/04	28/04 Seoma	29/04	30/04	1/05	2/05 Sinanga
3/05 Nalolo†	4/05 Kama	5/05 Lukona	6/05	7/05	8/05	9/05
10/05	11/05 Angolan Border	12/05	13/05 Ninda †	14/05	15/05	16/05
17/05	18/05	19/05	20/05	21/05	22/05	23/05 Muié †
24/05	25/05	26/05 Kutsi Camp	27/05	28/05	29/05	30/05
31/05	1/06	2/06	3/06	4/06	5/06	6/06
7/06	8/06	9/06	10/06	11/06	12/06	13/06
14/06	15/06	16/06	17/06	18/06	19/06	20/06
21/06	22/06	23/06	24/06	25/06	26/06	27/06
28/06	29/06	30/06	1/07	2/07 Kunjamba †	3/07	4/07 Kunzumbia
5/07	6/07	7/07	8/07	9/07	10/07	11/07
12/07	13/07	14/07	15/07	16/07	17/07	18/07
19/07	20/07	21/07	22/07	23/07	24/07	25/07
26/07	27/07	28/07	29/07	30/07	31/07	1/08
2/08	3/08	4/08	5/08	6/08	7/08	8/08 Cuito River
9/08	10/08	11/08	12/08	13/08	14/08	15/08
16/08	17/08	18/08	19/08 Menongue	20/08	21/08 Cwelei †	22/08
23/08	24/08	25/08	26/08 Kaiongo	27/08	28/08	29/08
30/08	31/08	1/09	2/09	3/09	4/09	5/09
6/09	7/09	8/09 Cwelei †	9/09	10/09	11/09	12/09
13/09	14/09	15/09	16/09	17/09	18/09	19/09
20/09	21/09	22/09	23/09	24/09	25/09	26/09
27/09	28/09	29/09	30/09	1/10	2/10	3/10
4/10	5/10	6/10	7/10	8/10	9/10 Kamundongo †	10/10
11/10	12/10 Silva Porto	13/10 Benguella	14/10	15/10	16/10	17/10 Lobito
18/10 Mossamedes	19/10	20/10	21/10			

Travel by train or vehicle   River journey; sea trip   Overland on foot/by *machila* (*tipoia*)

Stationary – in camp or at mission stations   † Mission Station

THE DIARIES
Volume One

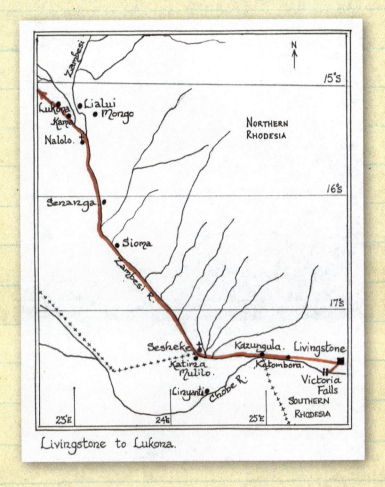

Livingstone to Lukona.

Rondebosch, Cape to Ninda
(April 6 to May 13, 1925)

'...it is such a fresh lovely morning and the flowers are lovely, the grass is short and besides the waterlilies there are huge rose-red *Convolvulus* blooms, the trails with slender slightly sagittate leaves float on the water and the huge trumpets rise from the surface.'

April 10. Victoria Falls with the Zambesi River in flood

April 18. 'Wa-Wa' birds, low flying waterfowl and Fish Eagle

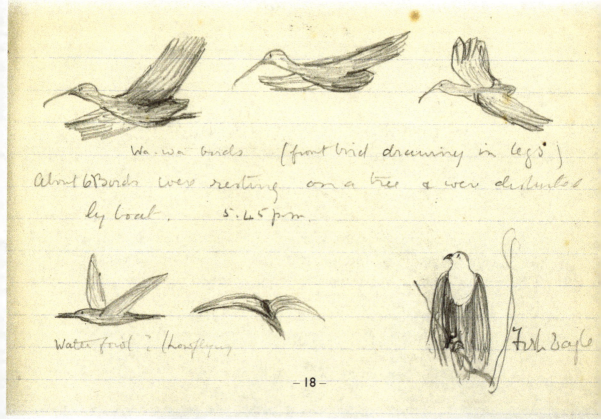

Wa-wa birds   (front bird drawing in legs)
About 6 Birds were resting on a tree & were disturbed
by boat.   5.45 pm.

Water fowl? (low flying)   Fish Eagle

JOURNEY BY TRAIN FROM CAPE TOWN TO VICTORIA FALLS

April 6th 1925 (Monday)—So the great trek has at last begun! I wonder what it will be like – somehow don't feel very excited about it.

In the morning I took my three packages (sack with stretcher etc., kit bag and paraffin box containing collecting material and various odds and ends – I hinged the lid on for convenience in getting necessaries out) and picked up Miss Bleek's at Mowbray. The tent and *machilas* were waiting at the station booking office. The Wellington store goods, however, were not ready and I had to wait some time for those. Eventually, however, I got everything and booked them (nine packages) for Livingstone.

I caught the 11.35 back. After lunch and packing my *padkos,* I had a final clear up and bath. We started at 3.20 – just as we reached Alma Road, remembered my *padkos* and had to go back for it, so after all it was past 3.30 when we finally started. Got to the station just before 4, found all the Bright family, various other friends and Miss Stephens, at the station. Miss Burrows (of Rustenburg) shares the carriage with us. After supper (pasties and rissoles etc.) a friend of Miss Burrows (Mrs Heath) and her son came in for a game of bridge. At Worcester we bought apples.

Tuesday April 7th—Went in to breakfast, otherwise *padkos* sufficed us, eked out by avocado pears bought at De Aar, where Mrs Heath and son (who had again been in for bridge) left us. In evening reached Kimberley where Miss Wilman met us. I invested in some more avocado pears which added greatly to the food supply.

Wednesday April 8th—Miss Saltmarsh joined us at Mafeking – coach next to ours. At Mahalapye Road we passed the down train (5 or 6 hours late) and W. Bolus was in the carriage which stopped just opposite ours. Was half asleep (3.30) when I heard a voice call 'Hallo Mamie!' and there was W.B. looking very well but much bored by her experience of wet farm life. *Karosses* at Artesia very fine.

Thursday April 9th—Reached Bulawayo early (just after breakfast) and spent the morning walking about the town. Saw Miss Burrows off (for Gadzema) on the Salisbury train at 12.30. Then collected our luggage for the Falls. Found we (Miss Bleek and I) were booked with two others – fortunately one did not turn up and the other who had friends elsewhere on the train shifted and we had the compartment to ourselves. Went into dining car before the train started, fortunately, as it is very full (Easter excursions). Rather annoying that books of meal coupons are not available. Finished the *padkos* – bread and butter, biltong and avocado pears for supper. The country is most beautiful – green parkland. Tried to collect at several halts including Gwaai – most beautiful flooded vleilands with sunset reflected.

EASTER WEEKEND AT THE VICTORIA FALLS. ZAMBESI RIVER IN FLOOD

Friday (Good Friday) April 10th —Reached the Falls about 7.30, an hour late. After breakfast got into our rooms – one 2-bedded room with a smaller single one opening

April 12. Zambesi River gorge below the Victoria Falls

off it in the first annexe – then went to the Rain Forest. Started at the Bridge, then to Danger Point and then right through the forest. It simply pelted, and the mist was so thick that only occasionally could we catch a fleeting glimpse of the Falls. I had no conception of what the volume of water, and the clouds of spray, could be like. At the Devil's Cataract the spray was less and we had a fine view. In the afternoon we had a long, much-needed rest – then heavy thunder showers prevented our going out. The country looked magnificent, the heavy thunder clouds most grand. After dinner we three and one of Miss Saltmarsh's travelling companions, Mrs Perrin, played bridge.

Saturday April 11th—Spent the morning collecting and pressing, with help from Miss Saltmarsh and Miss Bleek.

Oh, on the way to Danger Point yesterday I used my George green-lined parasol as a walking stick, whereupon it gently snapped in two, at the handle end. This morning it refused to close, so was taken out open and served excellently to carry the specimens in.

In the afternoon, donned bathing dress and mac (only down to my knees!) and went through the rainforest again – was better than yesterday but we were rather late (4.30) so did not see any rainbows. Met Mr and Mrs McNeillie (friends of Miss Saltmarsh) and their friends the Fauls on their way back. All got soaked but rather pleasant than otherwise.

After dinner played one rubber of bridge, then, just as we started the second, the moon rose and we went down to see the lunar rainbows. There was a faint one under the bridge, and better ones on the Eastern Falls. Mr and Mrs McNeillie and the Fauls went down too. We were very reluctant to leave, and stopped several times to admire the bows in the gorge, particularly a very fine one above the Boiling Pot.

At the hotel there was a big dance on, with a band from Livingstone.

Sunday April 12th—In the morning walked down to the Fifth Gorge, then back to the hotel and down to the Eastern Falls which were very lovely. The volume of water in the river (channel) above is most impressive.

About 3, we three and Mrs McNeillie paid our last, and most enjoyable visit to the Rain Forest – each time it has been better. This afternoon the bows were lovely, first on the Devil's Cataract and then, as we walked along the edge, each had her own rainbow. Finally at Danger Point, just as we were leaving, the mist lifted and we had a glorious view of the main falls, the end of the bow resting on the water in the gorge, the upper part curving over the falls.

After this strenuous day we were all tired and intended to go early to bed but the moon rose so bright and clear that she tempted us out again, and Miss Bleek and I walked down to the Eastern Falls. The bows were even finer than last night, the one right on the end Falls, curving back over the Falls being particularly magnificent.

I should like to spend several weeks here and watch the changes in the Falls as the dry season progresses. Every day there has been a difference.

BY TRAIN TO LIVINGSTONE

Monday April 13th—A day of waiting. We waited in the morning to see the excursion train leave for the South. We sat in the lounge and watched it cross the bridge. At the far side it stopped, the front engine and tender were uncoupled and crossed slowly, then the second engine and train crept across at a snail's pace, stopped as soon as the engine was over, re-coupled and then proceeded. We walked up to the station, waited about for some time, then returned to the hotel to lunch, as the train was so slow in leaving.

In the afternoon we packed, then learnt that the goods train for Livingstone would be in not at 4, but 5 or thereabouts. So we sat on the stoep and again waited. The view was superb; as the spray was blown back under the bridge, a glow of colour – not a bow, but an extended spectrum, spread over it, now the blue, now the yellow, and now the red end predominating. As the sun moved westwards, the spectrum moved slowly back up the gorge, rising higher and higher above the banks as the spray grew thicker.

A little before 5 we went up to the station, only to have another long wait, till the goods train slowly drew into the station. Then about 20 of us, with a good pile of luggage, all crowded into the guard's van. There followed another wait and finally at 5.40 we started with many creaks and groans and bumps. As we crossed the bridge we had a last lovely glimpse of the Falls in the evening light. At Palm Grove we stopped to pick up a dozen or so more people – Easter Monday picnickers – how they all got in I do not know, but they did!

The setting sun caught the spray from the Falls turning it now to molten gold, now to fiery red. The sky in the distance was golden, the rising ground to the south a rich blue – a golden evening. It was interesting to learn that the river – and consequently the Falls – is exceptionally full this year – fuller than it has ever been known to be. By the time we reached Livingstone it was quite dark and another wait followed till a taxi, kindly sent by one of our fellow-passengers, arrived to take us and all our belongings to the North-Western Hotel.

Still another wait! The hotel being full, the manager was doubtful if he could take us but finally managed to do so, by using Miss Saltmarsh's stretcher. We had dinner, then after waiting for half an hour or so, managed to get into our room and to bed, very tired.

A DAY IN LIVINGSTONE

Tuesday April 14th—A Mr Westhoven met us after breakfast and told us he was to take us out to Katombora. He had brought a party (Mr and Mrs Jalland) in from Katombora on Saturday. They had left Thursday and arrived Saturday afternoon (36 miles) owing to bad roads, shortage of petrol, punctures etc. He intended to send the boys on with half the stuff and follow tomorrow morning early with the (Ford) lorry. We spent the morning interviewing Mr Gow (of Shelmerdine's), buying stores, bank business etc., which we finished in the afternoon.

Then we presented the letter of introduction which Mr Phillips (Rhodesia Railways) had given me to Hon. Richard Goode, Chief Secretary. The latter was not in but we saw the Assistant Secretary, Mr Carlton, who was most kind. He gave us letters of introduction to various magistrates en route, and to Rev. Roulet (French Mission), and rang up Mr Jalland (to whom we had already spoken about his three little Sealyham terriers). We called at the Rev Roulet's house but he was away at Broken Hill. When we got back to the hotel, Mr Jalland came and had a long talk with us. They have just come through from Mongu where he has been stationed, and which is just north of our destination Nalolo. He gave us a good deal of useful information and Miss Saltmarsh had a talk with Mrs Jalland who was ill in bed with fever. At dinner time, came a note from Mme Roulet asking us to tea with her tomorrow – unfortunately impossible.

JOURNEY BY LORRY TO KATOMBORA

Wednesday April 15th—Up at 5.30, dressed and packed by 5.55 but no sign of tea or coffee. At last we got coffee and bread and butter, and by 6.55 we were on the lorry and ready to start. The first 12 miles were fairly good going – we left Livingstone to the south, getting a fine view of the Falls cloud, then turned westward along the Katombora road. About 12 miles out, on a piece of road recently 'prepared', i.e. with hard road metal on the surface, but a quaking bog beneath, our left hind wheel went in up to the axle. The boys with the goods, however, were just ahead and soon pushed us out. Then followed a series of bog holes, in some of which we stuck – others we passed. Several times we had to unload and have the things carried for several yards. The longest bit was about 7 miles from Katombora, not far from Westhoven's farm. At the latter we stopped for a rest and

Mr Westhoven gave us lunch – roast fowl and home-grown vegetables. He is growing cotton and the young cotton bushes looked very pretty. We were vociferously greeted by three dogs – two wire-haired bull terrier cross and one great dane × bull terrier.

THE SUTHERLANDS, KATOMBORA

After leaving the farm we had to leave the road and take to the bush. Passed over a black mamba (five or six feet long) in the road – front wheel went over it, but it wriggled back into the bush. After getting badly bogged once, we finally got to the Sutherlands' house where Mrs Sutherland kindly gave us tea. Finally one-and-a-half miles brought us to Katombora proper – on the way we met Mr Sutherland, an enormous man (6 ft 5 in) who sent one of his boys along with us to show us the camp. At the camp Mr Coulson's boy, Kandu, of whom Mr Jalland spoke, met us and we engaged him to act as cook boy etc. for us up to Nalolo. Miss Saltmarsh unpacked her delightful canvas bath and as soon as we had the tent up (the paddlers' work) we had much welcomed baths and got to bed, I in front of the tent and the others inside.

FIRST CAMPING EXPERIENCE

Thursday April 16th—We did not tuck in the nets properly and as a consequence got a good deal bitten – most unpleasant. There is a very heavy dew at night. We hoped to get on to Kazungula tonight, but the boys with the luggage did not arrive till past three, so we had to spend another night there. Mr. Sutherland came down in the morning and we fixed up things with him, arranging to start early tomorrow. Mrs Sutherland came down in the afternoon to spend some time chatting with us. She sent us down some milk and cream.

April 15. 'All push together'. Lorry stuck near Katombora

START OF THE RIVER JOURNEY

Friday April 17th—At last we have really started. We left camp, everything packed at 7.55. (As the boys opened the *machilas* to roll them up with the tent, a small snake wriggled out of them, much to the boys' consternation.) Got down to the boat (we had to be carried across to the landing), left about 8.30, Mr Sutherland seeing us off. After about 10 minutes however, the Induna discovered that we had forgotten the tie rope and back we had to go. Finally we really started at 9am.

DESCRIPTION OF THE FLOOD PLAINS AND RIVER

Most of the way we were in the flooded flats. We were in the main river for only a couple of hours, and then we hugged the shore. The flowers were lovely – water lilies pink, white and blue, all sizes but I think only one kind, a large red convolvulus trailing on the water or climbing in the reeds, a delicate mauve *Commelina* among the water lilies, beautiful red trails of bladderwort, with a circlet of five or six floats below each flower, and a white ?*Limnanthemum*. At midday we stopped for an hour at Kazungula. Now (4pm) we are out at the edge of the main stream, here flowing very rapidly, after a very difficult detour through particularly long grass to avoid some rapids. It is very lovely here – large trees and palms on the shore and islands – further back we passed a succession of papyrus clumps rising like rounded kopjes above the surrounding vegetation. Now, after poling all the way from Kazungula, the men are once more paddling.

Took to vleilands to avoid rapids below Mambova, which we could hear roaring as we passed. Reached Mambova (a small collection of huts) about 5, having had some difficulty in effecting a landing. The swift stream was only a hundred feet or so from

April 17. Miss Saltmarsh (standing) and Miss Bleek (seated in boat) waiting for porters at Katombora Camp

the shore, the latter being edged with a shallow strip of flood water, too shallow to float the boat which had as a consequence to be dragged over the mud. Took some time to get out things needed, as I had to have the crate and my box opened. Kandu killed and roasted a fowl quite successfully – nice to have some fresh meat again after two days bread and biltong! Did not finish our very welcome baths and get dinner until long after dark (7pm). Mosquito net arrangement much more successful – Miss Saltmarsh slept in Sutherland's bell tent, which is more roomy than ours. At 8pm the Great Bear was clear above the horizon to the north, Southern Cross to the south. Also Orion, Auriga and Capella, Gemini, and the ship's masthead light, Canopus, the Dogs and others – a lovely starry night. A man came to have his foot doctored.

Saturday April 18th—Left camp 7.45 on a beautiful cloudless day, with a cool breeze early. We went out into the main stream at once, crossing to the right bank (this is the east branch – there is another branch to the west, Induna says.). The river was very rapid at first, and we soon returned to the left bank. Further up, past a very large bend (where the other stream branches off) the river widened considerably. Have just passed a man in a dug-out, going downstream. We stopped for a little conversation. We saw a good many birds – pied kingfishers, dark grey herons, white egrets, black and white fish eagles, turtle doves, the 'hoo–hoo' dove, a kind of bokmakierie, etc. and the 'go-way' bird. (One told us repeatedly to 'go-way' when we returned for the rope at Katombora. He and his mate were sitting in a tree near the landing stage and we could see the crest and long tail well.) In camp last night, a cat from the kraal visited us and enjoyed chicken bones. This morning there is a very 'ancient and fish-like' smell, but the backs of the six paddlers in front are interesting to watch.

April 18. 'The backs of the paddlers in front are interesting to watch'

DESCRIPTION OF THE BOAT

This is our boat: a long, flat-bottomed structure with floor boards at intervals across the bottom. The centre is floored and covered with matting too, and the sides nearly straight up. The whole thing is some 30 ft or so long. In the front stand six paddlers, one in the bows is the leader of the front six, and directs the others, steering and sometimes chanting or making rhythmic calls to cheer them on. Then under the tilt we sit – there is a grass mat on the floor and on it we put holdalls etc. to sit on, leaning against the luggage. At the back of our stuff is the boys' food, on which sit our boy, Kandu, and the Induna (the head boy) who oversees everything but does no work. There is also a coop of fowls as provision for the way. At the back are seven more rowers. At noon, stopped at the Ingueri River for lunch. There met the Barotse schoolmaster, a Scot (from Aberdeen) his wife and little boy on their way from Mongu to Livingstone. A very wet, hot, uncomfortable spot. This evening's camp is said to be worse.

THE FLOOD PLAIN

Now (2.50) we are poling through acres of long grass mixed with a tall, delicate yellow-flowered leguminous plant which looks perfectly lovely against the blue sky. There is a good deal of thunder and for the moment it is cloudy, a pleasant change after the hot sun, with a strong following wind – probably will rain in a few minutes. The vleiland is really wonderful – heaps of water lilies, here mostly the small-leaved kind, bladderwort (here yellow, in the main stream some mauve and some yellow) by the yard, with bladders an eighth of an inch or more across, and *Vallisneria* (?), the male flowers from which are blown over the surface in clouds of white fluff, besides several flowers I do not know, *Commelina* (blue, or mauve), *Convolvulus*, and *Azolla*. Bird life, too, is getting much more abundant: purple and white herons, egrets, duck of various kinds, coot and a black waterfowl and we have just seen two darters sitting on top of a tree. In one spot were some large *Blechnum*-like ferns growing among papyrus.

HIPPO ISLAND

A camping spot was difficult to find. Finally we got to a tiny rise with two trees and some bush where there was a little dry land, so little that we decided to stay in the boat and let the boys sleep on shore: it was only about twenty by thirty feet in extent. At the back, in the mud, was the spoor (two hind feet) of a hippo, quite fresh. We had a light supper of chicken bones, spread all the rugs and eiderdowns on the floor, hung up two nets (one for Saltmarsh and Bleek, the other for me) and retired to rest (!) at dark.

A FIELD OF WATERLILIES

Sunday April 19th—9am We have just passed the Kasaia and are now in the midst of a field of waterlilies – all white, huge and very sweet-scented. We left Hippo Island at 7.30. Our leader is changed and our assistant cook boy has taken his place. I wonder whether it is because he dropped his paddle yesterday or just to give him a rest? Second cook was very quiet at first but is now working up to a fine dance of energy accompanied by cries of encouragement. He has just cracked his paddle and another was pushed across the tilt to him. Now the front boys have changed sides, I suppose they get stiff with the one attitude.

ATTACKED BY INSECTS

To go back to last night: such a night! I doubt if I slept more than half an hour in all, and never a deep restful sleep – Miss Saltmarsh the same. Miss Bleek fortunately fared better. To begin with, I had picked up a crop of tiny grass ticks, and their activity was even more irritating than the mosquitoes. I've got dozens of huge bumps on my legs and body, many more intermediate ones from mosquitoes on arms and neck, and a third crop of smaller ones, caused by I don't know what, chiefly on one shoulder.

However, these are all details. We were all glad when light came and we could dress and get out of our nets. It is extraordinary how little one feels the want of the sleep – it is such a fresh lovely morning and the flowers are lovely, the grass is short and besides the waterlilies there are huge rose-red *Convolvulus* blooms, the trails with slender slightly sagittate leaves float on the water and the huge trumpets rise from the surface. Then the large mauve *Commelina* (we found a much smaller-flowered one, otherwise very similar, at our stopping place this morning) forms fairy-like patches at intervals. A *Lagarosiphon* (?) is setting free its male flowers to drift over the surface in a delicate tangle of white fluff. A tufted plant with black heads (the larger ones at any rate are seeding) and tangled linear leaves is abundant, and the beautiful white, scented unisexual flower we found yesterday is poking up white buds which are just opening. A little while ago we crossed the open stream of the Kasaia River, and now we are getting to the yellow pea type again, nearer the shore. There is a line of trees some way off to the right. Apart from that, as far as the eye can reach, there is just the vlei with grass, lily etc.

SESHEKE MISSION

Stopped there to write to Mother, then we sighted Sesheke and for nearly an hour we were passing the long, spread-out settlement. Finally, right at the head we saw the neat little brick church built next to Livingstone's fig tree – a mighty old tree, or rather two trees grown together, under which Livingstone sat to talk to the local people. Now under its shade are benches for school children. We wandered past the pastorage (shut up in the temporary absence at the island garden a mile or so away of the pastor and his wife), one or two other buildings and finally came to a house with a wide, netted verandah, set among orange trees, flowering shrubs and gums, with indigenous trees in the background. This proved to be the headquarters of Mlle Dogimont and Mlle Giuglier, both of the Paris Mission. The former, an energetic capable woman, partly of

The Great War

In 1925 'the war' which Mary refers to would be understood to be the First World War, 1914–1918, a recent cataclysm overshadowing the lives of everyone at that time. The 'Union' was the Union of South Africa, formed in 1910 and succeeded in 1961 by the Republic of South Africa.

French peasant stock (her old mother, aged 76, whose photo we saw, comes of peasant parents) is 'Institutrice missionaire' of Sesheke where she has been for many years – she has been working in Barotseland for 13 years, a great part of the time in Sesheke.

Mlle Giuglier (Italian from near Turin) is a trained nurse who served during the war – was wounded in the wrist and nose – and doctors all the neighbourhood. Many come to her who are afraid to go to the government doctors, the latter as a rule not knowing their language. They even occasionally perform operations, e.g. opening and scraping the bone of the finger of a man whose wife had bitten him! Another case was a man who had been mauled by a lion in leg and arm. Mlle Giuglier came out with Mlle Dogimont when the latter returned from leave five years ago. Their chief work is training young native girls, for the most part in domestic work and handicrafts such as basket-weaving, mats, hat-plaiting, native pottery – in short, in Mlle Dogimont's words, training them to be good wives and mothers.

The natives don't mind making an effort to educate the boys, but the girls do not matter, so all the work is gratuitous. Many of the girls come from outlying parts and are boarders. They keep in touch with their old pupils and with few exceptions all have remained faithful to the principles in which they have been trained. They are monogamous and take trouble in bringing up their children etc. There is also a school for the native boys, and besides the trained native teachers, there is a Swiss craftsman, M. Barraud, who trains them in woodwork. We saw some of his work – one beautiful French-polished table with shaped legs, all done by hand. He, his French wife, a bright-eyed little thing who speaks excellent English (was six or eight years in England) and little girl, Gabrielle, had tea with us.

We were received with the greatest hospitality, given dinner, delightful hot baths, most welcome after our night in the boat. By the way they advised us most strongly never to sleep in the boat again, for various very lively reasons! Then after a rest, we all had tea in the garden under the trees, including Mr Reilly, an elderly trader from St. Helena, who is at Sesheke for treatment by Mlle Giuglier. He comes in from a trading station out in the bush. When it grew cool, we all went to see the settlement – the dispensary, surgery (with stretchers, surgical table etc.), two hospital huts, carpentry shed, and the Barrauds' house, the little cemetery, school, church, etc.

Both Mlle Dogimont and Mlle Giuglier, when they return from their leave (due next July – once every five years) are anxious to leave Sesheke and go to Bototela on the Machila, where the country is much better, and start an industrial school for girls there. Sesheke soil is poor and the climate not good – very dry seasons with an occasional exceedingly wet year make agriculture difficult. They say conditions are far more favourable and the position more central.

Unfortunately the funds of the mission are low (the exchange among other things being against them) and at present at any rate there is no government grant for the girls, so that it is very doubtful whether they will be able to carry out their plan. They insisted on our staying with them, gave us a bedroom and the delightful airy covered stoep where Miss Bleek and I put up our stretchers. We saw some of the boarders, bright faced girls, but most are away for their month's holiday. Also two or three of the teachers in the boys' school – they are also particularly intelligent of countenance. Mlle Giuglier is kept busy – among other things she says there is a good deal of leprosy, probably brought up from the mines.

HOSPITALITY OF THE MISSIONARIES

Monday April 20th—Up early, packed, had breakfast 7.30, did some re-arranging e.g. got out supply of dried fruit for ourselves and some for Mlle Dogimont's household. Also gave them most of the remaining apples – they had not seen any for two years. Their kindness was without end – in addition to all their hospitality they gave us lemons, guavas, butter and milk, and finally Miss Bleek and I were each presented with a dainty little basket made by the girls. Miss Saltmarsh will stay there again on her way back in ten days or a fortnight's time.

We left just after 9 – the place looked very pretty in the morning light. We went up a backwater for some way, then open river but soon back to vlei – very uninteresting compared with yesterday's country, far fewer flowers. At noon we stopped at a nice dry sandy spot with scattered trees near a native village (N'Manjanga) and now (2.15) are still poling through uninteresting country. The line of low sandhills to the west is beginning to appear. 3.15. After passing through seas of grass we have just come to patches of open water literally white with the fluffy flowers of (?) *Lagarosiphon*. They are apparently dispersed by the wind more than by water. In some parts they form little heaps on lily pads. Have also passed several patches of *Marsilea* (leaflets large and entire, with longitudinal brownish marks on the back, arranged fanwise) and a floating *Aponogeton*-like plant, with forked spadix of yellow scented flowers, and a hollow-stalked trailing yellow pea, the whole stem covered with roots.

Rather difficult to find a camping place. We finally met several men in dug-outs who directed us due east to a small, but delightfully dry 'island' – a rise in the vleiland with one large tree, some bush and two antheaps not far from a village which we could not see, but where they had an all-night dance, the sounds of which roused us at intervals during the night. Again rather late in making camp – past five.

Tuesday April 21st—While breakfast was preparing the Induna asked for the gun and had two shots but got nothing. After leaving our island we had to make our way almost due south and to be towed for some way before we could get into our proper direction. Later we went up a backwater with a lot of *Rhus* growing actually in the water, and finally came out into the main stream. In the backwater we saw an enormous chameleon sitting on some reeds and changing its skin. It was about 3″ high and 10 or 12″ long, bright green, a beauty. While the boys were towing they killed a big snake – could not see the length but it was as thick as my arm.

The trader at Katonja

Stopped at Katonja for lunch where a trader, Chipman, (25 years in the country) came down to greet us, sporting a several-days-old beard. I went up to his store to thank him for some eggs and sweet potatoes. The store was interesting as he had an enormous lion skin, about ten feet from tip of nose to black tuft at end of tail, with a fine mane, which they had poisoned nearby. He was quite interesting, and told us they had been three partners and had a large herd of cattle which they had driven up to the Angola diamond fields via the Caprivi Strip (formerly German territory) from which cattle are not allowed to be brought into this country. The whole trip, overland going, back by river, had taken them 16 months. They had then dissolved the partnership, and two of them, he and a man called Moody, had started keeping stores – he at Katonja since last November, Moody first at Sesheke then at Barotse. This Chipman came originally from Toronto, has been in this country for 25 years and was last in Livingstone 20 years ago. He originally walked up from 'Wankie's', and first saw a motor car in Angola a year ago! He does not intend to stay long at Katonja as it is getting too civilised! We gave him a *Punch* and he gave us various papers including an *Overseas Mail* with some crossword puzzles, one of which we have just (4pm) been doing.

After leaving Katonja we tried back-waters again but got into a kind of *cul de sac*, much to the boys' amusement! We backed a little, then were towed across some shallows and so into the main stream, where we have been ever since. Here it is a fine broad stream with bush right down to the banks – on this side a good firm sandy bank some 3-4 feet high. As always in the stream, we hug the shore. The country is beautiful, but the river from a botanical point of view rather uninteresting. However it is rather nice to have a rest from collecting. This is a lazy life not conducive to hard labour.

By the way Chipman says the river has risen again, 1″ since yesterday, probably owing to the upper tributaries having come down. He has been waiting three months for sacks in which to put mealies, etc. He had an old condemned boat outside his store full of mealies waiting for sacks and hoped when he heard our boat arrive that we were the transport barge with his long-awaited sacks. Also a hut full of the beautiful silky *Muhoha* fibre (*Sida cordifolia*) which he had bought hoping to find a market for it. Failing that, he was having it made into rope locally. It is beautiful stuff and the fibres are over four or five feet long. It must have been rather a disappointment – he said we were the third lot of ladies who had arrived at his station in the last three weeks, the others being Mlles Dogimont and Giuglier, and someone else who sounded like Mr and Mrs Kannemeyer. Among other things they had done (he and his partners) in the past was boat building. They had built most of these boats now on the river, but the war had put a stop to all that. There are such a lot of kingfishers here – they fly out of holes in the bank, two kinds, small and large but both pied. As we were having lunch two fine big black and white birds flew past, which Mr Chipman said were Egyptian Geese and very seldom seen in these parts.

This morning, not long after we left camp, a man in a dugout came along and offered two wild duck for sale. We bought them for 2/- probably four times what we ought to have given. He was still talking to the boys when there was great excitement. He darted off in his dugout, but his hat blew off and he stopped to get it – and one of our boys dashed off with the spear (the water was shallow there, only 12″ or 18″ or so). We asked what all the excitement was about and were told that there was a buck. However the boy returned, and nothing more was seen of the buck.

THE PADDLERS' ATTIRE
Our energetic leader of the first two days, with the shaven head (except for the top) is back in the bows again, and cook no. 2 is back at his original position of no. 2 on the right. His attire is interesting: in camp he wears a vest, a proper woven vest, and in the morning he starts paddling in the vest. Soon, however, off it comes and till evening he is attired in a fringe, red and (sometime) white, tied tightly round his loins with a tightly twisted roll of material and wearing at the side a small cheap pigskin purse! Most of the paddlers are comfortably and efficiently clad in a small piece of sacking tied tightly round the loins with a piece of string! For warmth at night some of them don an old sugar sack with holes cut for neck and arms, end split a few inches up the side, to make a pukka shirt. The Induna wears trousers, originally what colour? Possibly khaki, but patched artistically with navy serge, and other colours. He also has a vest and a neat little *kalabas* snuff box hanging at his waist. Kandu is quite civilised – khaki shorts and white cotton jumper. Over the latter he dons a long white cotton shirt!

SOME USEFUL SEROTSE PHRASES
Wednesday April 22nd—We had rather a good camp, a sandy knoll (again near a village) with very few mosquitoes – in consequence a much more restful night. We are trying to pick up a few words of Serotse. So far:–

> *Kosiami*: it is good/all right
> *Dumela*: greeting, to one
> *Dumelang*: greeting, to several
> *Sala hantle*: rest nicely = goodbye
> Reply: *Tsamaia fore*: go nicely
> *Ta kwano*: come here

The camp was on a backwater, so on leaving it we had first to go slightly southwards, then across a reedy bit and so into the main stream. In the evening a man came with

Why the Lozi speak Sesotho
During the years of Shaka's military expansion, the Difaqane (forced migration) affected much of southern Africa. In about 1830 the Patsa, a displaced group of the Sotho-speaking Bafokeng people invaded Barotseland and gained dominance over the Lozi (Barotse) people, with Sebitoane, leader of the Patsa, becoming ruler of the Lozi. This domination of the Lozi lasted until the 1860s. During that time the Lozi adopted the Sotho language of Sebitoane's people and it remains their language to this day.[1]

milk of which we bought half for 6d – he also offered us fowls and pumpkin which we did not want. Camped rather earlier (5.10pm), made bread ('doodgooi' proper) and Kandu roasted the duck. It was most delicious – tender and a lot of flesh on it, very good flavour. Started earlier too – 7.30, the Induna having sent Kandu to say that he wanted to start early as water 'very hard', i.e. should have to be on river working against a swift current most of the day.

UNEXPECTED PASSENGERS

For some time we kept to backwaters and then we appeared to be making straight for a low hill, near which we could see a hut with a couple of figures moving about it, one very much clad, near which to our surprise we put in (9.30). However the reason soon appeared. Kandu interpreting for the Induna, explained that the white clad figure (cotton shirt and fringed skirt) was 'Big Induna', and with his wife and boy wanted a lift to Seoma! Evidently a pre-arranged meeting. It seemed rather cool, but eventually we said they might come if they sat at the back. The Induna wanted to put the woman in front but we said no, there is quite enough in front as it is. However, the boy has joined the paddlers in front, so he at any rate is working his passage. Before they left they had to collect a sack of meal, etc. so that we were delayed practically an hour. It was very interesting watching them. On meeting, the Induna appears to introduce the new-comer to the crew, and the response is a soft clapping of the hands – where possible in a squatting position apparently.

SAYING FAREWELL

The 'Big Induna' was carried aboard (only the bows were on shore) seated on a sack of meal placed across two paddles resting on the shoulders of four men. This was far preferable to the 'grannie's chair' arrangement we have had to put up with, our arms clasping the dusky shoulders of our reluctant bearers, of whom almost invariably one is tall and the other short (making the transit even more uncomfortable than necessary). The woman had to wade out by herself! The farewells were interesting. An old grey-headed and bearded man came down to see the departure. He and the 'Big Induna' kissed the palms of one another's right hand, and then each softly clapped hands. The woman and the old man squatted in front of one another, gently kissed hands, clapped and then the woman came aboard. Just as we were about to leave one of our paddlers ran up to another woman, dropped on one knee and kissed the palm of the hand she held out to him. As we left the Induna called out 'Salang hantle!' (Rest well) and the reply came 'Tsamaiang fore' (Go well).

After collecting our unexpected passengers we made our way back to the main stream and wound in and out through the trees and bushes near the side – all at present deep in the water. These include a species of *Rhus,* a yellow *Laburnum*-like shrub, while among the trees one of the most abundant is a beautiful grey smooth-trunked one not unlike a poplar in appearance, with glossy elliptical opposite leaves and beautiful red fruit, pear-shaped and about the size of a small plum.

About 11.45 we stopped for lunch at a small opening among trees. (Barometer 26.15″ – 3400′) There we lunched off cold wild duck in the forest. Since lunch we have been in the main stream, here very beautiful, wooded right down to the shore, with a variety of trees, some towering above the others with dark twisted limbs and grey-green foliage, others a silvery grey.

This part of the river is marked as a series of rapids, but owing to the fullness of the stream we row all the way until we get to Seoma. The boys much prefer poling to rowing and do it wherever they can.

This morning the sky was cloudless and very pale. Towards noon, however, the clouds gather round the horizon and gradually spread, the sky becomes a deep intense blue with white clouds scattered at wide intervals all over it. That is the usual program – sometimes towards evening there is a little distant thunder. At the moment the white clouds and blue sky are reflected deep in the comparatively smooth water (1.50pm).

NEGOTIATING THE RAPIDS

4pm. We have just ascended our first real rapid (*Manye' Kanga*). We have passed several but each time we have avoided the swiftest part by going into flooded side country. This time, however, it was evidently impossible to avoid them altogether. We heard the rush of water and across our branch of the river was a line of rocks. This had to be negotiated on the far side where the stream was deepest. Even so our bottom bumped and scraped a good deal, and practically all the paddlers had to jump overboard and pull and push. One advantage gained: practically all our paddlers had a good bath, at any rate up to their waists! The leader ducked right under and seemed reluctant to come out. The Induna's son did not get out but stayed in and made great play with his paddle. I was sorry, as I think he stands in more need of a bath than any of the others, judging by the added bouquet from the bows since his addition to the number of paddlers.

We have just seen two boats shooting the rapids far over on the other side of the river, which with some difficulty Kandu explained were two of 'Mister Suverland's', and 'Miss can go' – meaning, we eventually discovered, that Miss Saltmarsh could if she liked return by one of them.

Barometric readings, elevation and time

It is possible to estimate altitude using barometric readings taken at two sites at different elevations. Mary was a mathematician so would have known how to use the formulae and do the calculations involved. Similarly she regularly corrected her watch against the rising sun when it was visible:

We were up at dawn – 5.30 – sun, unfortunately, again invisible, so that I was again unable to correct my watch (August 15).

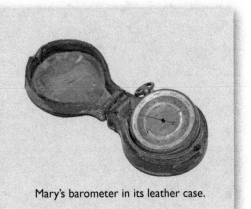

Mary's barometer in its leather case.

Now we are approaching another and much larger rapid. Here, at the side of the bay below them, there is quite a swell. We are however avoiding them by taking to the reeds at the side – no, not quite. We avoided the first part, then had to go into the strong current. For a time it was quite exciting – both Indunas got out of the boat on to a tree at the side, came to the front and then pulled. Our Induna got right into the water in front and pulled with the rope. The worst past, he stood among the paddlers in front directing them, and finally, past the actual rocky barrier but in strong water, he took the paddle from the other Induna's son (who was only too pleased to sit down on the tow rope and rest). He went to the bow and stood behind the leader, and then they two set the pace for the other men and kept them at it too, till the camping place was reached – a beautiful spot high above the river, with one or two trees but cut off from the actual bush by a grassy hollow containing water. There the other Induna superintended the pitching of our tent.

Thursday April 23rd—Woke as usual at daybreak (6am), packed and finished breakfast by 7.30. Going again hard all morning – no actual rapids but current at times so strong that the men would have to paddle for all they were worth to hold their place. Then the pressure would be raised a bit and we would progress slowly past the bad bit, then a comparatively easy spell would be succeeded again by strong water. Had lunch in a pretty glade with spoor of hippos and buck all around.

THE NGAMBWE RAPIDS

About 3 the water began to get particularly strong and a distant roaring to become audible. Then in the distance we saw a shining barrier across the river – the Ngambwe rapids. Here we had to put in. The boat was unloaded and we took to the shore. It is a

April 22. Campsite on the Zambesi

April 23. Boat being towed up Ngwambe rapids

lovely spot – the water swirls down, then drops some feet and swirls on again. At low water it must be an even grander sight. The tow rope was fastened to the bows and the paddlers manned it – one right ahead, then most of the boys near the boat in which were the two Indunas in the bows, and three boys at the stern (which by the way is square not pointed as I had thought). Thus pulling and poling the boat was edged round the end of the rapid where the rush of water was not quite so great. It was most interesting to watch and we took several photos.

Above the rapids (4.30) we made camp and while one boy baled out and cleaned the boat, taking out all the flooring boards (which looked as though they badly needed airing), the other carried the things up – including the Induna's 200 lb. bag of meal! It is to be hoped he is paying them for all the extra work they have to do. To our disgust his son, instead of going into the water with the other boys, stayed on the shore with Mama – he is a dirty little pig and smells worse than any two of the other boys! Most of them (particularly the leader, who simply wallowed) enjoyed the chance of a good dip. We longed to go in – there were lovely sandy-shored little bays with the water swirling through, and not a sign of a croc. I don't think they are found near the rapids – at any rate the boys show no sign of fearing them. What they *do* fear are hornets! Again and again in passing a bush they all duck down into the boat. Then there are guffaws of laughter – these hornets appear to cluster on the leaves of trees, making their nests on them.

ARROWHEAD BAY
Near the camp, on the shore of a tiny bay, found several small quartz (?jasper and cornelian) arrow heads etc., and part of a grinding stone. The 'Big Induna' asked for the

gun and went off with several cartridges. He came back without having seen anything, however, and arranged to go out again early tomorrow. A woman and her child came down to Arrowhead Bay, to fetch water – she saw me and called out something. I of course could not understand but gave her greeting and beckoned. She ran up laughing and evidently asking for something – salt, I thought she said, and so it proved. We gave her a little and asked for milk, which she promised to send in the morning.

Friday April 24th—The Induna left early with the gun and has not yet (9.25) rejoined us – probably will meet us at midday. An old man brought a pannikin of milk (about one-and-a-half bottles) for which we gave him a cup of salt. Our own Induna had evidently been up to the village drinking and had not recovered properly – was much less dignified and more noisy than usual. However he got things packed up all right and we got away not much later than usual (7.45). The water was rather broken at first but we have kept close in, usually just among the reeds, and now the water out in the wide river is gliding swiftly and smoothly down. Palms are becoming much more abundant, particularly where there are islands.

The men caught a dainty little bird, a fledgling – green, with a long beak. They handed it to me, and we made them put it back to the nest – however we could not make them understand and, before I could prevent it, they had torn down the nest, which was built on a large hanging leaf. Unfortunately, the little thing slipped away and flew to the reeds on the river side – got startled, flew again and fell into the river and was carried down the river. The mother was piping agitatedly in the bushes nearby.

11.30. Another rapid – the rapid of Lusu (*Rapide de la Mort*). There we watched two boats (one apparently a grain barge) shoot the rapids on the far side. The boys thoroughly enjoyed their dip. After surmounting the rapid, I happened to glance back and saw some of the stern paddlers having a good swim. Here we stop for lunch.

A SUCCESSFUL SHOOTING EXPEDITION
We had just finished lunch when a hail came from across the creek just below us – Induna Nasilani had sent for some of the boys to bring in the result of his shoot, two buck (*pala*) and six guinea fowl. The buck are beautiful creatures with rich brown coats and rather long tails, white along the sides with I think a black line. Miss Saltmarsh asked him to cure one of the skins for her. Our share was two guineafowl and hindquarters of the smaller buck, the rest going to the Indunas and crew.

We landed at 11.30 and did not leave again till 2.30, as it took some time for the boys to bring in and then skin and prepare the animals and birds. The guinea fowl are very small. While waiting I tried to do a sketch without much success – the cloud effects in the sky were much too complicated for me. When I started the sun was behind a cloud and the lighting soft, so I used Coxes paper – before long, however, the sun was out and I was simply grilling!

CAMP AT MONOMÉ

Did not camp till past 5, then crossed to the far side to a spot called Monomé – rather beautiful but covered with long brown grass about two-and-a-half feet high and as, soon after landing, Miss Bleek saw a snake in it quite near the tent, we did not feel too happy. Then started a great cooking of dinner and drying out of meat in the smoke. We had a couple of guinea fowl, very good and plump but rather tough. Kandu roasted them very well. The hindquarter of buck went to spend the night in the tent with Miss Saltmarsh, who expected terrible dreams but merely dreamt of house buying in Johannesburg.

The 'merrie sound of feasting' arose from the boys' camps till late at night, and from the Indunas', I fear also of drinking, judging from the raised tones of our Induna Sanpiere, who seems somewhat addicted to the local beer. I rather suspect the big calabash Induna Nasilani brought with him. He, however, seems all right, but both at Sesheke and above Ngambwe Falls ours seemed to be rather above himself. It is a pity, because he is a smart capable boy. However the effect seems to wear off very quickly.

Saturday April 25th—Meat for supper seems to make these boys wakeful – they were stirring long before light. After a sumptuous breakfast of kidneys, liver and bacon we got away at 7.15 – our earliest start so far. We crossed back to the left bank of the river and soon passed the entrance of the Njoko River. The scenery was beautiful beyond words for the first hour – the water (here very wide) like glass and the reflections wonderful. It is almost the prettiest hour of the day. I tried to do an impression of the glimpse I had just before sunrise – my bed last night faced the east. The evening was cloudy but most of the clouds cleared off. We could not see the Great Bear, as a fine tree a hundred yards or so north of the camp blocked it out. Now (11.25) we are passing a part wooded right down to (and at the moment, into) the river, where palms are unusually numerous. These are rather like our Rusdon ones and many are submerged right above the crown.

I have been trying to do some sketching but it is next to impossible in the boat. The view in front is completely blocked by the rowers and to get the side view one has almost to lie down, at any rate to recline. We have been doing crossword puzzles. The *Daily Mirror* two were good, but those in the *Daily Graphic* very feeble on the whole – perhaps we are getting too clever at them! Anyway, we have now exhausted our supply and I must divert the *Daily Graphic* to its intended use – i.e. as a receptacle for pressed specimens.

HUTS BUILT ON PILES

About 12.30 we saw a couple of dugouts crossing the river to our side, and we put in at the same place. Traffic as a consequence was somewhat congested! The river bank here was sandy and some 15-18 ft high, crowned by some rather fine old trees. The ground sloped back behind the bank down to a patch of marshy ground with a shallow vlei, about the edge of which were half a dozen or so houses built on piles, some 6 ft above the ground, and with branches crossed between the front piles to form a primitive kind of ladder. We took snaps but the light was not good so I doubt if they will come out. The huts seemed to be deserted. One or two had fallen down. Still, so far as we could make out, they are used as dwelling houses.

We tried to get our overnight's washing dry during lunch but it was rather overcast so our success was not very marked. We were glad to get away, as the spot was not of the sweetest, and when the sun came out it was very hot and steamy. It is often a relief to get into the boat and start moving after the midday halt. We made Kandu start cooking the two buck haunches as we feared they might not keep. The sky was very cloudy towards midday and for some time we could see a rain storm over to the north-west. Then came thunder in the distance and finally 'crash!' right overhead.

A DELUGE

Then followed the rain – such a deluge! For a time we went on, the boys singing and dancing at their paddling, till finally it simply pelted. The tarpaulin was pulled over, her nose run ashore and all seven bow paddlers squatted down on the tow rope in front (we were glad to see 'Son and Heir' pushed to his rightful place – i.e. least of the paddlers and therefore nearest the bows!). There we stayed for the best part of an hour – the atmosphere, what with warm paddlers in front and warm meat behind, growing thicker and thicker. Miss Saltmarsh and I tentatively poked up the tarpaulin on each side but the heavy rain made the water splash up as well as down, and until the rain abated we had to endure.

As soon as the rain had nearly stopped we started again. By the way, no sooner were the boys under the tarpaulin than they all started gnawing the large fleshy, nutty roots which were given them at last night's halt (Monomé). The roots look rather good – they call them something which sounds rather like manioc. Soon we came to the Bombwe

April 25. 'Traffic was somewhat congested'

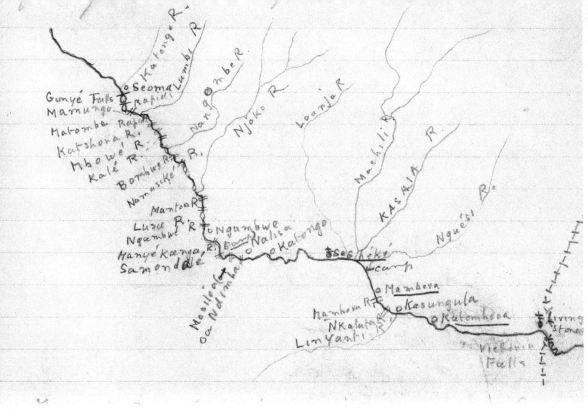

April 25. Mary's sketch map of the Zambesi from Livingstone to Seoma

April 25. Huts built on piles some 6 ft above the ground

Kamashi plant – Xyris sp.

Xyris sp.

{P 3   A 3} G(3)
...ll... yellow hair between perianth leaves
...n... punctate; seeds numerous.

Rapids – very fine to look at but so full that apart from the extra swiftness they did not present much difficulty.

Sunday April 26th—To make up for the time lost owing to the rain we went on till nearly 6pm, the latest yet. We had both tents put up, so could have the luxury of two baths going on at once. It was well we did so for various reasons. First, our tent was very mouldy and badly in need of an airing (at present it is spread on the roof of the barge, but as the tarpaulin has since been put over it I doubt whether it is deriving much benefit from its 'airing'). Secondly, soon after we started dinner the rain began again, so in my bed had to come, much to my disappointment.

DOSING MISS BLEEK

Miss Bleek has developed a cold, in her throat chiefly, which is making her feel rather wretched and has almost deprived her of her voice. Apart from quinine and hydrogen peroxide, we none of us brought any remedies for colds! It is so long since I had one that I had almost forgotten the necessity! It rained a lot in the night – fortunately there were few mosquitoes, but we have been a good deal bitten by other things, partly grass ticks, I think, judging from the large uncomfortable bumps, but partly also by other more obnoxious things that nibble! We had suspicions after what Mlle Dogemont said about the boat and today they were unpleasantly confirmed by a discovery: last night after my bath I washed vest and stockings as usual and today at midday halt hung them out to dry – and found a beast clinging to my vest! I realized what the daily examination of their flannel shirts implied to the men in the trenches! However, it was after all only one.

We broke camp late – it was still very damp although the rain had ceased. We started at 8am, the latest for a long time, and crossed the river to a village on the other side, where we got milk and eggs. The former had to be boiled, and while we waited, Miss Saltmarsh and I got out and took snaps of the group of women with their babies, evidently just very well fed, and gave them bits of chocolate. These the babies promptly handed to their mothers, who proceeded to eat them very doubtfully. Here again one or two of the huts were on piles (but here apparently only storehouses) while others were merely shelters – thatched roofs on logs as supports with mats below – in one or two there was a fire (usually two large sticks crossed) burning at the entrance and a man (or as in one case a woman with a child) sitting on the mat inside. The children were all very dirty.

A GIFT OF VEGETABLES

On leaving this village of only some half a dozen huts, we crossed back to our side of the river to another group of huts where we picked up our Induna, who had walked up from the camping place, armed with his crocodile spear. He had with him a clean, respectable looking woman and well cared for child of four or so, the former carrying a flat basin of something, which she handed to him as he came aboard, then squatted down and clapped her hands. The bowl evidently contained vegetables (parsnips?) as he presented us with a plate of the roots at lunch time. They will be very welcome for dinner. It is still, and has been all day, very cloudy, but looks rather lighter now, and I do not think we shall have rain tonight, but we may. Tomorrow we ought, I should think, to reach the Gonye Falls below Seoma. I don't know exactly where we are, there have been no very apparent rapids but several times the water has been very strong. We are now again on the right (west, or rather south-west) bank.

Monday April 27th—The 'parsnips' turned out to be sweet potatoes! And very good ones too – I've never seen any so white outside. We sent Miss Bleek to bed early with a large cup of *Maizena* laced with (oh, dreadful!) a dessertspoonful of brandy!! I hope it did her good. We started about 7.45 this morning. The western sky early was lovely – tiny grey clouds which suddenly flushed pink as the sun rose. Last night, too, we greeted the new moon which was hidden by clouds the night before. Today again it is cloudy. It is rather pleasant than otherwise and does not look like rain. It was good to sleep outside again last night. We camped a bit above the Kalé rapids – a beautiful spot – wide open glades among large trees.

KATSHORA RAPID

About 11am we neared another rapid (Katshora) and edged our way up, bumping and scraping on the rocks at the edge, till we reached the actual drop. There we stopped and the front paddlers got out and while Induna Nasilani gave instructions to the stern paddlers, attached the tow rope and began edging it along the shore above. ('Son and Heir', we noticed, pushed his way out first with his paddle and jumping ashore without wetting his precious toes, betook himself to land) When it was well

April 27. Boat being towed up Katshora rapids

out (one man goes right ahead with the end, then the others man it), Induna Sanpiere and one of the paddlers got right into the water on the inner side of the boat. Induna Nasilani directed from the front and thus we got up the two little shoots of water – it is always an interesting manoeuvre. The men know the river wonderfully well. There was another very well made strong-looking tow rope lying on the rocks at the side of the rapid. I wondered whether it was left there purposely or had been dropped by some boat on ahead. I tried to get a snap of the men hauling us up but am afraid it won't be much good.

Have just finished 'Life and Erica' (Gilbert Frankau). He is a clever writer, but this does not touch 'Peter Jackson'. 4.45. Have just towed up another rapid – Kaloonga according to Kandu – not marked on our map. We passed the entrance of the Lumbi River an hour or so ago. It was one of the prettiest parts of the river. There was a wide bay to the right and at the head of this two rushing torrents of foaming water creaming over a network of black rocks, higher than an ordinary rapid, yet not sheer enough for a fall, on both sides and in between the two mouths wild bush country with grand trees.

Since then there have been occasional high banks of pure white sand, or mounds of tawny rock to diversify the bank. The sun has come out too, and though that makes it at times unpleasantly warm, there is no doubt that it does add to the beauty of the scene, particularly at this time of day. Twice this afternoon in strong water a paddle has lodged between stones and been torn from the hands of the paddler, much to the amusement of the rest. The last time this happened was in a very narrow bit, where the water rushed between a rocky island and the shore. For a time the current was too strong and we were swept back, the paddle being left standing quivering in mid-stream. However, in a few minutes we got through and the paddle was rescued.

Tuesday April 28th—We camped in another lovely spot, the isthmus of a long sandy peninsula jutting outward into the river on the right bank – only the second time we have camped on this bank. Our tents were pitched on white sand, in which was a sparse growth of six-foot-high grass which was pulled up before the tents were pitched. Miss Bleek's cold was a bit better but she is still very hoarse, so we put her in the small tent again. Back of the sand ran a ridge of higher land with big trees and undergrowth. Apparently there was a chance of prowling animals, as a very large fire was made just behind the tents and at bed time three men brought their grass mats, rolled themselves in their blankets and lay down between the fires and the tents. The Indunas had another fire in their camp just behind. My bed faced due north and I watched the Great Bear slowly circle upwards as I lay in bed. Needless to say the arc I saw him describe was not a very large one! Being dry and sandy and with no standing water (the current here is strong) there were no mosquitoes, so we ventured on a game of cut-throat bridge. We could not finish the rubber, and went to bed at the unheard-of hour of 9.15! It was a perfect evening and day.

April 27. The view ahead (above Katshora Rapids)

April 26. Thatched shelter

April 25. 'The scenery was beautiful beyond words...'

April 27. 'A paddle was left standing in mid stream, much to the amusement of the paddlers'

More rapids and the Gonye Falls porterage

This morning we started at 7.30 and after doing the Matomba Rapids, those just below the entrance of the Lumbi River, and the Mamungo Rapids, we turned into a little bay to the right and there disembarked for the Gonye Falls porterage. We landed from the boat about 11, and after airing my bedding in the sun for half an hour or so (there was a heavy dew last night) and having watched the boys leave with one load of things, we started on the walk through open woodland – several sterculias, one of the bush forms, in full flower, looking like a pink-tinged white *Azalea*. Half way the boys passed us, running back for a second load! They really are a willing, as well as a happy, good-tempered crew.

Soon (after rather less than a mile I should think) we came to a spot where the wagon track and our path converged and dipped down into a sheet of water, to emerge on the rise on the far side. Here, by the water's edge, we found our things and settled down in the shade of a tree to wait. It was too early for lunch, so we did a little washing and mending, read and lazed a bit and then had lunch. Just about then the team of oxen which the Induna had sent for (to haul the boat across) appeared on the far rise. They looked fine as they were driven down into and across the water. In the middle it was deep enough for them to have to swim.

Some time later we heard in the distance the peculiar harsh, raucous cries which are apparently the indispensable concomitant of oxen driving. Gradually these grew louder and louder till finally the first span appeared round a bush and gradually the whole team appeared in the water dragging the barge after them. Thereupon followed a leisurely repacking, after which we poled up the vlei to the left for some way.

April 28. 'The whole team [of oxen] appeared in the water dragging the barge...' round Gonye Falls

April 28. Perilous crossing of the Seoma Falls. Our first experience of travelling by dugout

Gradually the water got narrower and soon banks began to appear, showing evidence of man's handiwork. These became higher, the stream narrow, deep and very strong and we were in a real cutting. Up this we poled with difficulty, tacking from side to side of the narrow way, partly forced to do so by the current, partly so that the men could get purchase for their punt poles. Finally we reached the broad river, our canal cutting the village of Seoma into two parts. Here we were told that if we wanted to see the falls we must get out and walk back along the river. With us came Induna Sanpiere, armed with his spear, one of the paddlers and a tall, thin, elderly Induna of Seoma.

SEOMA FALLS

It was quite a long walk – some one-and-a-half miles, zig-zagging round little square millet patches, past huts (built inside a small enclosed courtyard), along ditches where a small boy or girl was busy taking small fish out of little grass traps, till finally the way was barred by a fairly deep stretch of water some 30 ft wide. There our guides called to a bashful old man (he tried to hide himself against a tree as we passed him later), and he pushed a dugout from his bank so that it slipped across to us. Then, one at a time, Induna at one end, paddler at the other, we accomplished the dangerous passage.

The other two crouched down, Miss Saltmarsh resting part of her weight on her sunshade which (minus its handle of course) projected three inches on each side! I decided to stand, holding my sunshade crosswise as a tightrope walker holds his pole. At first it was all right – then a lurch first to right then to left nearly upset my balance and precipitated me into the water! It is not easy to keep one's balance. The dugouts (this small size) are frail little things and very narrow. The paddlers stand at the two ends and either pole or paddle. They look very graceful and pretty gliding past on the water.

After crossing we still had to walk some way, now hearing the roar of the falls, now catching a glimpse of the spray rising from them. In the end, we did not get very near, but could only see part of the falls. They are extensive – part horseshoe-shaped, different sections of the curve divided from one another by groups of trees. I hope we shall see them better on our return when the water is lower.

We were quite weary by the time we got back, after again traversing the pond in the dugout. We wondered whether we should find the camp made, tents pitched etc. The boat had gone, so we had to wait a quarter of an hour while a canoe was fetched to take us across the cutting. This canoe was twice the size of the other dugout, but none too stable. Miss Bleek was taken across first, then Miss Saltmarsh and I together, sitting on our sunshades. The water was deep and strong, and we were by no means sorry to get safe to the other side. Five minutes' walk brought us to the camp, where we found everybody *very busy*, just putting up the tent. They must have waited till we were in sight and then scurried round getting the things out, and were rather reluctant to put up the second tent.

We finished the useful buck at lunch and as the satsuma plum jam, which had been masquerading as red-currant jelly, had already been finished we found it somewhat dry – so for dinner we had soup, and very good it was too – four onions, three *Oxo* cubes, a plateful of small tomatoes that an old man brought and a little pea flour. An omelette followed, then coffee – and what more could one want? My third batch of bread was very successful, two white and two brown loaves baked in the pot. I do the mixing, Kandu the baking – quite a successful division of labour.

Induna Nasilani came up to Miss Saltmarsh with Kandu to explain that he was going to await the transport boat here at Seoma, and suggesting that she should wait too. After much discussion (much hampered by mutual ignorance of the other's tongue) she decided to do so, as so far as we can make out, the boat should arrive tomorrow or the next day. As he and his wife (and presumably malodorous little Son-and-Heir) will be camping just next to her and will look after her and her things, she is very glad they became our unexpected passengers. The evening ended with bridge – bed at 9.15 again.

Miss Saltmarsh leaves

Wednesday April 29th—We decided to stay in camp for the morning to keep Miss Saltmarsh company and to do some much needed washing, sorting etc. Among other things I boiled my sleeping bag sheet, pillow case and towel – all much in need thereof, the benefit being enormous. Miss Bleek did a most extensive and exhausting wash. I've kept on doing odd bits all along so had less to do. Then we rigged up a clothes line – our mosquito net rope tied to two paddles – and got everything including bedding thoroughly dry.

The boys, meantime, put themselves in the hands of the barber and now all, except Monomali who was kept busy all morning heating water, plucking fowls etc., have shaven polls with little black woolly patches of various shapes, with one or two exceptions who are completely shaven. The men are awfully well-built and very muscular – wide strong shoulders, slim straight hips, shoulders are almost too wide in proportion to the hips, and straight muscular legs. Of course their special type is to a certain extent at least, the result of their life. Standing paddling tends to the development of the shoulder, chest and back muscles, and magnificent they are too.

Our assistant cook, Monomali, is one of the tallest and best built of the lot, and his skin is a warmer, more chocolate brown, while others are more coal black. There is a great deal of variety in their faces, both in feature and expression. One is evidently a humorist, with a funny broad-mouthed face. I have quite revised my opinion of Egyptian art so far as figure drawing is concerned. Their broad-shouldered, narrow-hipped figures are seen here alive, attitudes and all. Well to return: we made Kandu kill (or rather, have killed – it is beneath my lord's dignity to kill or pluck a fowl – or to fetch and carry anything if he can avoid it) two fowls and roast them – one for Miss Saltmarsh, the other for dinner, which we decided to have at midday. We had vegetables too – cabbage, beet and tomato.

Finally, after a good sorting out of things, we left about 1.30. I hope Miss Saltmarsh won't have long to wait for her boat. We started in the river, but soon left it for a backwater, in which we kept as long as possible, then back to the river, where we could hear the roar of the falls. Since then we have been sometimes on the river, sometimes in vleis. Now (4.15) we are waiting in the middle of some vleiland while some of the boys are fetching mealie meal. Near us three boys are disporting themselves in as many dugouts, racing and crossing from side to side. In their little red kilts they look most picturesque.

April 29. Miss Saltmarsh and Miss Bleek at Seoma camp

All this time we have seen never a hippo, nor the sign of a croc! The fiercest things we have either seen or heard are one or two snakes in bushes as we passed by. A large wicked-looking spider with black shining jaws, a grey body marked with black, and golden brown thorax and legs, has adopted my fishing net as his home. He has spun his snare across the mouth of the tube and sits at ease near the bottom. If I can get my killing bottle I must pickle him. At Sesheke they told us there are plenty of scorpion spiders in summer and that they are poisonous beasts. I am not anxious to meet any!

Hippo sighting

5pm. – We have only just started – the boys took a terrible time to fetch that sack of meal. Tomorrow we shall have to get on. 5.20. Our first hippo! The boys in front said 'Kooboo', so we stood up, and away to the right front there was a black splodge on the water between us and the setting sun. Soon the splodge disappeared, and as we passed on the far side nothing was to be seen save a patch of troubled water where Mr Kooboo had submerged. When we were well past it, however, came a great 'plooh!' and cries of delight from the boys, and there away behind us was a great wet brown head with a vague impression of red eyes and ears. The latter he waved at us, blew, and submerged only to come up again in a minute puffing and blowing amid renewed cries and laughter from the boys, whereupon, giving another snort either of scorn or breathlessness, he submerged again and we did not see him anymore. What a pity Miss Saltmarsh missed him!

Thursday April 30th—We made camp in a rather mosquito-ridden spot – an old millet field near a village. Fortunately the mosquitoes proved less worrying than at first seemed likely, as they were apparently mostly in the old stalks. In the evening all the village came out and our crew held a reception, with much clacking and jabbering and calling out to distant friends, varied by the hand clapping, a gentle diminuendo, accompanying the introduction of one to another. I think some of our boys must hale from this village, Matina (or Imatina) so far as I can make out.

We left at 7.30 and all day have been in a kind of backwater – the Induna gives it a name but I don't know whether it is a tributary or only, as I say, a backwater. The stream is quite swift in places, in others deep, too deep for poling. It has been an uneventful day. I finished Miss Saltmarsh's sensational '*Upstairs*', and in the afternoon dozed a bit. The afternoon has been hot and trying. Then we did some Portuguese. We stopped at a high bank above the stream for lunch – rather a beautiful spot – not many trees, and very long grass between. About 9am too we stopped to get milk at a village.

Friday May 1st—11.50. Last night, after a long, uneventful day (7.30-12.30, and again from 1.30 to past 5) we camped in a pretty open grassy patch on a high bank above the river – a much better camp than the previous one. This morning again we started at 7.30 and have had a rather monotonous morning, chiefly through backwaters and vlei land.

The odour in the boat this morning is far from pleasant. I wish we would come to some more rapids, since a good strong rapid means a good wash for the front paddlers at

May 1. Travelling up the Zambesi, chiefly through backwaters and vleiland

any rate! It has been rather hot, too, though just at present there is a nice breeze coming in at the side. I have been reading '*With my wife across Africa*', (Statham) which is quite interesting and well written. Then we did some Portuguese and played three games of Piquet – I am rather uncertain about some of the scoring, which is annoying, still it is enough to get along with.

SINANGA

Saturday May 2nd—About 4.45pm we (apparently) suddenly headed for a tree-covered knoll, just opposite a very open flat expanse of green grass and blue water. As I was getting my things together ready for going ashore, Kandu and Sanpiere (the Induna) came round and the former, evidently acting under instructions, informed me that this was Sinanga. As I had understood that we should not get there till the following morning, I was agreeably surprised, but no doubt I had understood wrongly. There were several huts down on the green flat, but from later sounds the main village was on the hills (or rather ridge) where we were camping and we had evidently come past it without seeing it.

We were just making camp when two boats came sweeping round the bend to where ours was lying – the transport boats for which Miss Saltmarsh is waiting at Seoma! It is annoying – she could just as well have come on with us and trans-shipped here at Sinanga. Instead, she has spent two-and-a-half days all alone at Seoma and will have to spend another before the boats arrive. I hope she has gone shooting or something! There was a great chattering among the three crews, but the newcomers did not stay long, as they were evidently going down to the village to camp. I sent a note to Miss Saltmarsh by them, together with two letters for the post – one of Miss Bleek's, the other to E.L. Stephens to notify her of the parcel Miss Saltmarsh is taking back for her.

A BEAUTIFUL CAMPSITE

The camp was about the most beautiful we have yet had – our hillside was dry, covered with rather far-apart trees with grassy glades between. We looked due west, across a fairly wide stretch of river to a flat green grassy expanse, broken here and there with bands of water, and to the right by a few trees, and here and there huts were dotted about. Beyond, the green ended abruptly against a narrow band of bluish purple (treeland in the far distance), and this in turn ended flat against the sky. Above the horizon showed the tops of cumulus clouds. The sun set in a blaze of gold and crimson – most lovely, and the early morning colouring was almost as beautiful. I just had to try a quick sketch, though the bath was waiting – result, a lurid splodgy thing which, however, if one gets far enough away, does give a slight idea of the scene.

This morning, instead of gold and red, the sky, perfectly cloudless, was pale turquoise above, then shading down through palest green, yellow to red and finally low down (the western sky of course) to purplish-blue and the water perfectly still and iridescent. We left early and have evidently reached homeland for at least some of the crew, judging from the frequent stops to converse. At lunch the Induna collected two lady friends (query, wives?) with whom he conversed at great length.

VLEI FLORA

Most of the day we have been going through vlei land and have got back the varied vlei flora of the first few days, with additions – i.e. a delicate white flower with an enormously long, fine stem – then a stout up-standing plant rather like arrowhead, with male and female flowers. Bladderwort is again abundant. Here there are at least three species – a very pretty mauve-flowered one with a long horizontal forward-projecting spur and very abundant fringed floats arranged in a rosette, often bright red in colour. The bladders are medium sized, sometimes reddish, or when older purplish-black. The other two are yellow. One has very long slender peduncles, coming up from the bottom and most beautiful, finely branched bladder-bearing parts, often red. The third is a very stout plant, without floats. Usually near the surface the bladders are very large and dark and the branches bearing them stout. The flower, yellow with brownish guides, is very similar to no. 2 but on a short stout peduncle and only one or two flowers on each, whereas no. 2 usually has a more or less elongated spike with several flowers.

Then the delicate dioecious, sweet-scented white flower which we found at first with the water lilies is again most abundant, and dainty *Azolla*, red and green. Just here there are fluffy heaps of flowers from the male plants of *Lagarosiphon* (?). They are blown along the surface of the water, rather than floated, until they reach the female plants and fertilize them. In many cases, however, the little staminate flowers collect on the lily pads and never reach their mates. Water lilies and the lovely purple *Convolvulus* are again abundant. The latter with its large sagittate, dark green leaves is most handsome growing among the green water grasses.

The sky remained clear until nearly noon. About 11am tiny cumulus clouds began to show above the horizon and gradually they gather, till although still hot and sunny, the sky, a deep blue, is dappled with cumulus clouds. The perspective in the clouds is very attractive – they are all about the same level with flat bases which show one behind the other getting smaller and smaller in the distance. A little while ago (5.15) I heard distant rumbling, which grew gradually louder until it was unmistakeably thunder. Looking out to the east there was a great thunder cloud, dropping rain, and to the right gleamed the arc of a rainbow. Another greeting! Their speech is punctuated by *Wa! A-ah!* and similar un-reproducible sounds with varying inflections. I do wish I knew the language – it would add so much to the interest. Have had a busy day going through pressed specimens, changing papers and putting in and labelling fresh specimens, varied by a little sketching, a little Portuguese, a little sleep and a lot of diary! Of this last, enough.

A CAMP IN THE VLEILANDS

Sunday May 3rd—Our camp last night was yet another type – imagine green fields as far as the eye can reach, which however turn out to be not solid land but water with long grass, mingled with water lilies, *Convolvulus* and other flowers. Suddenly, a slightly raised bit is seen and the boat's nose runs against it till she refuses to move further. Then, out jump the paddlers, arrange themselves on each side of the bows and first raising an antiphone of call and answer, they pull the boat in as far as possible. Last night this meant that there were still several yards of water between us and dry land. Across this we had to be carried sitting on two oars and holding on to two brown shoulders. This 'island' was a fairly extensive roughly circular patch rising a few feet – four at the outside – above the level of the surrounding flats, but quite dry and hard. No trees, but a few bushes to the east, while to the west was a swampy patch from which rose a mighty humming, apparently flies, though they did not bother us. The rest of the island was grassy.

Moonlight again and the night was lovely. To the south lightning lit up the sky at intervals, while to the north, a quarter of a mile away rose the light of a fire in a nearby village which was hailed by our headman in the usual fashion. Standing on the island one had a perfectly circular horizon of enormous extent, the line being broken only at rare intervals by an occasional distant tree. Our journey by river is nearing its end. We should reach Nalolo tomorrow. At lunch time today we hit such a hot overgrown island, without a patch of shade, that we had to have lunch in the boat, and soon started on again.

NALOLO MISSION

Monday May 4th—About 5pm yesterday we were surprised by the sight of a house, brick-built with a proper roof and chimney, and a nearby flagstaff, while on the same island were several well-built, brick huts. Upon inquiry we were told it was Nalolo, and the Native Commissioner's residence! We were surprised as we had not expected to reach it till today. We did not stop as we thought we should be camping nearby and could present our letter of introduction after preliminary tidying – a mistake as it turned out, as we went on for another hour before stopping, passing several clusters of huts on the way.

Finally we stopped at a particularly neat collection of houses, huts etc. standing in a neat, sandy compound, with hedges and flowers and several trees. It proved to be

beautiful.

May 2. Sinanga campsite

May 2. 'Bladderwort is again abundant'

May 2. 'The lovely purple *Convolvulus* was abundant'

May 2. Water lilies and *Convolvulus*

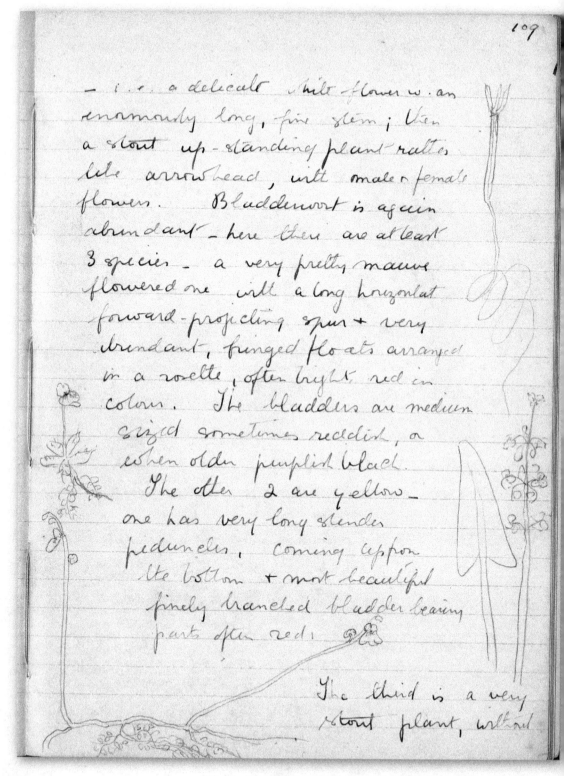

— i.e. a delicate white flower w. an enormously long, fine stem; then a stout up-standing plant rather like arrowhead, with male & female flowers. Bladderwort is again abundant — here there are at least 3 species — a very pretty mauve flowered one with a long horizontal forward-projecting spur & very abundant, fringed floats arranged in a rosette, often bright red in colour. The bladders are medium sized sometimes reddish, & when older purplish black. The other 2 are yellow — one has very long slender peduncles, coming up from its bottom + most beautiful finely branched bladder bearing parts often red. The third is a very stout plant, without

May 2. A page from Mary's diary, showing sketches of plants

the French Mission station, but the missionary was unfortunately away, and at first we found no-one who could speak English, so we had to make shift with Kandu as a very inefficient interpreter. Our camp was made in the centre of the compound and we got bananas, milk and butter, the latter a 'plesent' from one of the head boys. By and by appeared an intelligent young man, probably chief teacher in the school from his look, dress etc. who spoke a little very fair English, and he told us their *Morena* (missionary) was away at Sianda and would not be back for some days.

I am writing with a tiny green frog, half an inch long, sitting on my hand – there he goes to my shoulder. The toes are slightly knobbed and he has a white line running from his nose right down each side, and lovely gold and black eyes. I think they are 'tree' frogs only they live on the grasses! Now he is off on to my shoulder – now to the side of the boat. They are perfect little jewels – translucent jade green.

We were very late in getting our dinner – the last chicken roasted with a little butter (bacon, potatoes, onion are all finished and we have no fat) and gravy and rice, followed by stewed apricots and fresh plantains. We must try to get some fresh supplies. I expected to do so here at Nalolo, but the store is some way off and we do not wait here for Mr Dempster as we thought but go on to Kama opposite Lukona, another day's journey. The sky looked very threatening, heavy storm clouds over all, and a good deal of lightning. We did have a few drops of rain in the afternoon, but I risked sleeping out again. Apart from half a dozen drops or so had a lovely deep sleep right through till early morning when I was roused by Miss Bleek saying something to which I was far too sleepy and comfortable to pay attention. After a few minutes, however, it began to rain and we had to move my bed in. Whew! The tent did seem stuffy and airless after my outdoor sleeping. However, there was not much rain, though I should not be surprised if we get it later. The sky is very heavy and threatening. Most of the boys have colds, Miss Bleek's is still very troublesome and I have the beginnings of one which I am trying to ward off with hydrogen peroxide.

END OF THE RIVER JOURNEY

All morning we have been making our way – usually paddling, only occasionally is the water shallow enough for poling – through endless 'fields' of grassland with here and there an open channel of water. There! I've just seen two of my little green frogs sitting at ease on grass leaves – there are heaps of them! An occasional one gets brushed into the boat as we pass. They really are charming – most interesting little fellows. I don't think the feet are webbed, or only slightly so, so probably they do not swim much. Here we stop for the refreshment of a nice, fresh, ripe plantain (9.50am).

The luncheon spot was awful, a wretched little island mostly inundated and the part above swampy. We inspected it, then lunched in the boat. We had just finished when it began to rain, holding us up nearly an hour – boys and all under the tarpaulin. We played Picquet much to the entertainment of the front boys who were absorbed in watching us. As a consequence of the rain we did not reach Kama till 6pm, almost our latest camp, and not a nice one – too near the village. However it might have been worse. Miss Bleek wrote a note for Mr Dempster, which was given to a boy to take across to Lukona first thing in the morning.

START OF THE OVERLAND TREK

Tuesday May 5th—Up soon after 6 as usual, and then started a good old wash, not so extensive as at Seoma however, as we did not know how soon we should hear from Mr Dempster. About 9am a native boy, speaking a little English, and very neatly clad in red print kilt and grey flannel coat, arrived with a note (written 4 May), saying that bearer was to be at Kama that day to watch for us and send a message when we arrived. Simonkanga, as he was named, explained that he had been there at 5 the evening before, but further along the bank at the proper boat place, that he had prepared a place for a camp and that he had not heard of our arrival as our boys had gone to 'wrong place' – the old landing stage. As our messenger had gone long before, this boy stayed and gave some help about the camp – e.g. it soon got very hot, so he prepared a nice little camping spot under the shade of a small tree, cleared it and carpeted it with cut grass, and then moved our chairs into it. We then proceeded to look through the stores and repack them.

About 11 another boy arrived with a second note, saying that the boys with the *machila* poles etc. would be there by 12. This boy was Mubukwana, who is to be our cook for the trip. When we opened the tent at Katombora we found that we had been let down as regards the *machilas*. We have the hammock parts only, without any awning, so we must try to fix something up at Lukona. Soon after 12 the *machila* boys and porters arrived, and the next hour was occupied in arranging loads, fixing *machilas* to poles etc., while we had a hasty lunch. At 1.15 we started – *machilas* first, with blankets over the poles (the stalks of palm leaves – about 10 ft long, – and three or four inches thick – but quite light).

FIRST EXPERIENCE OF TRAVEL BY MACHILA

They had not slung them very well so that we were flat on our backs, head and heels raised to about the same level – off they started at a kind of jog trot – ugh! It was like lying flat on the back of a jogging horse – not at all comfortable. An hour's journey brought us to Lukona, at first through most attractive open woodland, big trees with very little undergrowth, then across vlei ground – grassy of course – where the water grew deeper and deeper, and nearer and nearer. The poles were shifted from the shoulders of the bearers to their heads and still I felt as though I were lying on top of the water.

LUKONA, MR AND MRS DEMPSTER

By and by, on the far side of the watery flat, a ridge of treeland appeared and on its slope, just above the juncture of grass and treeland, a number of scattered villages. Through several of these we threaded our way (this border line apparently is known as the *Mafoola*) till finally we reached a very neat compound hedged with *Euphorbia* – at the far side of the front was the store where we were welcomed by Mr Dempster (short, broad, bald-headed, whence his native name *Sitenda,* meaning bald head, and with spectacles). He took us through the store, which he locked, to the house at the back, where Mrs Dempster repeated her husband's welcome. She is fair, rather plump and looked German, but proved to be southern Swiss from the Ticino. They placed a nice

May 5. Mr and Mrs Dempster, with Miss Bleek trying out a *machila*, Lukona

little room at our disposal and suggested that we should spend a day there with them and leave the following day, Thursday, and this we decided to do. Their abode consists of two or three separate buildings, one for sleeping, near which is our room, one a bath room, and a third with living room, mosquito-proof stoep etc. It was rather nice to sleep quite dry, without even dew, but I did miss the free air! We had supper with them then sat talking till about 10pm.

Wednesday May 6th—The morning was ushered in at 7.30 by coffee, followed at 8.30 by a most substantial breakfast. I had the luxury of a hot bath morning as well as evening. After breakfast we packed things so that Mr Dempster could weigh out loads and get an idea of the number of carriers needed. Then I made my *machila* cover while Simonkanga sewed Miss Bleek's. Mine is composed of my old Cambridge hammock, lined with pale blue sateen (*selike* as the natives call it), while Miss Bleek's is store calico with red sateen lining to the head part. It took some time to sew on tapes etc., and by then it was warm and we were thirsty, so we made lemonade which we had just finished when Mrs Dempster came along to ask us to come and have tea. It was nearly 1 o'clock so we supposed that 'tea' was instead of lunch – not a bit of it, they had only not realized how late it was and lunch soon followed, although we did not really need it.

VISIT TO M AND MME BEGUELIN

Then Mrs Dempster in her chair machila, Miss Bleek in Mr Dempster's and I in my newly finished blue and white one set out to visit M. and Mme Beguelin, missionaries about an hour's march off. We were rather late in starting (3.40) and a storm was coming up. The road ran through the bush skirting the mafoola. Just after starting, I saw a head of the beautiful scarlet thistle Mrs Dempster had spoken about and a little later, the large white 'lilies' (Amaryllidaceae?). Unfortunately I was not able to get either and did not see them again.

After an hour through the bush we reached the mission station and spent a pleasant hour with M. and Mme Beguelin and their two little girls of five and six-and-a-half. In their courtyard were two magnificent pointsettias, all ablaze of blossom. I should have liked to go round this place, but the time was too short, especially as the thunder was growing nearer and nearer, and the sky heavy and livid. Soon after 5.30 we started back along a different road, lower and rather sandy, winding in and out of the Lukona villages (Lukona is really a district). The lightning was magnificent – at one moment the whole of the north-western sky would be lit with a violet light, then there would be a vertical flash, like a crack in the floor of heaven. Half way the rain started, but fortunately only a light shower, and then a boy met us with coats sent by Mr Dempster.

By the time we reached the store it was quite dark. It really was rather a stirring journey. The *machila* boys kept on shouting and chanting. Mrs Dempster said they were saying 'Hurry up! The rain is coming. We must not let our people get wet!' and so on. My *machila* is very much more comfortable now and the awning quite a success. After tidying up a bit I took some writing along to the house, finished my letter to Munnie and Uncle John with enclosures (account of our visit to Falls and river trip). The first was hurried and much more must be written about Nalolo onwards. After supper we sat chatting till past nine, then to bed.

First long trek

Thursday May 7th—Again a frantic packing up and sorting into loads, which proved more numerous by one than Mr Dempster thought, as he had counted a two-man load as one man's. Finally at 10am we got started with our cavalcade of 35 boys – six for each of the *machilas*, cook, waiter, four carrying meal for the boys (these go back after three days) and the remaining 17 porters carrying tent, stretchers, bedding, food for the journey, stores, clothes, miscellaneous camping paraphernalia, books, collecting apparatus and so on. We got sweet and ordinary potatoes, paw-paws, lemons and bananas (the latter two a gift from the Dempsters) at Lukona, all very welcome additions to our larder, as fruit and vegetables are scarce and our Sesheke lemons were just about finished. Then Mrs Dempster kindly let her boy bake us a fine big loaf and gave us a bottle of yeast for our next baking.

We trekked till about 12.30, first through open woodland, then swampy grassland. At the edge of one such change we stopped for lunch for a little over an hour, then trekked on again till 3.30, about five hours in all, and camped near a village where woodland abutted on grass (the water is in the latter). The village was full of Mouwika women with their peculiar headdress, hair in little tags done with clay and castor oil, looking rather like a curled wig – name something like 'Kata'. They made quite a nice camp.

Friday May 8th—Up with the sun – I stupidly placed my bed facing due east, so that I

May 4. *Pleiotaxis ambigua*, Nalolo

273

Pleiotaxis ambigua Sp Moore
Nalolo, Barotsiland.

Herb. Rhodes University

No. 6790

PLEIOTAXIS AMBIGUA Sp Moore

Loc. Nalola, Barotseland.

LEGIT M.A. Pocock.   DATE May 1925

G. & G. GTN.-2,000/10/83

May 9. *Barleria ramulosa.* Dried plant and painting mounted together as a single herbarium specimen. Kassassa, Nalolo

Barleria

M.A. Pocock

had the nearly full moon in my face most of the night. We rise at 6, dress, pack, breakfast and start at about 8 o'clock. The trek was much as yesterday in character. About 11.30 we passed an extensive village, or series of villages. All the women with their wig-like headdresses and brass bangles, and the children ranged themselves in a band and as my *machila*, which was leading, reached them they started clapping and a shouting chant with a dropping cadence, making a simply deafening noise. After preceding and following me for some time they left me and attacked Miss Bleek. Evidently considering her the more important personage, kept it up for a good quarter of an hour – I could hear it rising and falling in the distance.

Meeting with the magistrate
Not long after, after passing another village, we stopped at the edge of the woodland (the villages are all in the grassy flats) for lunch, and we had just started when behold! A white man, in a chair *machila*, with two dogs and a large escort! Our visitor, for such he proved to be, was Mr Warrington, the magistrate for the district (in other words, Native Commissioner, Barotse District, Kalabo), very plump and pink, nice and clean, in khaki shorts etc. He was holding court in the village we had just passed and very kindly came along to see if there was anything he could do for us. His retinue including a 'Messenger' with four stripes, in their queer uniform of grey-black coat, pink vest, belt and elongated grey-green fez lined up and greeted us and then faded into the background. He stayed chatting for half an hour or so and then returned to his work.

Messenger added to the retinue
Half an hour later just as we were preparing to go, up came another 'Messenger', with two stripes, and presented half a dozen oranges and a note, both from Mr Warrington, the latter saying that he was sending the messenger to go with us as far as the border. He could keep the boys up to the mark, see that we got milk, eggs etc. en route, and though he did not say so, see that no harm came to two lone females wandering through Kalabo district! It was very thoughtful of him, and certainly the presence of the messenger made the boys buck up considerably. He superintends the pitching of the tent and provides the boys with entertainment – he is apparently a good raconteur as his deep bass voice (quite different timbre from the local boys) is heard narrating something punctuated by hearty laughs from the boys. The Messengers (really a kind of native police) are not Barotses but come, I believe, from the north-east.

We had done a good three-and-a-half hours in the morning so made a short trek (one-and-a-half hours) in the afternoon. Mr Warrington warned us that there would be a good deal of water and the Messenger made the boys sling the *machilas* higher, so that a recumbent posture was a *sine qua non* – not so comfortable but less likely to get wet. However, we had no deep water this afternoon. Again our camp was at the edge of woodland looking north-eastward across grassland. The full moon rose, a great golden ball, as the sun set and soon it grew really cold, our first cold night. Hope it will quiet the mosquitoes – they were rather troublesome through the grass yesterday and today, so today I put on my 'fly catcher' made of Bessie's green net. It was a great comfort as it just prevented the odd half dozen or so mosquitoes getting at my face. Although not the malarial kind (these only come out at night) the day ones are every bit as annoying in their immediate effects. I feel like the contents of the meat safe inside it, and probably look like nothing on earth, but comfort first!

Saturday May 9th—The morning was cold. I woke with the eastern sky blazing with gold right in front of me and, just as I was dressing, the sun's disc, molten gold, rose above the horizon. I was quite glad of my thick coat. The quinine bottle got broken yesterday, and I had to unpack my little case to rescue the quinine, scattered throughout its contents, from destruction. This delayed my packing considerably. However, by 8 o'clock we had started. The first part of the trek was simply lovely. To begin with, we usually walk a mile or so (the 'so' sometimes is zero or even a minus quantity). This morning we walked for over half an hour through the woodland, and then we got into the *machilas*. It was really quite chilly – there was a delicious cool breeze, and the woods looked quite autumnal.

One of the most abundant trees is a leguminous one with pinnately compound leaves. The ovate, inch-long leaflets turn brown (rather mottled) and drop separately. The whole tree is not unlike a silver birch at a distance, though the bark is grey, not silver, and it was these that gave the autumnal look. Most of the other trees are evergreen. The undergrowth was not very thick, and the ground carpeted with brown leaves. I enjoyed the first half of the morning more than any other part of the trek so far. Then we got to a fairly dry grassland where the flowers were particularly interesting – two ground utricularias, a minute yellow and ditto violet with narrow leaves underground like *U. capensis*, a small *Nemesia*-like plant, the rose *Dilatris* and blue *Salvia*-like plants appeared again. This grass strip went on and on, bounded in the distance on all sides by woodland which never seemed to get nearer. In parts it was dry enough to walk and soon will be quite dry.

At 11.30 we came to a part where a tongue of woodland came down into the grass, and stopped there for lunch. I had taken out a fresh roll of film and to my annoyance managed to lose it at this stop, perhaps while I fished in a lily pool just beyond, full of yellow *Utricularia*. After leaving this spot we did have to go through deep water, but not much of it. We only trekked for an hour-and-a-half and made camp here, at the Magistrate's camp of Toowa, or Tuwa, just near the village of Kassassa where the meal boys go back. Now to transfer today's specimens to the press and then for my bath (4.45).

Meal rations
Sunday May 10th—Such a fuss – the Induna, Muyé, who is evidently new to the job, had to give out the meal for the next half of the journey. Mr Dempster gave the boys food for three days, and sent a sack (200 lb.) of meal for the rest, carried by four boys who are to go back from here. At our last stop they wanted to take on another half-sack of meal and asked who should carry it. Mr Dempster had said nothing about it so we disclaimed knowledge thereof and consequently it was left behind. Well, Induna Muyé

appealed to us to know how many 'cups' (a large two-pint measuring mug) each boy was to have. After calculation we said four. So he started and ladled out the meal, evidently with a lavish hand, for by and by there was a great outcry and there was Muyé, wild and woolly, powdered with meal, an empty sack in one hand, mug in the other, and five or six grieved-looking boys. 'There!' he said (in Barotse, but one could see what he was saying, besides hearing what Mubukwana interpreted), 'I told you you ought to have brought that other meal. Now there is not enough, three, four, five, seven boys no meal.' After a good deal of chattering, Miss Bleek solved the problem – she made all the boys bring back their meal, and put it back in the sack. Then she assembled the whole crowd round the sack and made Muyé give out meal to each individually – mug full – but not heaped up, first two cups each, to all the 34 – then another cup all round. Still plenty in the bag – so round went another half cup – still more. So a final half cup made up the four cups, amid peals of laughter from the boys.

Muyé is rather worth a description – wild and woolly is the best phrase. His hair is woollier than most, and tufts of hair seem to sprout promiscuously over his face and body. He is clad in a ragged piece of sacking which dangles from his waist, and altogether

May 10. Crossing the Luati River. Porters and loads waiting their turn

is as wild and disreputable as the wildest. Another character is 'Grandfather', who was evidently intended to be put upon – easily the ugliest, with practically no chin, only one eye and altogether a beauty. If the pot is behind and we are waiting for it, 'Grandfather' is sent running back for it – something else to be fetched – off he goes with his good natured, foolish ugly face.

CROSSING THE LUATI

This morning again we set out at 7.40, passing through part of Kassassa's domain, two rather prosperous looking villages with bananas and paw-paws. In the first I saw a particularly neat-looking woman kneeling beside her meal basket and mill – beautifully posed so I snapped her. We walked for nearly an hour. It is delightful in the early morning. Soon after 11 we came to a wide shallow river with a village, Sityonamba, on the bank – the Luati River which had to be crossed by *makoro* (dugout). First we had lunch, receiving several deputations – two large groups of women and children who came up and greeted us, and three men who brought a present of milk and eggs. By the way, Kassassa sent us a basket of millet meal and a very indignant little black hen, protesting vigorously. Then very leisurely we proceeded to send the 'goods' across first.

There were three *makoros*, one a good one, long, level and fairly wide, about two feet. The second was long and narrow and had a tip tilted nose, so that when it had more than one man and a load, or two men, its nose went down and the water came in at the bend, and if the load were shifted further back, it came in over the stern. The third was quite a small one, just capable of taking two men, so the process of taking across 34 boys, 20 loads or so, the *machilas* and us, was slow to say the least.

The river was fairly wide and except near the far shore, shallow, with bushes and grass growing here and there on sand banks – most beautiful. After all the loads and most of the boys were across, the *machilas* went, one at a time in the largest *makoro*. The *machila* was slung on the shoulders of the two steadiest bearers who stood in the boat, with the paddlers, dodging the projecting end of the *machila* pole, at the stern, paddling very gingerly. First one, then the other passed safely over. Then the remainder of the boys, except Simonkanga, Induna, Messenger and one *machila* boy (the fat one) who was left to bring milk. Then a pile of dry grass was brought down, placed in the one large *makoro*, and we were invited to enter, which we did, kneeling on the grass and then sitting back on our heels. Incidentally one of my feet got into an awkward position and before we reached the other side I did not know what to do with it and daren't shift. Then the Messenger took the stern paddle himself – another boy tried the bows but water entered so he got out – and very carefully paddled us across.

By the river bank just below our camp was a long pole (?totem) with part of the upper half of a crocodile skull on top. Mubukwana said it was because they had killed a crocodile and put it there 'to show people'. After crossing we made a short trek (2-3.15) across one grassland and through a bit of forest at the far edge of which we camped. I did some collecting in the swamp on the far side, got the lower part of the twining blue flowers, which proved to be a *Utricularia*, and several other rather nice little things. Collecting,

May 10. Crocodile skull on pole, Luati River

washing, labelling and putting away specimens took till past 5, then I had to hurry to put my bed ready and have my bath, etc. We finished dinner at 7pm and went straight to bed as there were a lot of mosquitoes. It has been much warmer this afternoon and this has brought them out, unfortunately. Near the village were large fields of manioc. I asked what it was and by and by the messenger and a headman of the village brought a plant to demonstrate. The roots are thick and white inside. They peel them and then eat the nutty centre. I asked for some to try – it is not at all bad, nutty but with a starchy taste and very juicy.

PORTUGUESE BORDER

Monday May 11th—A much warmer night, but the mosquitoes did not worry us as I feared they would. The boys were very merry, chatting and laughing and singing till fairly late. The cook and waiter are, I think, mission trained boys. The 'wild animal' seems to have only one tune, with variations – a falling cadence rising towards the end and then again dropping. Last night these two started whistling, in harmony! It sounded very well – I think it was *Far away, far away*. Then they sang *Jesus loves me*, followed by *Oh my darling, oh my darling, oh my darling Clementine!* And snatches of *And we go marching on*. A truly catholic selection! They were in a hurry to start this morning – said we had a long way to go – but in the end we started at the old time –7.50.

The early morning trek was again delightful, through most beautiful grassland, in the centre of which were one or two clumps of palm, a new feature of the landscape since we left the river. In the west hung a pale, cloud-like moon above banks of cirrus, really all in one stratum but seen from below they appeared as if in several layers. The lower, more distant ones were varying shades of softest grey while the higher ones were just

May 11. 'The messenger and his satellite attendant left us at the Portuguese border'

tinted with a warm tone between pink and gold. The sun slanted behind us and a fresh breeze stirred the grass now and again. Round the horizon stretched a belt of trees, up to which the tawny grassland extended. In the distance the grass is brown gold. Near at hand the water glints between the tufts which are green at the foot shading upwards with purple, green and gold stalks, to the golden heads – tawny gold. The longer heads show some purplish, others yellow against the sky, which is palest turquoise low down, shading softly upwards to a clear pure blue.

Among the grass grow many flowers, of which the most beautiful is a delphinium-blue *Utricularia,* the flowers rather larger than a violet. This twines round and round grass stalks, the beautiful two-lipped flowers, with darker centre, hanging from the grass. The stalk below is a fine thread and buried in the soil are the leaves, threadlike green tips resting on the surface, while below are delicate little traps to catch and use for food the teeming microscopic life of the swampy soil. There is also a yellow-flowered twining *Utricularia* a little smaller than the blue one. Then there are other bladder plants standing erect, some only one inch in height, others reaching as much as six. There is a tiny white one, the upper lip just tipped with yellow, a pure yellow one a little larger, another yellow, fringed with brown at the back, and a small purple one with white and yellow markings. Near the fringe of the grassland is a bushy plant with *Dilatris*-like flowers of a peculiar pink–vieux rose, really – and a very handsome *Salvia*-like bush with spikes of deep blue flowers. Then in the wet places are a straggling plant with large pink flowers, and another with starry white blooms.

It is most tantalizing to dash past in the *machila* and see these all out of reach. Every now and again I get out and walk, but the 'road' is non-existent, merely a narrow one-man footpath. Etiquette is that the *machila* follows its passenger if she is walking, the four spare *machila* boys follow it, then the next *machila* outfit, then all the porters, headed by Messenger, Induna, Cook and Waiter.

Now my *machila*'s place is at the head of the procession. Hence if I stop to pick a flower, unless I just grab it in passing, the *machila* stops on my heels, and all the procession is held up. If I stop several seconds, the second *machila* passes me – if longer still, the rest of the procession. Then when I get back in – 'Hurry up, we must get back to our place. *Ah we! Ah!*' and off we go at a racing pace announcing our coming, each one passed having to step sideways from the path to let us past. No rest, till panting and streaming with sweat, my *machila* boys are once more in their place. Naturally, apart from the laziness induced by *machila* travelling, one thinks twice before upsetting the procession! They go at a good round pace – porters a good three miles an hour, *machila* boys more. When I lead I have to put my best foot forward, and my short legs have hard work to keep up a pace which I feel is as good as theirs, and is too, but an hour of it is about enough at a time.

Today I've walked quite a lot – quite two hours in all, I am sure! If a branch has fallen across the path, it is not moved. No, either we step over it, or if it is too large for that we make another path round it. All explained largely by single-file travelling in large parties. Soon after leaving camp we met a man, woman, boy and two goats. The latter turned sideways on at the sight of me 'in the blue', *maa'ed* and had a good look. We saw their tracks in the firm (wettish) sandy path for a long way, then passed their camp, a freshly burnt-out pile of logs, and near by a hole scooped in the sand, about a foot deep and full of clear water, an easily obtained water supply.

At 10.30 my *machila* stopped where a few trees cut across the grassy plain. Under one of them was a conical grass hut – the Portuguese border. Here we stopped for midday rest, as the Messenger had to go back from here. We gave him a present of some millet meal (Kassassa's gift!), some salt and a shilling 'to buy more food'. I took his photo with his attendant satellite, Miss Bleek gave him a note for Mr Warrington and he departed. At 12.10, we started again and did a long trek, till 3.45, when (after passing our first village in Portuguese territory at 3.30) we reached the second, Kalumbeya, rather extensive, well-built and keeping pigs! The first we have seen, some *red* with long rufous hair, others black. On our arrival all the inhabitants, men with fluffy hair, women with their queer headdresses of hair done in knobbles with castor oil and black or red clay, collected, ranged in two groups, men and women separate. Our boys and the women squatted, and then the women greeted us. All clapped gently – I responded, and then they clapped again, a greeting for each of us. Thereupon they gradually dispersed, and some of the women brought meal etc. to trade with our men.

By the way the 'Luati' has been with us all day, in the form of extremely badly dried or smoked fish which our boys have brought along with them. At lunch time Simonkanga, very indignant, brought the store box: 'The boy who carry sis box, have fish in his pot, put it upside down on box. Run all over.' I should think it did – it simply stank, talk about 'a very ancient and fish-like smell'! Fortunately, practically all the contents were protected by the bread tin at one end and Miss Saltmarsh's old chocolate box, containing our temporary supply of dried fruit, at the other. So out came all contents, tins and box had to be washed by Simonkanga (the delinquent being made to fetch supplies of water). As for the chocolate box, I took the fruit out, turned up my nose and threw it away, to be eagerly caught up by the cook at whose feet it had fallen, with the words 'For me?' I said 'Yes, for you or anybody so long as you take it away'. Since then he has carried it proudly under his arm.

Tomorrow, they say we reach Ninda, where I suppose our whole cavalcade will be re-arranged, these boys going back to Kama. Now (5.20) the crate containing my press has at last arrived and I must put away my specimens, then make bed, bath etc. Our menu today may be of interest: Breakfast – fried sweet potato (left over from last night) and eggs, coffee, marmalade, toast (butter finished two days ago). Lunch – whole small pawpaw and lemon each. Tea. (Yesterday we had Ovaltine made with milk – a fine drink). Dinner – the perennial 'chicken', this being the first day, roasted in the pot with sweet potatoes, plantain or oranges (Mr Warrington's gift). Coffee. Not a starvation diet, is it?

THE NINDA RIVER

Tuesday May 12th—This morning one of my good *machila* boys, one of the best, went sick, with a bad leg. I think he strained it yesterday as he stepped down into one of the swamps. It is hard luck as he is a good steady boy and as we have had to engage another boy from the village it probably means a shilling off his pay. Miss Bleek· had the new boy, a bearded ruffian, and I had her 'fat boy'. As a consequence my *machila* was much slower than hers and had to take second place. About 10.30 we reached the river Ninda and from there on our way more or less followed the beautiful little plain of this winding river which flows almost due west. By the way, South Africa publishes a most excellent map of central and southern Africa, which Mr McGill has – the best I have seen. We crossed the Ninda by a rough bridge – poles laid across supports – at a most picturesque village, quite the prettiest spot we have seen on the whole trek, and well built. Then we kept up the left bank – chiefly at the edge of the plain.

The river winds in and out of the wooded ridges and every now and again the path leaves the edge of the plain and cuts through the bush where the spurs are specially long. The plain itself is most attractive – all shades of green and yellow with touches of bronze, gold and purple, and here and there a splash of blue in the grass, where a tall, dark *Salvia* grows. The river winds in and out, mostly hidden till one comes close. On both sides are ridges of bush country, the dark green of which stands out behind the plain. Although the character of the country remains much the same, the detail varies constantly. To me it is a country of infinite variety, and it is sufficient just to do nothing all day but enjoy the changing panorama – nevertheless I have managed to do a certain amount of Portuguese as well as a good deal of collecting, and pressing – though under difficulties. We made a long day's trek – 8 to 12 in the morning, 2 to 5.30 in the afternoon. Then we camped in the bush at the edge of the Ninda.

A lone paddler in his dugout canoe on the Zambesi

'Lazy boy' was two hours late at the midday halt, arriving with my bedding at 2, just when we were ready to start. Consequence – a scolding which we instructed should be given him, by a combined force of Induna, Cook and Waiter, after which his soft load was taken from him and given to the 'man with the wire comb', instead of the crate which was left for Lazybones. Further consequence – at 7pm when we had finished our dinner and it was pitch dark, Lazybones had not yet arrived. We spoke to the Induna, whose business it really is to see to stragglers. He told two boys to go back and look for him. They refused, saying it was 'full night' so he had to go himself armed with lantern, matches etc. To our relief in half an hour or so he returned with the boy whom he had found, 'going so slow as a snail', says Mubukwana. Mosquitoes very bad just after sundown, but by 7.30 all had disappeared. Boys want to start <u>very early</u> tomorrow so as to reach Ninda by midday.

NINDA MISSION. MR AND MRS McGILL

Wednesday May 13th—Up at day break, dressed, packed, finished breakfast and started by 7am (by my watch which is I think about 15 minutes slow). To my annoyance found that one strap from my press in the crate had disappeared. Couldn't possibly have come off without considerable effort on someone's part – I do not think the Lazy Boy took it himself, but probably he stopped to talk somewhere and the strap was taken by someone. Hope many of my specimens have not dropped out. Walked a little way in the delicious fresh air of early morning, but not far as we did not want to keep the boys back. Such a lovely trek, particularly the first cooler part – river and bush again. We crossed the Ninda again and got into bush, then up the right bank on sandy ground, partly through bush, till suddenly we came to a wide sandy clearing or road, cut sharp off – 'Portuguez' said my carriers. Half an hour up that and we reached the first building – the Fort, we hear. Then another half hour and at 12.30 we reached the mission station, where Mr and Mrs McGill welcomed us. Here I am writing this and shall send it back by the boys for Mr Dempster to post, as the mission mail bags are, they say, usually very full. I hope it will arrive safely and interest you all.

End of Volume I. Ninda 3.30pm. Wednesday.

THE DIARIES
Volume Two

Angola, Ninda to Kutsi
(May 13 to June 22, 1925)

'Then began the ticklish job of getting us and our belongings across – most nerve racking at times – e.g. when all my specimens in the crate, precariously balanced on one edge, went wobbling perilously across in the dilapidated old canoe.'

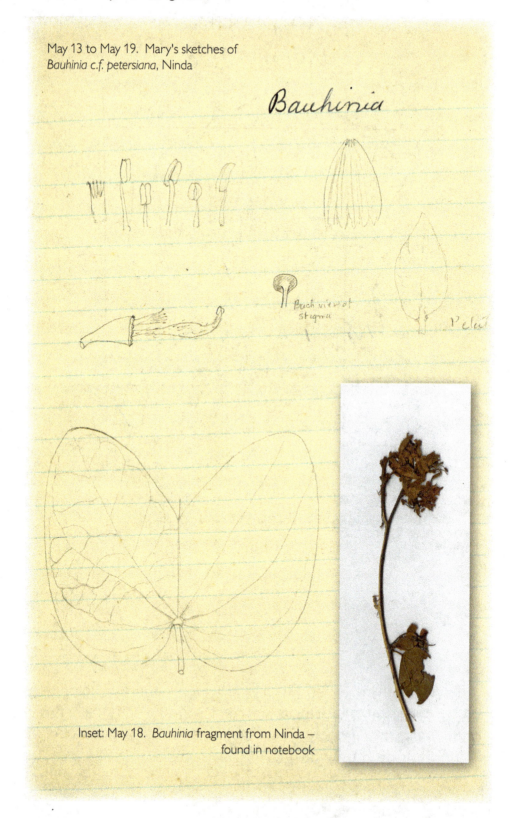

May 13 to May 19. Mary's sketches of
Bauhinia c.f. petersiana, Ninda

Bauhinia

Back view of
stigma

Petal

Inset: May 18. *Bauhinia* fragment from Ninda –
found in notebook

AT NINDA MISSION

Wednesday May 13th—Well here begins the second volume – the first having gone back with our carriers to Lukona for Mr Dempster to post. The last porter struggled in a couple of hours after our arrival – Mubukwana made us hot water for our baths. Mr McGill gave them food for the return journey and two bags of fish as an extra relish, and Miss Bleek tipped the Induna and the two personal boys, and they departed to the other side of the compound for the Induna to give out the food. Judging from the sounds, their extent and loudness, he got into as great a snarl over his measuring out as he did at Kassassa. However about 5, the drum sounded for service and they evidently thought it was time for them to depart which they did, and silence fell around. Mr McGill came up to ask if we would care to go to service. Miss Bleek went but I unfortunately was in the midst of an extensive wash of myself and my belongings and could not go. Hope to tomorrow if we are here.

Thursday May 14th—It was quite strange to be sleeping in a house again – don't like not being *en plein air* and could not get to sleep for some while. Mr and Mrs McGill suggest that we should stay on here till Monday and then go on to Muié. It is very good of them and the stay of a few days will be quite acceptable. I hope it is not too great a tax on their hospitality. They are very busy re-plastering the houses (with a mixture of black clay and yellow sand) after the rains and making sun-dried bricks. The women are bringing loads up in baskets either on top of their heads or carried by a strap around their foreheads – many have their babies on their backs as well! A lot of small boys are carrying as well. These Wambundu have very little in the way of clothes – mostly merely a skin hanging behind and a much shorter fringe in front. Calico has been an awful price – what costs 1/- in Barotseland has here been costing 4/-.

COMMENTS ON THE PORTUGUESE

They give the Portuguese a very bad name here. There has been a stream of emigration into British territory, not due, as Statham suggests, merely to economic reasons. The Portuguese, a degenerate Latin race, are by nature cruel, and do not treat their native dependants well. Of course financially things are much against them – the escudo, formerly nearly equal to 4/- is now at 100 to the £1 and goods very expensive. Then they use forced labour, i.e. make the natives work for them, willing or not, and pay them

Money and measurements

The monetary system of both Britain and its overseas possessions was based on the pre-metrication British system of pounds, shillings and pence, abbreviated: £.s.d. A shilling was often indicated thus: 2/- and an amount of shillings and pence would be written thus: 2/6d. Once they reached Angola, a Portuguese territory until 1975, the monetary unit was the escudo ($) with coins of very little value being referred to as piastres or centavos. Measurements are given in feet and inches, the imperial measures of length before decimalisation. 1 foot (ft or ') = 30.48 cm. An inch (ins. or ") = one twelfth of a foot.

nothing and in addition there is a great deal of taking of women by the soldiers in charge of the forts, either Portuguese or half-castes, a very poor lot. Near the mission stations of course things are a bit better but even here at nights the soldiers go out and seize the women. The traders are evidently a poor lot. They (the Portuguese) are very much afraid of the Union which they say wants Angola – a very good thing for Angola if they could come under British rule – Union rule I'm not so sure of! The Portuguese seem to have all the best harbours of Africa, which means that they are practically of no value. Lobito Bay they say is a magnificent natural harbour, very deep and almost landlocked. Beira and Delagoa Bay too are very good.

The commandant of the fort (about equivalent to an under sergeant) has been invited in to supper this evening to meet us so that he can have a look at our passports, if he wants to, endorse them, and generally 'pass' us, also incidentally so that Portuguese Rule shall not be insulted! It saves us going up to the fort at the entrance to the village.

INFANT MORTALITY

A sick baby was brought to the house about 9 o'clock to be doctored. It had been taken ill in the night and was very bad – with flu, Mr McGill said. By noon the poor little mite was dead. There has been very high mortality among the babies the last two years, what with pneumonia, flu and smallpox. Most of the small babies died, and now most of the women have others of 4-6 months old. They, both men and women, are very fond of their babies, the McGills tell us. Miss Bleek had the two elder children here in the afternoon and later 'Bebe' as Jessie calls her, joined them.

Then there was another event – they killed a pig. It made a horrible noise, much too close here for my liking! Everyone is glad to have a change from chicken occasionally but the chief reason for killing a pig is for the sake of the fat – it is apparently about the only source of fat up here. After tea we got our hot water – had baths and then went to evening service – tonight the funeral of the baby who died this morning. The people collected in front of the hut, with pigs, chickens and dogs running about in the background, then walked up in single file to the graveyard, just a piece of bushy grassland, where the service was held.

MEETING THE CHEFE

On the way back, just at sunset we met the *Chefe* with two traders (one of them coloured, the other curry powder yellow!) who had walked down with him, to all of whom we were introduced and with all of whom we had to shake hands. After a somewhat one-sided conversation, we left the others, the *Chefe* (Sr Joachim) going on with us to the house. We sat on the stoep for a time, then had supper after which the *Chefe* who was evidently rather shy, and simply dying for a cigarette (not of course allowed in a *Christian* household!) was made happy with the gramophone, having first inspected our passports and made a note of our names, callings etc.

Friday May 15th—Spent the morning mending Mr McGill's camp chair and amusing the children. The latter occupation considerably retarded the former, in fact it was not finished till after dinner when Baby had gone to sleep and the other two gone to 'the other house' with Miss Bleek. Then I indulged in a rest – quite miss the *machila*! After tea went to the evening service in the schoolroom – quite interesting. The 'school' room is a large building with a high thatched roof. Walls of poles placed rather far apart so that it is nice and airy, and an earth floor on which the congregation sit, on small logs, or their own stools which they bring along on their heads.

Saturday May 16th—Again spent the morning with the children. Forgot to say on Thursday that we went up to the dispensary and saw Mrs McGill (who is a trained nurse) give out various remedies. The people are very poor physically and constantly ailing. Colds and itch seem to be the most common ailments, but bad eyes, bad gums (pyorrhoea?) and many other ailments abound. After tea I went out to photograph a flowering shrub outside the compound. I thought it was a *Sterculia* (lovely pink and white *Azalea*-like, have been finding it ever since Seoma) and was somewhat taken aback to find it was a leguminous plant! Then took a view looking up the valley – hope it turns out well, it looked lovely!

Sunday May 17th—Our program – Church in the morning for everyone including the babies, who were given flowers to play with and were really very good. Afternoon – Miss Bleek again played nursemaid, I put my papers to dry, had a good long laze then took the children off her hands.

Monday May 18th—Had arranged to go out to a village about an hour's march on the other side the river to try to get in touch with some Bushmen there, so had early breakfast and started out in the *machilas*, but before we reached the gate of the compound, word came that the Bushmen had got wind of our coming and had flitted, so that was off. Instead we went down to the river and collected and photographed. I expended one of Mr McGill's old rolls, and took exposures (plates) of 1) a mauve-flowered bush, possibly scroph, 2) China-blue *Salvia* (new), 3) Delphinium-blue *Salvia,* 4) *Dolichos* with enormous pods. It was very lovely down along the crystal clear, fast flowing stream, which is full of small fish.

Arrival of post

Afternoon and rest of morning was spent as follows: – 1) putting away, labelling and sorting specimens, 2) a short rest, 3) packing, 4) writing letters, 5) bath. Then we went down to supper and found the post had arrived and was being sorted. My share was large – three letters from Munnie (April 14, 21 and 25), one from G.V., one from W. Bolus and a couple of others, 3 papers (*Weekend Argus*), and specs, so I did not do badly. One from Oudtshoorn brought a snap of *Conophytum* Kopje – such a contrast to this country where I have not yet seen a stone since we left the Zambesi! We left early as the McGills had their mail to see to, and we letters to finish. Another baby died last night – taken ill in the night – apparently lungs – dead in the morning, the 4th or 5th who has gone suddenly.

DEATH OF A PRISONER

A man was brought to the dispensary this morning, very thin, his chest in a dreadful state, all cut up – spitting blood and so ill that he was 'stupid'. Mr McGill made enquiries, found he had been a prisoner and one of the soldiers had beaten him with his sjambok. The man had as a result got ill, and seeing that he was very bad, he was turned loose. The old Chief is taking care of him.

Tuesday May 19th—The man died in the night. No good saying anything to the authorities – the *Chefe* was spoken to once in a similar case and his reply was 'Well I did not do it. One of the soldiers did.' If higher authorities are informed, the letter is ignored.

DEPARTURE FROM NINDA MISSION

Morning went in packing and arranging loads, complicated by the arrival of three Bushmen! They had heard that we would not hunt them but give them salt so two men and a boy came along – one typically Bushman, the others with more traces of admixture with their neighbours. Miss Bleek set to work asking them names etc. and got a good deal of material, and I photographed them with her camera and my Vest Pocket Kodak. Finally by noon the carriers were all given their loads and instructions, and we went to dinner. Then, while Miss Bleek settled accounts with Mr McGill, I snapped the children. Just after 1 we started – nine *machila* men and twelve porters – we have to get

May 19. Three Bushmen arrive at Ninda Mission

three more *machila* boys on the way. Two boys were sent on with loads (store box and sweets etc.) on Saturday. We got two of the extra boys this evening from the village near which we stopped on the Ninda.

Wednesday May 20th—Half an hour brought us to the old Portuguese huts now in ruins, after which we crossed the Ninda. A well-built (for this country) causeway led across it, and then we got on to a Portuguese 'road' – a wide track cleared through the bush. As Miss Bleek was crossing the bridge the third extra *machila* boy came along, so now we have our full numbers. The first part of the trek was delightful, the last very hot and we both felt the heat. I had a headache which however did not last long after we stopped. We trekked through woodland all morning, first going up 100 ft, then dropping 150.

THE LUATI VALLEY
About 2pm we came to the valley of the Luati River, on the far side of which on the hill is the Portuguese fort. We stopped for some time to let the porters catch up, then started to cross the river. (Here I found 2 spikes of purple *Gladiolus*) A causeway led half-way across the valley and then we branched off and trekked up along the river for some way, finally crossing it by a bridge where a crowd of women were at work. They stopped to sing a greeting. My *machila* men stood still and beat time till I got tired and told them to go on. They carefully avoided the fort! We camped in the bush above the river, but below the villages, one of which had a very fine chief's house. The camp was not high enough to escape the mist which was heavy along the river.

Thursday May 21st— Left camp 8.05 – heavy mist last night, bed very wet – especially the mosquito net. The river bed (Luati) looked very lovely with swirls of mist, while blue smoke was rising above villages on the high ground beyond us. The way was up over woodland, partly along a broad road which stretched before us. Then it became much overgrown and degenerated into a narrow path and it was difficult to get the *machilas* through. There was a long thirsty stretch, where we stopped once to rest, then went on again. By and by 'Moma!' was said pretty often, everyone hurried up encouraging one another with varied cries and down a steep little hill we rushed.

There was a spring of sorts, in a kind of swamp – beautiful crystal clear water. They said it was the Luar but later I found that the river itself was well across the flat. This apparently was the source of a tiny tributary. Here I found two kinds of sundew – one with a flat rosette of leaves which catches those spotted clear-winged slow- flying butterflies and the other the stalked form already found; also another small-flowered climbing *Utricularia*, yellow *Polygala* and a *Lycopodium*. A short trek along the river bank brought us in under half an hour to our camp in open bush some way from the river to avoid the mist. We do about five-and-a-half to six hours actual trekking each day.

THE WAMBUNDU
Friday May 22nd—These Wambundu are very different from the Barotse (Gengi people

= easterners here!). They are not so well built physically, and do all their carrying with their heads or shoulders. This shows well in the *machila* boys. Usually when they want to shift the pole to another shoulder or the head they call in the assistance of one of the others! Head takes regular turns with the two shoulders and they have a plaited grass pad. Their arms are weak and lifting power small. Manaza is our personal boy and bosses up the carriers. Many of the latter are oldish men and quite a lot have small beards. They have brought four or five small boys to carry their personal gear, one of whom is carrying my green rucksack – that, with my bed etc. did make a very heavy load and I am not surprised that the old chap found it too much, though he has not complained. This little boy shall get some salt as a reward.

Today has so far (1.10) been delightful, with a fresh breeze, though with hot sun. Last night was very cold. We trekked through wood, then steeply down to the Luar, here a deep clear stream like the Ninda, and up again on the left bank past a deserted Portuguese post. Then the sandy road led up through a belt of wood and into an open grassy stretch which took about an hour to cross. I stopped at the bridge to get two yellow waterlilies (first I've seen) and then tried to take a snap looking down the stream where there were masses of dark blue *Salvia* – only to find my shutter was jammed. I do hope I shall be able to put it right – it probably has some grit in it and needs oiling.

We stopped in the middle of the plain in the shade of a tree for half an hour's rest then came on here, another 55 minutes. The plain ended and we are on rising ground in open wood country once more. The wide road continues more or less all the way but sometimes very much overgrown, otherwise very sandy. Occasionally the way is varied

May 22. *Nymphaea sulphurea*, Muié

May 22. *Nymphaea sulphurea*, Muié

L ikande
mankante

Nymphaea sulphuria
Gila,
[404] Muié, Angola

May 26. *Ottelia lancifolia, now O. ulvifolia,* Kutsi River

Tianamema
Ottelia lancifolia Rich.
R. Kutsi, Muye.

401

by a little friendly rivalry between the two *machilas*. If they are near to one another when two fresh carriers go on they often have a race, lasting 5 or 10 minutes until one has to stop to change the pole, when the other forges triumphantly ahead for the time.

Saturday May 23rd—The afternoon yesterday was much cooler than the day before, though still hot enough. We made a short trek to the Chikoluwe (really spelt Cikoluwe). At the top of the hill my carriers, in advance as usual, stopped and invited me to get out. Just beyond I saw some trees with masses of red 'bloom', so I invited my old man to climb up and get me some, which he did forthwith. The 'bloom' turned out to be fruit – a hard nut-like fruit with five large bronze-red sepals, colouring much like that of autumnal hydrangeas, and very handsome – the leaves are large and tomentose on the underside, the young flowers small. The boys call it *mulangandombe*. After gathering this, Miss Bleek's '*tipoia*' (pronounced *chipoya*) or *machila* (Barotses call it *machila*) came up and we went on to the tune of a vigorous chant past a Portuguese trading station where the trader, hearing the noise of our coming, came out and had a look – I just caught sight of his yellow face – and then disappeared into his house.

PUTTING AWAY SPECIMENS
Passing the post, we turned to the left and went a little way along the river bank to a spot where two or three trees cast a little shade and there made camp. I got out my press – fortunately the boy with the specimen box was always one of the first in – changed the papers of all in the press and put in the specimens collected on the way from Ninda to Muié much to the interest of all the carriers and some native girls who came along to sell meal – damsels of 14 or so with red clay headdresses, most neat and dapper.

Then a couple of boys brought offerings – each had a small bunch of very small zinnias and a vivid orange-red cluster of Thymelaceous (?) flowers. The former I rejected, but after consulting the 'blue book' asked '*Kuli?*' (where?), whereupon a shout of laughter went up from the surrounding boys, the tall one giggled and with a flirt of his skin train (goat I think) turned about and proceeded to lead me some quarter of a mile along the path where sure enough, it was growing wild in the bush. Then just next to my press, I found a *Protea* with a couple of dead flower heads, only the second I have seen (the first was way back the other side of Ninda). Altogether it was quite a satisfactory day. It took me a good two hours to get the papers changed and dried and the new ones put away. And then I had not touched those in the box, some of which are still quite damp.

APPAREL OF MEN AND BOYS
The boys have no calico at all. The men wear a whole duiker or goat skin back and front, fastened by a broad leather belt. Trekking they sometimes fold the skin up half-way, or just leave them dangling. The little boys have a single skin – some quite nice cat or jackal skin – fastened by two paws in front, and two paws and the tail at the back. The little tails sticking out at the back of the waist and wagging as they walk have quite a jaunty effect!

Many of the men stick a bare knife with a sharp pointed blade into their belts (handle down) in front, most dangerous looking.

We made a pleasant and leisurely trek – the morning was cold, though there was far less mist on this river than on the Luar or Luati, so we started energetically (8.10) – 5 minutes along the edge of the swampy plain brought us to the causeway. This was simply river soil lifted slightly above the surrounding level and held in position by short piles – logs some 4" in diameter driven in close together on each side. The pools along the sides of the causeway had small yellow waterlilies, very pretty, and a tiny white flower (2 specimens of which I collected in the Zambesi vleis but which are spoilt). I collected some of the latter and had just taken my hand out of the water when up swam a 2-inch brown leech, evidently hoping for a meal!

It took 20 minutes to cross the causeway and then came a long hill, straight up which the *strada* led. By this time the temperature had risen considerably and the sun on our backs was, to say the least, warm! However, with the exception of a brief spell in the *machila* to put away my specimens, we continued walking till the top was reached – well over an hour from camp.

Near the top on the right we saw a lot of grass-like plants, into which the boys plunged, coming back with a lot of red fruit (*litundu*), some of which they gave us. The *litundu* fruit is almost 2 inches long, rounded on one side, flattened on the other. On being split open the inside is found to consist of a soft but firm white substance in which the black seeds are embedded. It is very pleasant to eat on a hot trek – cool and pleasantly tart, most refreshing and not unlike the 'cream of tartar' fruit though nicer. We saw a lot more of the plants later on.

Then one boy plunged into the bush on the other side and brought out a branch of a shrub bearing orange chilli-shaped fruits which he ate in the same way as the *litundu*, and

Litundu

Aframomum is a genus in the ginger family, Zingiberaceae, represented by about 50 species, 5 of which are found in Angola[1]. The name *litundu* refers to one or more of these species. The fruit-pulp of these plants is edible and has a slightly peppery flavour. According to Stanley marijuana smokers in the region used to eat the fruit to mask the odour of the smoke from their employers[2]. Until the beginning of the 19th century, large quantities of *Aframomum* seed, variously known as 'melegueta pepper' or 'grains-of-paradise' were imported into England from West Africa. These spicy seeds were used for flavouring and medicinal purposes instead of cardamoms and as an additive to some alcoholic beverages[3].

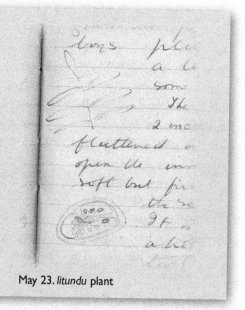

May 23. *litundu* plant

called *mulianhyma*. They have much less flavour and are not nearly so good though very pretty. *Litundu* is evidently monocotyledonous – *mulianhyma*, a dicotyledon with dark green leaves. Another bush with striking rough orange coverings round the fruits they said was *mundeemba* but whether that was its name or a descriptive term I do not know.

Muié Mission

About 11.30am Manaza pointed to a blue line beyond the next rise and informed me that was '*donga* Muié'. We went on for about an hour over the rise, then stopped in the best shade we could find and had lunch, after which we went on to Muié – down the long slope to the tune of very vigorous chanting, getting good views of the extensive settlements on the far side – post on the right, mission buildings and stockade centre, villages left and at the back, across the pretty little swift river, where we filled our water bag at a spring with a spout fixed below it (the mission's water source). The bridge is very dilapidated, as is often the case here – there we met the boys returning from school and most of these turned back with us (a couple helping with the *tipoias*) and escorted us to the mission station. Apparently this must be reached in style at the swiftest pace, so as the hill was steep, my boys at any rate were puffing and blowing. However they got there first by a few yards!

There we were welcomed by Mr and Mrs Bailey (both Americans) and Miss Jensen (Norwegian). They had expected us the evening before, it appeared. After a very welcome cup of tea and 'strawberry short cake' they took us to the nice little two-roomed guest house, with bed and stretchers all nicely prepared for us. Then with Mr Bailey's help we paid off the carriers (with the exception of Manaza who elected to stay till Monday) and gave them a present of salt each, as they wanted to start back at once.

Evening prayers

In the evening after dinner there were rather prolonged, but quite interesting prayers – interesting both from the sidelights they let in on the life and interests of the little community and the insight into the characters of the participants – each one took part both in 'quoting Scriptures' and in prayer, i.e. Mr and Mrs Bailey, Mr and Mrs Reinhart and Miss Jensen. The Wilsons and Lewises are away, the former with their family taking their holiday camping in the bush, the latter I think on account of Mr Lewis's health. In 'special subjects for prayer' Mr Reinhart included besides the subject of the permit for the removal of the Kunjamba Station, the case of a boy who had apparently accepted their teaching and was 'living a life of grace', when suddenly it appeared that, though quite young, he had two wives – 'I don't know how he got them' brought from Miss Jensen the remark 'Perhaps he inherited them', one older than the other. The boy was quite willing to have one only, but he wanted the young one who would have nothing to do with him, while the elder was quite willing. It struck me as humorous in the extreme, and in addition jolly hard lines to insist on an already much married man giving up all but one wife – not on the man but on the rejected wives! However, in the circumstances, one could not well argue the point!

EMPIRE DAY

Sunday May 24th—One of the hymns in church was sung to the tune of God save the King – which I did not recognize at first! Two-year-old Helen found the service rather long, and sang lustily in the wrong places, particularly when her father was preaching. In the evening it was altogether too much for her and her mother took her out. She is a sturdy little thing with a somewhat surprising habit of saying 'Dam!' when she means 'yes.' It sounds awful! Her father said she started it while he was away on trek. Her mother retorted that she was just beginning to say 'yes' nicely when he returned and she took to the other monosyllable! Presumably she picked it up from something the natives say or probably it is her attempt to say 'jam!' Mr and Mrs Bailey, Helen and we had tea and supper with Miss Jensen.

SHUTTER TROUBLE

Monday May 25th—Spent most of the day in a vain attempt to put my shutter to rights – interesting to see the working of its innards but no help towards the making of photos. I shall try Mr Reinhart and if that fails send it to Kodak's and hope to get it back in time for use on the return journey. Bother it.

In the afternoon, by arrangement, Mr Bailey took us up to the Fort to pay our respects to the *Chefe* and get our passports and all put right. The way lay along the pebble road recently built with forced labour. Most interesting river stones, (first I've seen since below Seoma!) agates, jaspers, cornelian, rounded quartz etc., of all colours, green, white, red. Mr Bailey and I made quite a nice little collection as we walked along, ostensibly for Helen's amusement but really from the collector's instinct whetted on a little geological knowledge! The *Chefe* said our passports had to go up to Kangamba, the head post, to be registered. We were reluctant to part with them but there was no help for it. We urged the need for haste and he promised to do his best to let us have them back in six days' time – also inspected our *guia* from Ninda and pointed out that it was said that the men were to be paid in his presence at Muié before returning! Neither we nor Mr McGill had noticed it, fortunately.

THE MISSION PRINTING PRESS

On the way back we inspected the printing press, where they are busy on St. Matthew's gospel – Portuguese and Mbundu on parallel pages. It was very interesting. In Mrs Wilson's absence the three boys in charge are running it themselves – setting type, taking proofs and finally printing all on their own, while Mr and Mrs Bailey do the proof reading. The translation is being done by various missionaries and several of the boys themselves, since though speaking the language well enough for ordinary use, the former have not really mastered the idiom.

I returned to my tussle with the shutter while Miss Bleek went to the evening service at which a boy preached and handed in a bundle of fetishes which his converts had given him. By the way, yesterday an old man got up and in a language which even Mr Bailey did not know properly, made a long speech. He came from a neighbouring tribe it appears, and his speech was to the effect that he had lived 'in sin' but now that he was old and weak he wanted to settle down and live a good life – query, attracted by the 'truths of Christianity' as expounded by the missionaries, or by the comparative material ease and greater comfort found in the neighbourhood of a mission station? After dinner Miss Jensen came in with her knitting and we talked, looked at photos etc. for some time. Waited, expecting evening prayers, but apparently they have those only on Saturday and Sunday.

MOVE FROM MISSION TO CAMP ON THE KUTSI

Tuesday May 26th—Miss Bleek was anxious to get to work so we decided to go to camp across the river near the cattle camp close to which they say the Mueri Bushmen have moved. I should rather have liked a day or two longer at Muié, but it could not be helped.

Miss Bleek went down to have a look at the collection of fetishes (baskets, bits of feather etc.) and then packed. I went up to the Reinharts' house, was directed up to the Dispensary for Mrs Reinhart and found her just finishing there. She sent me back to the carpenter's shop where I found Mr Reinhart up to his eyes in work – superintending three boys shoe-making and half a dozen others busy at woodwork – making window frames etc. – in a delightfully large airy shed. I stayed looking at wood, watching the work etc. for some while then went back to their house for tea. He had a look at the shutter but could do nothing with it, alas! But he undertook to grind my knife for me.

Then I went back, took meal and yeast round to Miss Jensen and had a lesson in bread making, leaving Miss Bleek to pack and sort all the household goods! Finally I had just half an hour to do my packing before lunch – quite inadequate! As a result, I held up the departure for half an hour afterwards i.e. we left at 2.30 instead of 2, with 8 *machila* boys, cook boy and 14 or 15 carriers.

CROSSING THE KUTSI

An hour through very lovely bush country brought us to the Kutsi, into which the Muié runs. We could see the blue line of the valley from the station. There we found one tiny dugout with no paddles. However, much shouting, poling across the river, and more shouting ended by producing two more dugouts, one with a large piece out of its side, a man and some paddles. Then began the ticklish job of getting us and our belongings across – most nerve racking at times – e.g. when all my specimens in the crate, precariously balanced on one edge, went wobbling perilously across in the dilapidated old canoe (which had already sunk once, fortunately when a boy only – Massovi – was in it and then only in 3 ft deep water!).

However everything went over without mishap. Then just as the mosquitoes began to get tiresome, the two *tipoias* went over, flat on top of one another on the best dugout. Then at last Miss Bleek went, after a couple more loads had been taken over. After that, the remainder of the loads and boys, all save two, went over in the other two canoes. The mosquitoes grew worse and worse as the sun sank, and I began to grow weary of waiting. At last, back came the best canoe propelled by the official paddler. In I got, and down I knelt on very wet grass. Then across over the deep pools, and round to the left into a narrow channel, scarcely a yard wide through the reeds. It wound in and out and round about, getting narrower and shallower, but at length we were poling over almost solid weed (broad-leaved *Potamogeton?*) and the end – black mud – of the channel hove in sight with my *tipoia* and four boys waiting. I was not sorry to get off my knees, out of the dugout and into the *tipoia*.

By this time the sun had set and the colouring over the grassy flats was lovely. A short stretch of ankle-deep swampy ground was succeeded by a comparatively firm path through the marsh. Then a difficult bit across a little tributary stream with deep clear pools, and firm ground again brought us in the gloaming to the stockade of the cattle camp (about half a mile from the end of the boat crossing) inside which I could just see the gleam of our white (comparatively) tent. There I found Miss Bleek and the rest of the boys. They had pitched the tent there before her arrival – not ideal, but it did very well for the night.

SETTING UP KUTSI CAMP

Wednesday May 27th—After an early breakfast, we found the 'back door' to the camp – i.e. a hole in the fence easily closed by slipping in 3 or 4 timbers, and selected a camping site on the fringe of the wood, a little behind and to the north of the stockade. Then we took Massovi and with the help of my note book in which were written phrases supplied by Miss Jensen, I very quickly made him understand what we wanted: – '*Mu tingeni nsinge*' (build a hut) – marking roughly on the ground the shape and orientation of the shelter. '*Mu tingeni helu tia m'balaka na veti na mil a*' (Build a roof for the tent with sticks and grass). Delighted at last having received an order in an intelligible language, he assented smiling and we returned to the camp to shift to the spot we had chosen.

This accomplished, the next thing was to sign on workers, which had to be done in style – 10 for grass and 6 for 'sticks', some of them our yesterday's carriers, the rest men from the nearby villages, of which there are two, one on each side, about half a mile or less apart. These then disappeared in various directions. I corrected the orientation of our shelter by compass (the sun answered just as well!) and then Massovi set to work to clear and level the ground. The poles brought by the woodcutters were firmly planted in holes in the ground to form walls on three sides, with smaller poles laid across. Then the roof and walls were thatched with long grass collected in the river plain, with leafy branches in addition in the case of the walls. The tent was pitched in front and work finished for the day at about 2pm. Built: a roomy and shady shelter. Cost: 16 escudos.

Meanwhile the headman (*Capitea*) of the cattle camp had gone off in search of *Vasekele,* and about noon he returned with 2 men, 2 women and 2 little children, one in arms, the other walking. Miss Bleek set to work, though rather bothered by the noise of the building and numerous spectators, who 'assisted' at very close quarters. Various vendors of merchandise – fowls, eggs etc. came. I tried to buy a small wooden pestle and mortar to grind salt etc. in – '*U nji tungila cini na muisi ari vindende*' (make me a small

May 27. Kutsi camp

pestle and mortar). The result was not so good as in the case of the hut – by and by two women came up, one with a wooden meal mortar so heavy she could hardly carry it and standing about 2 feet, the other with the pounding stick as long as herself and two-and-a-half inches in diameter!

My next attempt or rather our joint attempt was to get a sieve to sift the maggots out of the flour. This produced a woman with a winnowing basket and a deep conical basket. We rather doubtfully turned some of the flour over to her tender mercies. She was given a seat in the shade outside the little fence the three boys (Telosi – cook, Fuelu – wood and water and Massovi – man of odd jobs – a luxury!) had constructed around three sides of the kitchen, made to wash her hands (by Telosi!) and then she started operations. The winnowing basket is held slantwise and shaken steadily – all the chaff and light stuff being gradually thrown into the conical basket, the heavy meal staying behind. It was quite successful until I found her pounding up in the mortar the remnants, including the grubs, and then resifting the results!

ARRANGEMENT OF THE CAMP SITE

Thursday May 28th—The *nsinge* in itself was quite all right, but the tent still needed a roof. So another lot of 13 boys was engaged to *Mu tingeni helu tia m'balaka* and meanwhile I set Massovi making two corner shelves in the *nsinge,* one for each of us and then a long one right across the back wall for cups, shoes etc. By 3pm we had the *helu* built and the tent put up a yard further out so that we don't trip over the guys every time we go in or out. The arrangement is now that of a large open shelter facing south while half of this side has facing it a similar lean-to, open on the south and east but walled in on the west – thus we have our back door to the west, while the south end is practically open except for the low wall below my window. I sleep near the south end, Miss Bleek in the tent. Altogether a very comfortable, roomy little abode – total cost 30 escudos, including shelves and carpet, i.e. 6/8d at the rate we changed our money. Now we can start getting to work.

THE BUSHMEN VISITORS

Today the Bushmen included the two men (Kavikisa and Golli-ba), three women – one small and yellowish with natural hair, one darker and with her hair dressed *a la* Mbundu with red clay and oil. Her 'man' Golli-ba also has the top of his head similarly dressed but in black. The third, an old woman (very light, a warm ochre yellow tinged with brown) and a little boy, Golli, son of Golli-ba (whence the latter's name, 'father of Golli'), Golli being evidently his eldest child.

While Miss Bleek questioned them in turn, I photographed the others with her camera, each in turn and then a group with all except the older man Kavikisa (yellow-red brown, small and unmistakably Bushman type of face) who was with Miss Bleek at the time. The photographing was superintended by Massovi and Fuelu who much enjoyed being allowed to look into the viewfinder afterwards at the group – then each of the *Vasekele* had a peep, much to their delight, but I doubt whether this is the first time they have been photographed.

CAMP KITCHEN

Friday May 29th—The boys have made quite a neat little kitchen. They cleared the ground and built a three-foot wall of grass and sticks round three sides. In the middle is the fire – two logs with the fire between their ends. The logs gradually burn away – being pushed closer as need arises. We have two guard fires, one outside the back door and the second on the far side of the tent, but evidently there is not much fear of intrusion by wild animals here, not even hyaenas – it is too civilised! Cattle camp just below us, a village five minutes to the north and another ten minutes to the south (the river here runs nearly north and south). These fires, too, are made of huge logs – usually four placed with one end of each in the centre from which they radiate like spokes of a wheel. At night the centre heads are pushed close, a little dry grass and sticks placed between them. The embers are blown upon by Fuelu and the fire starts. He attends to them once or twice during the night, and in the morning empties our wash basin of

May 28. Kavikisa with top knot, Kutsi

May 28. Four Bushmen, Kutsi

water over them, and usually pulls the logs apart so that they do not burn away during the day. The Bushmen were very late in coming – noon. It appears that they live one or two hours from here, on the Mueri. They have not moved, apparently, as Mr Bailey thought. We understood from the *Capitea* that they were coming to sleep here, but so far there has been no sign of them doing so.

Saturday May 30th—There was a good deal of distant lightning in the south-west and south last night. During the night it rained, not very heavily, but several showers. I woke with a start, at the sound of movements in the *nsinge*. I found that the boys had come in (two of them at any rate) and were lying on the far side near the door. Fuelu getting up to tend the fires had roused me. That is all very well, but it won't do, so this morning we told Massovi to get two more men, get sticks and grass and build themselves a shelter, which they have done. It is a three-sided lean-to, long side to the east, open to the west. That ought to stop further intrusions. Probably we shall not have any more rain, but it is as well to be on the safe side. The Bushmen did not come at all today, but the *Capitea* again says they will come to sleep nearby. Miss Bleek had plenty of work to do, however, going through the words already obtained.

Hardly a day passes without Noki bringing us a note and something or other from either Miss Jensen or Mrs Bailey – eggs, *makovi*, etc. Today is Mrs Bailey's birthday. Miss Jensen mentioned it a day or two ago, so we decided to send in some preserved fruit and a few of the boiled sweets. I spent most of yesterday morning making a cadurie for the latter and a little box for the former of white paper and a blue ribbon. Noki brought back a graceful little note of thanks from Mrs Bailey.

BUSHMEN ARRIVE TO STAY NEARBY

Sunday May 31st—A most peaceful morning – no crowd of sightseers or vendors of merchandise (thanks I suppose to the missionaries!). About noon, to our relief, the Bushmen, all but the old lady and the woman with two babies who did not come again after her first visit, arrived with the *Capitea*, who said they were coming to stay. They at once went a little to the south and started building two small huts of leafy branches – about 5 feet high and incompletely roofed over. That is a relief – now they are close at hand.

SYSTEM OF SUPPLIES

Monday June 1st—Today Noki brought a parcel from Miss Jensen – a 'taste' of the party, two slices of a most delicious birthday cake, four milk rolls, some marshmallows and a bottle of ginger beer – all most delicious. Our system of supplies – chickens, eggs and sweet potatoes, and meal for the Bushmen, is gradually getting organised. I have obtained a list of prices from Miss Jensen, which is a great help. At present supplies are pouring in (all except milk – there are no cattle, barring the mission cattle, and though there are plenty of goats they apparently don't milk them) and we have had to put a limit to chickens and eggs – latter are of course most useful. Market price: dessert spoon of salt per egg, or 1 escudo per 10 eggs. Chickens – two-and-a-half escudos.

MORE RAIN

It rained again last night, hard – and came through the roof all over. We were thankful we had made the boys build themselves a shelter! The morning too was overcast and things took a long time to dry up. Then we told Massovi to get a couple more men, gather grass and reinforce the roof. This they did, but I doubt whether it is yet rainproof. The rain is most unseasonable of course – it is a couple of months past the time when the rainy season should be over. This last week there has been a spell of mild weather – coolish days (I mean not scorching – some clouds and wind, just enough to be pleasant) and warm nights, very different to the cold spell a couple of weeks back, which is, I believe, normal for this time of year. I cannot get my specimens dry, which is a nuisance. There is a heavy cloud of mist along the river, the upper layer of which sometimes reaches us, so that even under the shelter, net and bed cover are dampish in the morning.

Tuesday June 2nd—A farewell gift of *makovi*, a delicious gooseberry pie, some guavas and two autograph albums for contributions arrived this morning. The Baileys and Miss Jensen leave today for their month's holiday. They are spending it at Ninda and camping.

The Bushmen are getting organised – they stroll round about nine, 'work' for about an hour or sometimes two, when they announce that they have a pain inside and want to go 'a *mmm* –' i.e. are hungry and want to go and eat. Then they are given meal and depart for a couple of hours. They show a tendency to think that is enough and Miss Bleek had to be stern and tell them 'no work, no food'. Since then they appear regularly, both morning and afternoon.

PLANT COLLECTION

I have started going through my plants — such a lot are mouldy, alas! — and have done a little collecting chiefly in the river swamps where I have found yet other *Utricularias* — one with beautiful little peltate green leaves which are just above the ground and covered with a hemisphere of clear translucent jelly; the flowers are small, white and only about 2–3" high. The second is one which has both the typically finely dissected floating leaves and the long simple linear leaves of the terrestrial forms, and large yellow flowers standing 4–6" high. I also collected a small *Chara* with bright orange fruits, and in mounting it found a second most beautiful delicate little one, also fruiting, of which I must try to find more. The river bottom must be thoroughly collected. There are, however, not many flowers about. One morning we walked up the hill at the back and got a few specimens — that tree with the red bracts was up there. I want to go up one day soon and photograph it.

KEEPING THE BUSHMEN ENTERTAINED

This morning Miss Bleek was driven nearly distracted by the whole crowd of Bushmen and two or three Wambundu sitting around while she tried to question one. So I sent off the Wambundu and set the remaining Bushmen round the remains of the south watch fire. Then I provided Golli and Baita-de with needle and silk and set them to threading labels for my specimens. Then when they were threaded I introduced Golli to a pair of scissors — tiny ones. He put his fingers right round the handles and tried to use the point as a knife, so after I had shown him once more, his father, Golli-ba, came to his rescue. A few minutes later I looked at them and this was their procedure: Golli carefully measured the necessary length of silk and held it up solemnly. Golli-ba, squatting in front of him, then opened the scissors (3 in. long!) with both hands and severed the thread, after which Golli carefully tied the knot. When she had finished threading, I gave to Baita-de a knife! She got Kavikisa to help her eventually. They actually moved out of the sun into the shade of the tent eventually! I gave Golli a plate to put his completed tags on, so Baita-de (= Baita's mother) thought she ought to have one too. Failing to make me understand, she borrowed one, a gay floral one, from Fuelu.

BAKING ATTEMPTS

By the way, on Saturday I set sponge (using the sifted flour) for a raisin loaf and put it in a hole with grass — no good. Much too cold and in the morning I had to put cinders underneath to warm it up. After doing this and mixing in a little more yeast and some baking powder to start it, it rose a bit, and about midday I mixed in flour and raisins, and left it to rise again, baking it late in the afternoon. It did not rise much but quite a nice light raisin loaf resulted. I shall have to try some other method of raising the bread — my hole is too cold.

Wednesday June 3rd—Today we finished the last of our bread. Miss Jensen's loaf and the four milk rolls (simply delicious toasted) had lasted us well, and there is only a small piece of raisin loaf left. So this evening I mixed wholemeal sponge and put it in the pot to rise – put cinders in the bottom of the hole and covered it up – hope for the best.

I started going through my specimens and writing up the descriptive slips for each. Such a lot have got spoilt, most of the water ones, and a good many of the others. It is most difficult to get the sheets dry and I have not changed them properly. I should have put in two or three days more work at them at Ninda. However, one lives and learns! Yesterday evening took our evening walk down towards the river. Seeing us start armed with collecting tin, Fuelu volunteered to accompany us and carry the tin. He showed great and awestruck interest in the collecting.

The magnifying glass

I amused a large audience in the afternoon with my little magnifying glass. The expressions and contortions as they tried to look through it were most amusing, and at times the competition to view was so keen that two or three heads met in their owners' efforts all to see at once. The most successful exhibits were firstly a piece of print and then a bit of an arm – the glass placed flat on the arm of the man who was looking. This brought roars of laughter. Finally I had to say 'Kunahu' (enough, finished) and put it away for another day.

Visitors arrive

In the evening yesterday I heard the sound of joyful welcoming in the kitchen (I was in the tent having my bath) and we were told that Telosi and Massovi's wives had arrived – there were four girls, one big and shock-headed like himself was Massovi's wife. Telosi's was small, quite a young girl. Who the other two were I don't know – one I think was Fuelu's wife (the second one apparently!). The Wambundu and Luchazi (our three boys belong to the latter tribe) of course all practise polygamy which is sternly put down by the missionaries as dreadfully sinful.

Fuelu

Fuelu was in the catechumen's class but backslid awfully – he had married a wife much older than himself, and had now put her away and married a young girl. Of course this is all right among the heathen, but for a Christian! So he was put out of the catechumen's class and frowned upon morally, although Mr Bailey gave him a character as a 'gilt-edged carrier'. He is certainly a good worker, very cheerful, and an accepted wit and mimic. Quite an asset to the camp.

Massovi and Telosi

Massovi, too, is full of humour and mimics well – he is a strong pillar of the church and very reliable. Both of them sing falsetto, soprano, but particularly Fuelu. Telosi seems to have less natural wit and to be more sophisticated. They have regular prayer meetings in the evening with intervals of prayer at midday and before retiring for the night. On Sunday they asked permission to go to the neighbouring villages to hold services.

Massovi, I think, does the preaching and is the leading spirit. We had given two of them leave to go to Muié if they liked, but they did not wish to. Then in the evening and again yesterday (Tuesday) they had service with sermon etc. round the kitchen fire – men only – the women sat around the watch fire and did not take part in the singing. Two of them have their hymn books with them. They are all anxious to learn English, and ask the names of various things. Telosi speaks some Portuguese. Today (Wednesday) Massovi asked leave to go away for two days and left with the girls going northwards – possibly to visit relatives.

BREAD-MAKING BY TRIAL AND ERROR

Thursday June 4th—Spent a busy day changing drying papers, doing bread (such bread!) etc. The sponge was cold this morning so I rejected this hole as no good, had it filled up and dug a smaller one in the sun near the back watch fire. After leaving it there till afternoon I made up the bread and put it back to rise while I tried a sponge cake. Unfortunately I made my oven (the pot in the centre of the four logs, with cinders on top) too hot and the cake had not time to rise properly and got burned all round. Such a pity as the inside was delicious though a trifle soggy. Then I stepped backwards and went into the hole on to the bread. I doubt whether that made it rise less – anyway I then proceeded to bake it and made real good 'doodgooi' It is quite nice toasted. In future I shall try setting the sponge early morning in the sun and see if that does better. We went for a short walk northwards about 5 and I got some nice specimens, not new, but ones I badly wanted, to replace others which had got mouldy.

THE DAILY ROUND

Friday June 5th—A cool, breezy morning, no good for pressing flowers, so I have been writing up this diary which had got much behind, having the water bottles (which were distinctly 'high') thoroughly cleaned, the cups scoured inside etc. I had just had a fowl killed when a boy arrived with a fine big fish (we had one, a different kind, one day last week and it was delicious), somewhat of the barbel tribe, which we bought for 2 escudos and a teaspoon of salt.

By the way we bought another table hen yesterday who promptly escaped from the *camba;* after being put back once, she escaped again, and stopped in the centre of the kitchen just before hopping over the log to lay an egg. Then both of them were allowed to run around the camp. A fierce and long chase was necessary in the evening before Telosi could catch and cage them. Now one is dead, and by the sounds I should think the little lady has laid again. This really is lovely weather, but I do want a scorcher today to dry my specimens. Now for lunch.

Various items omitted from the above occur to me:

THE PESTLE AND MORTAR

After repeated requests for 'Cini na muisi' (mortar and pestle), one morning I noticed Fuelu questing about with the axe among neighbouring stumps (a tree has no chance to grow large near the villages and heaps of stumps nearby attest to that fact) and by and by I saw him shaping out a 10 inch block into a *cini*. He spent most of his day at it, thoroughly happy. For hollowing it out he borrowed a curved strip of iron with a cutting edge he kept sharp by repeated rubbing with his knife. Part of the time he held it in his fingers, then set it in a handle made of a piece of wood. For finishing touches he borrowed my penknife, and then did some poker work with a spearhead. The *cini* took him most of the day. Next day he started on the *muisi* for which he used a harder wood. Finally towards evening he brought the completed outfit much larger and heavier than I intended (more overweight luggage on my return!) but most efficient for grinding salt for which I wanted it. The *cini*, 8″ high and five across top and bottom is of rather soft yellowish wood (*musese*) and the *muisi* 18″ of hard reddish heartwood of *musivi* sap wood – yellowish, rather similar to gum.

EGG CUPS MANUFACTURED

Then one day, two days ago, we had boiled eggs for lunch for the first time, and felt the lack of egg cups. We made shift with two blocks of wood with a hole roughly cut. Yesterday I drew a diagram and explained what I wanted. I told him to use the musivi wood which he did, but unfortunately used the light sap wood instead of the rich-coloured heart wood. Final result of a day's hard work, including several cuts on his thumb, two quite efficient little egg cups with our initials pokered onto them. I have been trying to put a polish on to mine.

MAKING RUBBER

The other day I was getting the names of plants from the boys and Fuelu informed me that from one they made rubber – the bush with fruits like a bird's egg, green (yellow when ripe) with brown spots – with latex. I told him to show me and he pulled up a bush, heated the root in warm ashes, pounded the root with a stick till the outer layers were loose, stripped them from the wood which he threw away, and then pounded them thoroughly. After this he washed the pulp first in cold then in hot water, then pounded again – after more washing the result was a tiny lump of rubber. It rubbed out pencil quite well. I told him to get me a bigger piece, but he was busy then and did not do so. This morning, after fetching water he was not given any work in particular and by and by I saw him hammering something with a block of wood he had shaped. Later he had a thin sheet of the rubber which he hammered and pounded, doubled, hammered, washed, etc. and finally brought me a nice lump of excellent rubber which I have docketed as a specimen. He has to be doing something – if nothing else offers he is twanging his 'piano' or cracking jokes with other boys. He may be a sinner but he is a decidedly engaging one!

A MEASURING POLE

With Fuelu's help I have got a comparatively straight pole, peeled and fixed erect in the ground at the back of the *nsinge* to serve as a measuring pole for the *Vasekele*. The sand

June 5. Curved strip of iron and knife used to carve the mortar and pestle

June 5. The completed mortar and pestle

June 5. The pestle, made of *musivi* (*Guibourtia coleosperma*) wood, used to grind salt

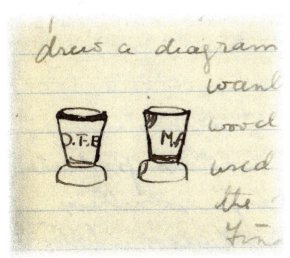

June 5. Fuelu's egg cups with initials pokered onto them

Bush Rubber

Although Mary wrote the name *kambungo* phonetically the correct Luchazi name for *Landolphia camptoloba* (and other species in the genus) is *mumbungo*[4]. The plant is a scrambling shrub or climber reaching about 3 m in height found in Angola, Zambia and Zaire. Raw rubber is produced by heating and beating the roots to extract the latex that is washed and beaten again a number of times. This was a labour intensive process requiring about sixty man-hours to produce a kilogram, worth about four square yards of cloth (stuff) at the turn of the century. From the 1870's 'bush rubber' became increasingly important as an export commodity from colonial Central Africa with Benguella being a port of transit for this raw material. The production of 'red rubber' (alluding to bloodshed, not the colour of the latex) was the basis for King Leopold's ruthless subjugation of the Congo. In Angola Bushman slaves were engaged by the Bantu speaking peoples to collect and prepare the product. It was traded to the Portuguese from whence it went to sheath the electrical wires then being strung all across Europe. Rubber production was a major industry around the confluence of the Cubango and Cuito Rivers in 1898[5] completely replacing money as local currency[6] thus explaining why plants were so scarce in 1925 and providing context to the tenuous relationship between Fuelu's people, the elusive Bushmen and the Portuguese almost thirty years later. In 1900 the German botanist Hugo Baum found that 'root rubber' (*Landolphia thollonii*) was completely extinct at the Kutsi and Kubango Rivers[7]. The Luchazi also use an infusion of the edible *mumbungo* fruit as eye drops in conjunctivitis[8].

Preparation of rubber:
1 Roots are heated in ash.
2 Pounded well till bark is loose.
3 Loose outer tunics stripped from wood.
4 Pulp beaten well, over and over.
5 Washed in cold then hot water.
6 Result - efficient pencil eraser.
Bushman name *Bungwé*

June 5. Rubber plant, Kutsi

at the foot I smoothed and levelled as much as possible, then Miss Bleek and I measured one another for practice, also because she said she was 5 ft 6 ins and I could not believe it! Our measuring brought me out at my old sixty-three-and-a-half inches and her an inch taller, at which I was surprised, as I felt I was the taller! Then I smoothed the ground and left all in readiness.

That afternoon I saw a group of women coming past the back of the *nsinge* on a line which would lead them right over my levelled patch, so dashed out to divert them – a most successful diversion. The leader, a buxom shock-headed damsel with a baby on her hip, leapt off at a right angle like a startled buck, with a shrill scream of terror from the infant, followed by the rest. I could not help it, but simply exploded, followed after an instant by the rest of the women! No. 1 was by that time a dozen yards off, but seeing everyone laughing came back looking somewhat sheepish but thoroughly enjoying the joke against herself. It makes me laugh whenever I think of it. They all came around to

June 5. Fuelu (second from right) playing the thumb piano, Kutsi

the front of the hut and had half an hour entertainment – i.e. watched us and our doings – ending up with having photo taken (Wednesday).

PHOTOS TAKEN

I finished my first dozen plates by taking Bushmen (Kavikisa, Golli, !Kõ) and then a group of the 'kitchen' buildings, Fuelu, Telosi, all the women and all the Bushmen. I tried to arrange them, but got a long straight (in parts!) line of very solemn faces and wondered how on earth I would ever raise a smile! However the anxiety of the women brought Fuelu's tongue out, and when by sign language I told him what he was doing, the resultant 'smile' was so broad that my whole line wriggled!

THAT FISH…

That fish stood us lunch, dinner and breakfast! The river fish are really very good and not very 'fishy'. The first had scales; this one was a very big, black thing, somewhat of the barbel tribe, with a huge mouth edged by a long feeler at each corner. No scales but curious eccentric markings on the head and red flesh – which cooked light, and very red blood.

'MADAME'

'Madame', our little hen, made herself a snug little nest in the grass wall back of the kitchen and there laid a neat little egg. She wanders round the camp scratching for food in a most homely manner. She is small, has a topknot and brown, partridge-marked plumage, a typical little Angola hen! I am not going to have her killed for a good long time if I can help it!

Measuring and drawing the Bushmen

This morning I measured some of the Bushmen, much to their interest. !Kõ (the small yellow woman), Baita-de and Golli-ba (who had to be checked in a tendency to stand on his toes – so soon do they learn tricks!). I don't think I mentioned that I tried my hand at portraiture – a tiny one of Kavikisa followed by slightly larger ones of !Kõ and Golli-ba, the latter not at all bad! I want to try larger ones; it would be very good practice for me. Some of my paper is horrid. I hope the sheets I brought up will be all right.

Before Massovi left I had him rig me up a rough easel. I cleaned up one side of a paraffin box to use as a drawing board. Now I am waiting for the inspiration! I started a sketch just before sundown this evening – a short way to the north, looking up the 'golden valley' of the Ninda which, lovely always, is specially so at sunset, when a shaft of red-gold light shoots down along the river on the far side, lighting it up against the blue of the wooded height beyond. Looking south-east, the blue of the Muié *donga* (= river) breaks the line of high ground. The river itself is marked by a strip of vivid almost emerald green.

More problems with bread

My bread will not rise – my latest attempt is true 'doodgooi', but I rather enjoy it toasted. I must try again tomorrow. Today I ventured on a sponge cake. Alas! my oven was too hot and before the poor dear could 'sponge' a crust so hard and rigid formed that the inside remained nearly solid – but very nice when the quarter inch of charcoal is removed!

Since Wednesday we have had the luxury of a pint (my big mug full) of thick white, but very creamy milk, most delicious, and it's wonderful how far it goes. It does us for lunch, morning tea, evening and breakfast coffee, not stinting either. I used Miss Jensen's nice little pie dish as my cake tin. I shall try another sponge, eggs being plentiful.

Trading with safety pins

Safety pins are at a premium – we have been using them for filing Miss Bleek's index slips. Fuelu came along with two eggs and asked for a safety pin. This we gave him in exchange for the eggs. Then *Capitea* with four eggs – since then there have been repeated requests. As mine cost 6d for 50, the profit is somewhat large – at the same time the supply is limited! What a pity I did not bring three or four bundles.

My cooking is improving – the other day I made a perfect light, frothy omelette (not much 'ome' – chiefly egg) and this evening, an equally nice tasting, but not quite so well-shaped, jam omelette, with some of Miss Jensen's most delicious strawberry jam. I am not progressing much with my specimens – I have been rather lazy, but am trying to get the wretched things dry.

Lemon syrup and marmalade

Saturday June 6th—Set sponge for bread yesterday, but it would not work – so after warming it in the ashes I packed it round with sacks in my specimen box and left it in the sun all day. I expect it will get cold again tonight! Then as our lemons showed signs of old age I made a bottle of lemon syrup. It seemed a pity to waste all those beautiful skins, and we have plenty of sugar and not much marmalade, so I sliced up the skins, put them to soak and left them overnight. They look quite promising. Then, to finish up the culinary account, I tried another sponge cake. This time I went to the other extreme and made the 'oven' too cold to start with (the 'oven' is always the cooking pot) and it took nearly an hour to bake! Consequence was that the bottom sat still, then when I warmed things up the top rose up and up into a beautiful light frothy sponge, a rich honey brown colour! Very delicious, and better than the first attempt but not yet perfect. Perhaps the third attempt will be better.

We measured Kavikisa this morning. About 11am Massovi returned, accompanied by a large black sow, and two families, presumably going on to Muié, all of whom, including the sow and half a dozen fowls proceeded to settle down as a matter of course, in our camp! The boys have given their shelter to the women of the party, and are sleeping round the kitchen fire, while the men of the party have made themselves a shelter of branches to the left. There are drawbacks to being so near Muié!

The afternoon passed – I went to sketch, Miss Bleek going on for a walk. I had not time to do much before the light went, and the moon (all but full) rose like a great salmon balloon above Muié valley – a lovely sight. The sky was blue above, shading through yellows and reds to a deep dusky blue low down, with sufficient light from the sunset to light up the 'golden valley' and the green trees in the foreground. I made a rapid pencil sketch and must try to put in the colours from memory. Then home to baths and dinner.

ROAST CHICKEN

At least I've got Telosi to roast the fowl properly - he will flood the pot with water. Now I boil the potatoes separately, roast the fowl, then make the gravy (very good thick gravy it is too!), put the potatoes in and let the whole simmer while we have our baths. The *makovi* is an excellent 'duty' vegetable. What a lot about food! But I am interested in cooking experiments. Also, it is rather enlightening as to the possibility of 'living on the country'.

BUYING CLOTH

Telosi had cleared away and we had just settled down to a game of Bezique, when he and Massovi came and squatted down with eager faces, and Telosi reeled out a long and very rapid speech. We stared and he reeled out a bit more. I still stared but Miss Bleek tumbled to it – they wanted to see the *tanga* (cloth). We have fixed the price at nine escudos (2/-). It cost a shilling at Lukona and we have had to pay porterage on it all the way. Besides we do not wish to make it too cheap. The Portuguese, I believe, charge something like 4/- the yard. Well, Miss Bleek got it all out – white, blue and just to see what would happen, red. The latter was quite out of it. The blue was much admired, but put aside for the future. Telosi, being the only monied man of the party ($19.00 to his credit) went 'bust' over the white. After much discussion and calculation – I finally got paper and pencil and put down: 1 yd. – \$9.00, 2 yd. – \$18.00, showing just how much one yard and two yards were. He said he wanted a little more, to the *donga* he said, pointing to the fold of the stuff. Then he did a calculation of his wages, on paper adding 12 to 7, but putting the 7 under the 1, getting a result puzzlingly disproportionate to what he knew he actually had. Finally, with our assistance he decided he had \$19.00 and would have as much as that would buy. The extra escudo would buy 4 inches, which brought him almost to his old *donga*! I cut it professionally, letting the scissors run, and the lines being crooked, landed on the far side of his *donga*, so he got more than his 2 yards 4 inches.

Finally they retired satisfied, after Fuelu had tendered his smallest safety pin (he has done quite a large trade in pins and had four or five) in exchange for a needle. This we gave him together with some thread, and told him he could keep the pin. This however he indignantly refused to do, saying '*Pone*' very emphatically, so that was that. Then we returned to our Bezique.

CANDLE MAKING

Our supply of candles is getting low and we must go gently. No more using of bare candles or reading at night. I took all the soft run-over wax from the lantern and returned to my old childish delight of 'making candles' – used some darning cotton as a wick and made a tall, thin, grubby-looking, knobbly object which shed quite a good light but required repeated snuffing to keep it in order. However it helped us out. We sent a note to Mrs Wilson to ask if candles could be obtained at the store and had a reply saying they could at the cost of three escudos each – over 6d each! She had three packets she could let us have if we liked. They will be most welcome. She sent us out (more food!) some guavas, roselles, Cape gooseberries (which we have been eating hard!), some *makovi* and other vegetables. We hope they and the Reinharts are coming out next Saturday to spend the day.

Sunday June 7th—We expected post yesterday, but it had not come, so we decided it must be late. The morning started with bread making. My sponge, after two nights and a day is still nearly flat.It looks as though it had started to work and then taken a chill. However, I made it into one raisin and one plain loaf, and tried 'raising' it on the hot ashes. This time I think it was too hot. I baked in the afternoon – result: 1) a good substitute for a cannonball, alias a raisin loaf and 2) a flat, semi-lunar affair, good old 'doodgooi' again. Then I put on the lemon peel chips to simmer, and sat down with a magazine to enjoy a little well-earned recreation.

SUNDAY SERVICE

Came Telosi, kneeled down before us and eagerly commenced to ask us to do something. It turned out to be a request to go to the neighbouring village and hold Sunday morning service. After some discussion we agreed but said someone must stay in the camp as the goats had already tried to eat my sack which was covering the bread. So Fuelu stayed behind. Massovi and Telosi led the way, each with one of our chairs. Then came Miss Bleek and I and behind Kavikisa, Golli-ba and Golli (who by the way had been given a little instruction by Massovi first in theology – the name Njamba (God) had cropped up several times – and then in reading, starting with phonetics – vowel sounds alone, then with 'b' in front – at which Golli eclipsed his elders.

We trotted off to the big village, a quarter of a mile to the south, wended our way in and out of patches of well-grown tobacco (grown on ridges of heaped-up sand about 6 ft long and one-and-a-half high) and picturesque huts, to the centre of the village where a fine tree (the one with the little brown and white flowers, all over twigs, etc.) cast a welcome shade. There our chairs were set down. We seated ourselves with our two dusky henchmen, the evangelists of the occasion, and the three *Vasekele*. A few villagers gathered round, and we started a hymn. Massovi and Telosi set the tune and we joined in as soon as we got the hang of it. The language is fortunately a very easy one to sing. We were accompanied by the hammering of bark for bark blankets by a man and a boy, whose hammering logs were nearby, and the dismal howling of a small thin dog of the whippet type for whose nerves our melody was too much. Quite a fair number gathered, and then after three or four hymns, Telosi got up and preached at a great rate. Then Massovi followed, one or other of them prayed and the meeting finished. When we first appeared, a small babe of one-and-a-half or thereabouts, thought the end of the world had come. His eyes nearly started from his head at the sight of such strange beings!

POST RECEIVED
When we came back it was about time to see about lunch, and then my 'oven' being freed, I baked my dismal bread and made marmalade in the enamelled billy. The latter quite a success, though it took a long time. Then a short walk and half an hour's sketching, and we returned to find the very welcome post – one from M. just before sailing with the snaps of the Falls, some not at all bad – one from Miss Heyworth, another from G.V. with enclosures from F.E., one which I had already seen, a request from the Grahamstown Library for my subs for 1925! A note from Mr McGill with stamps, as change from the 2/- I sent, my lens having cost 10d as I expected – and my missing strap, with a note for Miss Bleek from Mr Dempster! He says 'Lazy Boy' is not quite *compos mentis* and so is erratic, and that the head boy found my strap in the kit of one of the boys on the way back – good for old Muyé! He must have told which of us it was too, as it was expressly addressed to me. I wonder how he described me. Also, which boy it was – 'Lazy Boy' or another.

The evening has been spent reading letters and writing up this diary. Tomorrow I must start writing letters for Friday's mail. Letter writing is a trial! My sketches are not going well – the paper is horrid and is going into 'veins' in an extraordinary way. I've broken the top off the handle of my beloved fat brush, and the new ichneumon ones are disappointing. They will divide in the middle instead of giving a good point. Of course it may be the result of prolonged travelling! I must try to finish my portrait of !Kŏ tomorrow. She has rather an attractive little face.

Monday June 8th—Did very little – a few odd jobs. Sent Massovi, very reluctant, to fetch grass from the river to re-carpet our living room. The floor is very uneven and messy after two weeks. A new Bushman appeared, brought by the same old black scoundrel who brought the first lot – a tall, scrawny, knock-kneed specimen (taller than Golli-ba) with no front teeth and a *very* dirty face, who claimed the old lady Shové as his aunt. A message was sent to the latter, saying we wanted her to come again. They stayed a couple of hours (in the afternoon), then after being photographed and measured departed. Instead of sketching, went for an hour-and-a-half walk to the rise in the plain to the north, and then into the wood. I collected several bulbs etc. – some of a large *Crinum*-like thing with narrow leaves, a curious rhizome, a couple of *Gladiolus* corms, a large tuber with single elliptical leaves (to six inches long) – ?*Eriospermum,* and a couple of *Euphorbia*-like plants in the woods. Quite a tiring walk! So long for us!

A TABLE IS CONSTRUCTED

Tuesday June 9th—Miss Bleek had a brain wave and suggested we should have a table made after the style of the shelves, partly to be ready for Saturday's visitors if they come, and partly for our present comfort. So we set Fuelu and Massovi, after they had each fetched a bundle of grass, to work – result a firm, comfortable and reasonably level table. It is such a treat to sit up to a table of a decent height for meals and work. Of course 'level' is a comparative term as the framework is two 4 ft long, rather irregular poles, with split sticks laid across and tied down on each side under another thin split stick with bark. The whole thing is somewhat rough, and when one wishes to set down a cup for instance a level place has to be sought. But it is a great acquisition and I don't know why we did not have it made before. Now they are to make a bench along one side of the *nsinge*.

Used the last of the yeast to set sponge and put it in the billy in a hole, in which fire has been all day – last hope. I expect it will be cooked and the yeast dead in the morning. Walked up the hill at the back. Started late as I was busy with my specimens, so we only went a short way. Mrs Wilson sent out some fresh pork yesterday and a message offering more bedding if we needed it. A couple of pork cutlets fried for breakfast this morning were most extraordinarily good.

Wednesday June 10th—Fuelu made his first purchase last night – two yards of the white. He tried hard to get it for less than 18 escudos! My fears were well founded – the 'sponge' was solid porridge this morning. Must fall back on baking powder. However, by stirring up in water and adding a little baking powder I got quite a decent plain loaf and a rather more solid raisin loaf. The remainder of the afternoon and the evening were taken up with letter writing – to Munnie, G.V., Miss Heyworth, Grahamstown Library, Uncle John. To the latter I also sent eight pages – Travelling up to Date, part 4: By Machila. Don't suppose any of this will be of use for the beloved *Courant,* but hope it will at least please him and afford him some amusement. It was quite late when I finished, nearly 10, and the camp long since silent. Then when I had at last got to bed, my glasses got entangled in my mosquito net and I had to re-light the lantern and spend five or ten minutes disentangling them – no joke either!

Family and friends

In her diaries Mary refers to family and friends by their initials or by a nickname. ELP is her mother, Elizabeth Lydia Pocock; GVP, her sister Grace Vernon Pocock; FE, her youngest sister Florence Edith (Fania) and LGP, her brother Lewis Greville Pocock. She makes several references to 'Munnie' who, with 'B', sailed to England where they spent roughly the same length of time as Mary spent in Angola.

Apparently 'Munnie' was the siblings' nickname for their mother. One might hazard a guess that 'B' refers to their eldest sister, Elizabeth Dacomb Pocock, who was known as 'Bessie'. We are given no explanation as to why 'Miss Saltmarsh' joined the expedition for the first part of the river trip and then, by arrangement, returned. Nor do we know anything else about her.

Thursday June 11th—'Madame' started the day by laying an egg. I knew she was on the nest because the mother hen and chickens were scratching around nearby and whenever they approached her nook a fierce warning came from it. Last evening we suddenly noticed a line of fires in the grass nearby. I called Telosi who proceeded to put them out and informed us that Noki had lighted them *para nada, a n'goucho,* i.e. for nothing, just out of mischief, to which I strongly object. This morning Noki brought in a badger skin to sell – quite a big one but spoilt by three large holes. I suppose the animal was speared. It looks a typical ratel. Yesterday Fuelu and Massovi made the bench – not bad, but a bit too narrow. When they came in with the poles they brought a poor little 'bush baby' which they had killed – such pretty little hands and feet, with long hind limbs.

SUCCESSFUL BREAD!

Friday June 12th—At last I have succeeded in getting the 'sponge' and the bread to rise! Good for my raisin yeast! This, by the way, has been working hard, popping out its cork two or three times till at last I tied it down. It froths over the neck of the bottle in a vigorous way and a little of the top added to lemon squash makes an excellent drink. I got two excellent brown loaves baked in the billy and its lid placed inside the pot. The

June 11. 'Madame' outside her coop announcing the daily egg, Kutsi

raisin loaves were not so good so I made another baking powder one. Also a sponge cake – baked in two instalments – came up simply beautifully, but lifting the lid cooled it down a bit too soon, so that some of its pristine beauty had fled by the time it was finished. Altogether a very busy day – cooking practically the whole day.

We got two chickens, had them killed and dressed early. Then came a boy with a half fledged gosling – big but with gosling down still on its neck. We wanted it but the boy would not take what we offered, ten 'bits' i.e. five escudos, twice the price of a fowl, and hesitated so long that Miss Bleek gave it up as a bad job and returned to her Bushmen. My baking took so long that it was nearly dark before Telosi could start roasting the fowls, and they had to be left to finish in the morning. We had a note from Mrs Wilson saying that they were all coming and would start at 8.30, coming round by the ford. I suppose it will take them a couple of hours or more to get here.

VISITORS

Saturday June 13th—Directly after breakfast set Telosi to finish roasting the fowls (by the way, stuffing was onions, egg, stewed apple and flour) and the others to turn out the *nsinge*, tidy and level up the floor and spread fresh grass, then arrange the boxes round the table for seats, spread the hessian and some more grass on the bench – made my bed into a couch with cushions etc. and generally tidied up. Then I had to superintend the stewing of the peaches, and when they were done and turned into the only dish large enough, i.e. our small hand basin, make the blancmange – two eggs, condensed milk and maizena; this took such a long time on the boys' awful little fire, and just as it was finished a lookout on top of the framework of the cattle shed outside the camp signalled that the *machilas* were in sight.

Then a wild rush – turn the pudding into the other small basin, give directions to Telosi to eat up what was left in the billy, clean it and put on water, make the gravy in the big pot and leave it to simmer with the sweet potatoes in it, scrub my hands, change my dress, shoes etc., and just as I was giving the last touches, i.e. powdering my nose and cleaning my nails, the first *machila* with Mrs Wilson and Gordon (three years?) arrived, followed almost at once by the baby in a cage *machila* (ten months). By and by Mrs Reinhart arrived, then Gracie (7 going on for 8) in a straight-sided green canvas *machila*. The men, Mr Wilson (Augustus) and Mr Reinhart (Gene) did not arrive for half an hour or so, so we started tea, for which we were all ready. They had very thoughtfully brought their cups and plates and spoons, knives etc., Mrs Reinhart a tin of salted peanuts and Mrs Wilson a beautiful cake. My little sponge cakes and raisin bread were quite good.

The children are fine, the baby (Paul) such a jolly little fellow; Gordon at first was rather solemn, but both he and Grace soon became quite friendly. It turned out that Mrs Wilson had made a mistake about the ford and they had had to cross the river. It was past eleven before they arrived, and by the time the men had come and finished tea it was just about time for lunch. We gave Gordon and Grace theirs first on account of the smallness of our table. The menu was: 'Roast chicken assisted by Bully Beef'. Sweet potatoes stewed in chicken gravy. Mashed potatoes. Pumpkin. Followed by stewed peaches and blancmange. Then for dessert we had dried and crystallised fruit and chocolates, all of which were much appreciated. Paul had his bottle, which he finished on my bed-couch and then went to sleep for a short while, and woke rosy and very jolly.

We sat and talked – Miss Bleek got various items of information from Mr Wilson, assisted by Grace, who has more Mbundu than her father. Then we had tea and almost before we knew where we were, it was time for them to return. They left soon after 3.30 and we walked down to the first stream with them. About 5.30 Fuelu pointed out the last of them going up the bank on the far side of the river! The crossing must have been a very slow business. It was very nice having them all, and I think they enjoyed the day away from the station, though crossing the river with the three children must have been anxious work.

Sunday June 14th—A thoroughly lazy day after the last two busy ones – read most of the morning; in the afternoon lazed, then Massovi wanted some words written out and pronounced. Fuelu also, but after writing them once I gave them a piece of paper and said they could copy them themselves. Telosi asked permission to go to Muié for the morning so we gave him the day off. Then darned my stockings and painted two or three flowers rather roughly (two peas, red thistle and a labiate).

A DAY OF PHOTOGRAPHY
Monday June 15th—Slept late this morning – 8 o'clock – so we did not get breakfast till nearly 9. Then Telosi and Fuelu departed with the washing which they do at the river – the latter taking his piano along – and I took Massovi and went off up the river to photograph. Took close-ups of 1) the peach-blossom shrub which has fruits forming – hope to get some ripe. 2) The tufted red-flowered shrub, no leaves, with a fine *Polygala* – the blue one; then I took a nearer one of the red tufts. 3) Red *Thymelaea*. 4) *Clematis* with a small boy balancing a water calabash on his head. Then I went into the bog and found two new species or varieties of *Utricularia* – a tall stout chap with mauve and yellow flowers, standing 8-10 inches high, and occasionally twining, leaves rather short and spathulate – and a small-flowered deep blue, almost navy, twiner, possibly a variety of the ordinary blue. I photographed the former, of which there was a lot. I must go again tomorrow to collect and try to photograph others. I must try to paint all the Utricularias and write them up fully – wish I had brought some more books and particularly a cartridge paper drawing book for flowers. I think I shall have to cut up one of my sheets of drawing paper and use it for flowers – it would be worth it.

Meanwhile Miss Bleek had a delightfully quiet morning with the *Vasekele*. Golli-ba's other wife with her two rather fine little boys (one four or five, and the other about two, trotting about but still being nursed) had appeared upon the scene. !Nishe, the wife, the smallest yet, just over 56 inches, is !Kǒ's half sister – same mother but different fathers. My photographing took nearly all morning. I got back about 12, very thirsty! Put away my new utricularias, helped photograph the new Bushmen, then got lunch (boiled eggs and pumpkin 'fritters'), after which we both succumbed and had a good laze. The rest was interrupted by a hopeful soul who proffered 50 centavos and asked for a tin! Much too precious at this stage to part with.

At 3 I suddenly remembered something had to be done in case the pork Mrs Wilson promised to get us did not come in time for dinner, so threw some beans into the pot and put them on to boil. Then continued sorting, writing up and packing away my specimens. The Zambesi ones are no good at all but I am writing them up for my own edification. I want to get on to the local ones to get the native names of as many as possible. Noki brought a basket and two notes from Muié – six tins of condensed milk, *makovi,* peas, strawberries and the pork, and a big roll of English papers for us to look at, all very welcome. The milk was sent as our store is low and we asked if it would be possible to get some from the Portuguese store at Kunjamba with the salt Mr Wilson is getting for us. As a consequence we have just had a most sumptuous repast – excellent bean soup (beans, onions and a little of the pork), fried pork cutlets and sweet potatoes, followed by strawberries and cream, and a piece of Mrs Wilson's excellent cake and coffee. How's that for the wilds of central Africa in June? Now we are busy – Miss Bleek at her index, I at this diary and writing description forms, for a short time before our nightly game of bezique or piquet. Fuelu's and Telosi's wives departed this morning, to our joy!

BOTANICAL ACTIVITY

Tuesday June 16th—The boys asked for their week's salt today. As Miss Bleek was busy, I gave it to them – a cupful each, and to my surprise Massovi protested loudly that it was '*poco*'. I later discovered that Miss Bleek had been giving them the tin full! Took Massovi and Fuelu and went collecting in the bog – very interesting, found a fine *Lycopodium* and an insectivorous plant with spirally twisted trap structures, – very close to *Utricularia,* possibly *Genlisea* – I think two species. Took a few photos. Spent the rest of the day drawing, labelling and putting away specimens, but did not get half done. Set sponge.

Wednesday June 17th—Bread really very good. Spent the day finishing yesterday's specimens – another whole day's work. I've had Massovi and Fuelu making me a couple of frames for pressing. They are not very successful but will do to pack half-dried specimens away. There is a good deal more to be done about here, and I cannot get on with the packing away of the earlier material. Used the last of the yeast to make sponge – tried leaving out the hessian blanket.

Thursday June 18th—The 'sponge' missed the blanket and did not work, also possibly the yeast was a bit tired. 'Madame' has gone broody – I'm afraid that means her days are numbered. We are still eating the pork, and today a man brought four fish, of which we bought two, rather the size and shape of silver fish but darker in colour – excellent eating. They catch them in basket traps.

Went collecting with my two satellites towards the river, trying to get the finer charophyte. Got good material of the fruiting *Nitella*. Photographed the mauve figwort

June 14. Red thistle, *Pleiotaxis linearifolia*, Kutsi

Mukutassankuma (Kunzambi)

uyé

9299

June 18. Mauve figwort (*Gerrardiina angolensis*)

D⁰⁵ Gerradiina
    angolensis Engl.

and yellow water lily. Otherwise found nothing new, except two plants of the larger *Genlisea* fruiting – I think it a distinct species; it is quite glabrous and though there was only a bud on one shoot (all the others were fruiting) the flower is apparently larger, different in shape and colour from the first. There are also variations in size and shape of leaves and of traps. The afternoon was spent putting specimens away and mounting them, punctuated by bread baking and yeast making. We are going in to Muié on Saturday for the afternoon – really because I want a game of tennis!

PORTRAIT OF SHOVÉ

Friday June 19th—Announced the proposed excursion tomorrow and in less than half an hour the *tipoia* boys were all signed up. Then departed with Fuelu for the '*donga*' – went south, then across the plain to the water and back up the south stream – mosquitoes terrible! Did not get much – the smaller *Lycopodium* (with tubers) fruiting, and the small blue and green *Utricularia*. On returning, I retired with two basins and two billies of hot water to the tent, washed my head and then my feet, shoes and stockings – three latter all black.

In the afternoon Shové and !Nishe arrived and instead of putting away my specimens I spent the afternoon painting the former – drawing rather badly out and paper horrid to work on. Got a likeness but have rather lost it I'm afraid. It is really rather a fine old face, rather tragic in expression. She is very tall for her race, just about my height, although she is the lightest in colour of all we have seen so far. !Nishe looked quite different – like !Kõ she has washed her face since coming into our vicinity. It's wonderful what a difference it makes. She is several shades lighter (but like !Kõ is yellow-brown, not copper-yellow colour like Shové and Kavikisa) and looks years younger. I had just time to paint the tiny-flowered blue and green *Utricularia* before the light went. I hope the plants will be fresh enough to make specimens of them tomorrow.

VISIT TO THE MISSION

Saturday June 20th—Putting away the rest of yesterday's specimens took over an hour-and-a-half, then for half an hour I was writing up and packing away old specimens. A feeling of tension pervaded the camp – *tipoia* carriers began to roll up by 10, and before 11 one felt they thought we should be late. At last, at 11.15 I told Telosi to put on the kettle, we had a light lunch, tidied ourselves, put Telosi in charge and gave him directions for dinner. By 11.50 I was away, Miss Bleek having started ten minutes before. Twenty minutes brought us to the river, another twenty across. While my *tipoia* and carriers were being brought over we walked on up the hill as I wanted to do some collecting. At the top of the hill near the first village we got in and the joyful sounds of *tipoia* carriers at work recommenced.

Our cavalcade included the *Capitea's* wife (he being one of the carriers) carrying a tiny, sickly, squawking chicken, and a tiny puppy, complete with innumerable fleas, who shared my boat and cuddled close to me to get shade. He was rather a dear with a pretty head, well shaped and nicely marked, two white fore-feet, body yellowish, head dark.

I rather wanted him. After landing he was carried over the shoulder of one of the small boys we soon picked up – two little paws and pretty little frowning face looking back over the boy's shoulder, white hind paws maintaining a balance against little boy's front. A second small boy carried someone's axe, while a third took Fuelu's basket and a fourth the roll of papers Mr Wilson lent us.

The hour from the river to Muié is through as pretty a bit of woodland as any we have yet seen. The path is fairly wide and open – the *Chefe* did not like the old winding path so he had this cut. At 1.30 we reached the Wilsons' house. It was a good thing I was firm and did not start when the boys wanted to. Mr and Mrs Wilson, the latter with 'Paulie' in her arms, came out to meet us. The other two, Gordon and Grace, were resting, but soon appeared on the scene. We sat and talked and listened to the gramophone. I spun Grace and Gordon's *piastres* (50c) for them and looked at Grace's collection of pebbles (from the *Chefe's* road), then we had tea and it was past 3.30 before we made a move to the tennis court. Then we had to go first to the printing shop and get our pads of paper, cardboard etc. and my hymn book.

A GAME OF TENNIS

Then we collected the balls and racquets and went down to the court. The boy had put up the net for Mr Wilson – said he knew how to do it – resulting in pegs being placed nearly a foot from the centre line. However we soon adjusted it, removed the pegs and

June 20. Boys at Kutsi going fishing wearing bark cloth aprons

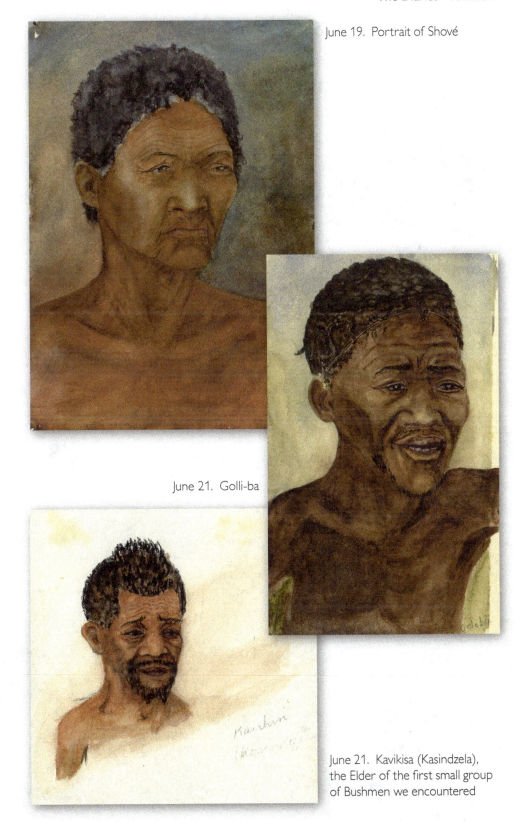

June 19. Portrait of Shové

June 21. Golli-ba

June 21. Kavikisa (Kasindzela),
the Elder of the first small group
of Bushmen we encountered

June 22. 'Found the bark tree (bark used for tying) in bloom... got plenty of the curious little, very sweet-scented blooms' (*Cryptosepalum exfoliatum* subsp. *pseudotaxus*)

POISONED

Herb. Rhodes University

No. 67

(CRYPTOSEPALUM PSEUDOTAXUS
= C. exfoliatum DeWild.

Loc. Kutsi : alt. 4200' : bark
for Tying.

LEGIT M.A. Pocock DATE 22 J.
G. & G. GTN.—2,000/10/53

30

hammered the centre pole to approximately the right height. The net is a local product of native-made string. The court is of the usual reddish clay which they use for building the houses. There was a rather large hole, so in the middle of a game Mr Wilson called to a boy who brought a large lump of red earth and they stamped it in. We had only time for two sets, one to each of us, and then it was 4.45. I hurried back while Mr Wilson looked for the other ball. Miss Bleek started off and 15 minutes later I followed – stopped while Mr Wilson talked to Massovi etc. The journey back was very pleasant, in the cool of the evening. I had to put on my coat, and by the time we reached the top of the hill it was dusk. I got cold, so walked a good bit before the last village (also, my boys, the cow-herd and another tall thin boy, were so slow I felt I could get along faster on my feet). Then I got in, and Fuelu and Massovi took their turn and went at a good rate.

I was afraid the two *watas* wouldn't be there, and my suspicions were justified. Instead of finding one or two boats waiting, there was only a solitary man with a paddle, and by the indignant conversation that followed I gathered that Master Noki had taken one of the boats. As a consequence we had to wait 20 minutes till Miss Bleek's boat got back, and it was quite dark except on the water. When at last the boat appeared, the reflection as it crossed was lovely. The men and the *tipoia* were soon put across and then with Fuelu in front and another man (boatman) behind, both paddling, I crossed – very wet around the knees. What with two men moving at different angles the *wata* takes on a kind of compound pendulum motion and it was distinctly difficult to keep one's balance. However the open river was soon crossed and then came the short poling up the tortuous channel through the reeds, while the *tipoia* carriers splashed along in the dark a few yards away, their course being roughly parallel to the channel.

Then came the final *tipoia* part, feeling step by step through the swamp, then firmer ground, once more a wide stretch of bog water, then my little gurgling stream and across that the fairly dry path with which by this time I am pretty familiar. My two boys (Fuelu went ahead to show the way and Massovi had disappeared) were evidently not accustomed to the work – one had gone in up to his knees once or twice, and one (if not both) was quivering all over, so I got out and with Fuelu leading we proceeded at a good round pace across the remaining marsh land, up the sandy slope and so to camp, the cheerful glow of our fires leading us straight home.

The sky was lovely – as I waited for the boat, the river gleamed gold in front of us. Away to the right (north) was the Great Bear, then across the river to the west was a row of four bright lights – almost in a straight line, Venus, Procyon, Sirius and Canopus – glorious. Then Venus dipped, followed soon by the Great Dog, and where he disappeared there was a wonderful light effect. It was quite dark, but there to the west-north-west was a kind of pyramid of light, or rather 'less dark' – a most wonderful sight. As it grew darker this grew more marked and it was clearest just as we reached the camp. I wonder if it was 'Zodiacal light'.

We find Massovi does not want to go on. We are sorry as we should have liked to have him for the trek. He wants to go back tomorrow, Mr Wilson says. Telosi had dinner and hot water ready, both very welcome. The post is very late again this week.

Sunday June 21st—Massovi asked to return this afternoon. In the morning he took a last lesson in English, getting a large number of words from Miss Bleek. When he departed, by way of a polite goodbye he bid us 'bye bye' – it sounded most ludicrous. Telosi asked to go into Muié to arrange about food for Kunjamba. Fuelu, too, does not want to go on, for which I am very sorry. We are debating offering an increase in his wages, now that he is alone at his wood and water job.

REACTION TO THEIR PORTRAITS

I thought I should have a good day at labelling and packing – not a bit of it. To begin with Shové appeared so I had to get on with her portrait – not much good, and I am afraid to work at it much more, as the paper shows signs of 'veining'. Then we had a show; Golli and !Kõ were there and they had to have a look. Then Golli-ba and Shové joined in and talked hard. I did wish we could follow. Miss Bleek gathered that Shové said it was not a pretty woman, was a young one, another Bushman. And they weren't quite sure if it was man or woman. Anyway *tone* said even more than words. Then Kavikisa came prancing up. (There was a dance at the native village last night – girls' initiation – and both Golli-ba and Kavikisa attended. The latter was, I fancy, a bit exhilarated.) Anyway he approached, dancing, and joined the sightseers. He recognised it at once for Shové by the white hair round her face. Then I produced Golli-ba's and held it up next him, to everyone's amusement. He acknowledged it was his portrait, but tried to hide his face. Then I turned to the small one of Kavikisa, who immediately went into fits of laughter, pulling his top knot and the tufts of his beard with his two hands – both somewhat accentuated in the sketch! As for the unfinished one of !Kõ – she would not admit it was meant for her, and certainly it is not flattering. Shové's does not do justice to her fine old face, nor does it make her old enough.

CULINARY TASKS

Then I made lemon syrup, and put the lemon skins (very dry) to soak to see if they would do for marmalade, and by then it was time to see to lunch. I had also made a fire in the bread hole to warm it well. My 12 lemons (the 13th was bad), the last of the Lukona lemons, made a wine bottle and the honey jar full of lemon syrup. After lunch Miss Bleek helped me cut up the peels which seem to be softening quite well. Then I set my sponge, put beans to soak for tomorrow's soup, got out prunes and rice, and apple rings for tonight, and started Fuelu at the dinner cooking, and then, about 4pm, started my specimens.

Massovi left about 3, so Fuelu had a dull and busy afternoon. Oh, for lunch I made pumpkin fritters – quite a success. Telosi did not return till after dark, about 7, and soon after came the long-awaited post, a very slender one. My share of it was a single letter, a very interesting one, from Miss Saltmarsh. Next post, three weeks hence at the earliest I suppose, I hope to hear from England. Miss Bleek's book, 'With my wife across Africa' arrived from Mr McGill.

I have just been changing my second dozen plates. I have to retire underneath

Mary's *sonneke* (drawings) delighted the Bushmen: (clockwise from top left)
Kavikisa, 'head of his group', Kutsi; Ndala with his mother, Kaiongo; Gandu Ndende, Kaiongo;
Bushman visiting Kaiongo

a blanket for the operation, a stuffy and tiring procedure. Tomorrow I hope to go photographing up the hill.

Monday June 22nd—Temperature wrong again, so my sponge did not work. However, made it up with rather curious results – first two did not rise but baked well. Third rose in baking but took hours to do. Took Fuelu and went up the hill, took several photos of trees including the fine timber tree *musivi* (*Sterculia*) of which there are some very fine specimens half an hour's walk from here. Found the bark tree in bloom, and Fuelu acted monkey for me, climbed it and got plenty of the curious little, very sweet-scented blooms – two large branches thereof.

A couple of days ago Miss Bleek bought five fowls, such tiny ones, for 48 inches of cloth – cloth at $9.00 the yard, chickens $2.50 unless very small. Today bought a 'red cat' skin – pure white with rufous markings on head, back and tail. Applicants for posts as porters to Kunzumbia are turning up – a dozen or so have already been engaged. The parcel of paper – scribbling pads, and larger pads of better paper, cardboard etc. arrived from Mr Wilson this evening. The bill looked awfully large – 48 escudos, but is less than 10/-!

End of Volume II.

THE DIARIES
Volume Three

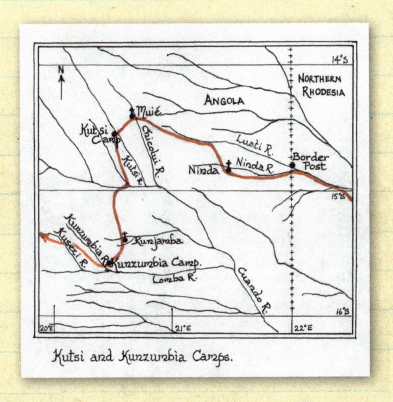

Kutsi and Kunzumbia Camps.

Kutsi Camp to Kunzumbia Camp
(June 24 to July 22, 1925)

'As we stalked the buck, I thought to myself what foolhardiness it was – stalking game in lion country, (armed with a pocket pistol!) and had visions of us being stalked by a lion.'

June 24. Yellow orchid (*Eulophia speciosa*), Kutsi

Kutsi, Angola

Ibulwa (Somthwa)
Eulophia sp

(= *Lissochilus speciosus*)
E. speciosa (R.Br. ex Lindl.) Bolus

A WOODPECKER

Wednesday June 24th—The other morning early I heard a 'Tap-tap! Tap tap tap tap!' in the tree behind my end of the *nsinge*, and saw the boys looking up. I went as quietly as I could, but not quietly enough and Mr Woodpecker took fright and departed. But a day or two later towards evening while I was busy with my specimens he re-appeared and I had a good look at him hanging on to the leaning trunk and tapping with all his might – a beauty – olive green back, brownish grey body and two bent black bars across the side of his head and a long slender bill. There are a good many about. Then there is a little black-backed chap, rather larger than a Cape robin, but with the same flirting action of tail and wings and a rather similar note, but quite unlike the robin in colouring.

This morning I took Fuelu and went to see the yellow orchids we found yesterday in hopes the buds would be far enough out to photograph. I missed the way at first and found myself right up at the large clearing where I photographed the big *musivi* (*Sterculia*). However, eventually I got to the proper clearing on the right, took the photo of the four-winged nut tree and then went down to the eulophias. They are not yet open so I left them for a couple of days. As we came down through the wood there was a flash of wine-red colour, and a lovely bird, white-bellied, black back and wings when still, but with a perfectly wonderful wine-red back when flying, darted across. It alighted on a bush and then flew on so I had a good look at it – a lovely creature about the size of a *spreeuw*.

In the afternoon I spent some time photographing the specimens of the yellow orchid (?*Eulophia*) I had collected yesterday, and also made some excellent biscuits, baked in the frying pan! – as well as baking powder loaf to carry us on till tomorrow. I started writing letters before breakfast and went on for a time afterwards. In the evening I was just about to continue when the boys gathered round our fire for their prayer meeting so we had to lend our voices to their hymn singing. After that I resumed letter writing and went on till nearly 10, very late for us! The second volume of this diary was dispatched by this post.

Thursday June 25th—Miss Bleek has only signed up 16 porters so early this morning she sent Telosi to try and find some more. He was away till past noon and came back unsuccessful, so Fuelu was sent on to Muié with a letter for Mr Wilson giving him the names of the men from across the river for the *guia* and asking him to get us 10 more men.

A very busy day bread-making – sponge rose with a little extra persuasion but the bread did not. Very disappointing, possibly due to the wind, or else my yeast is tired.

About 5 we went for a short walk up the valley. The grass was being burnt in some places, and over the largest patch of fire two black birds with swallow tails were hovering. I thought possibly they had a nest and went to investigate, whereupon they flew away. I think they were simply attracted by the flames – they were flying almost through them. A little further on I saw a young bird, half flying, half running through the grass, such a pretty creature, soft pinkish fawn with a glossy black head and rather long bill. I caught the little thing and put it in a bush – at first thought it was a plover but think it was a kind of wood pigeon or dove. I have heard some about and it was rather like a young turtle-dove in size and shape.

Got Fuelu to repair my *machila* this morning.

LAST DAY ON THE KUTSI

Friday June 26th—Our last day in this camp on the Kutsi; it has been very pleasant on the whole. For me it was a very busy one – baking bread and biscuits in the morning, packing up the rest of my specimens and personal belongings etc. About noon I went up to collect and if possible photograph *in situ* the yellow *Eulophia*. Only three spikes were left and those were in bud, so I only collected the tubers. On the way, had a game of 'I spy' with a woodpecker. I startled him from his tree and he flew to the next one, then recovering from his fright, looked to see who the strange creature was. He hid behind the branch and peeped over; I approached, he ran along still behind the sheltering arm, peeping over at intervals. Finally I outdistanced him and got to the other side of his tree, when with one more peep he decided to retreat and returned to his first tree – as he did so I got a good view of his back – the top of the head has a bright red streak; above the eye is one black bent bar, and below on the throat another.

There is a good variety of birds in these woods, some with very pretty notes – usually a whistle. One has a falling cadence of three or four notes repeated over and over, in a minor key. Another, a contented little song rather the quality of a blackbird but much more restricted. About 5 Noki (baptismal name Enoch!) came with a note from Mr Wilson telling of a hitch over the granting of a *guia* by the *Chefe*, and evidently expecting Noki to reach us about three. All effort to raise a messenger for Muié failed – even two escudos would tempt no-one to go. Finally we had to decide to leave it till the morning and hope for the best.

After dark we heard shouting and were told it was our porters; some half hour later they came in with the two bags of salt and the parcel of money. It was all in nickel so it took me some time to count it, while Miss Bleek wrote to Mr Wilson and made out the cheque. Finally it was 10 before we got to bed, very tired.

DEPARTURE FROM KUTSI CAMP

Saturday June 27th—After a somewhat troublous time with evasive porters, allotting loads etc., <u>we</u> were ready by 9am, and then were somewhat dismayed to see all the porters etc. sit down to eat! Finally at 10 to 10 we got started. Then there was some difficulty over the route. Miss Bleek depending on what she had gathered from the Bushmen expected to go west; instead we went south down the Kutsi and continued to do so all day in spite of protest. We were a bit short of porters, some not having turned up, so had to put *tipoia* boys on to porters' jobs and go short with *tipoia* boys.

The valley was lovely, and every now and then there was a splash of vivid scarlet where spikes of the deciduous bush with the clusters of red flowers waved among the grass – they were far the finest I've seen, some shoots over 5 ft tall, and numbers of plants

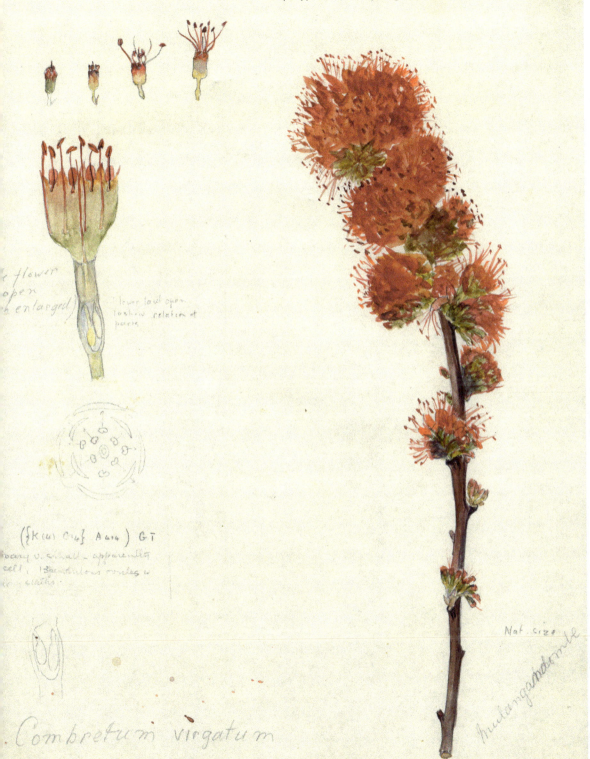

June 27. 'A splash of vivid scarlet' (*Combretum platypetalum* subsp. *virgatum*), Chikoluwe River

e flower
open
h enlarged)

flower laid open
to show relation of
parts

($K_{(u)} C_{(u)}$ A $a_{(u)}$) G Ī
ovary v. small — apparently
cell, 13 pendulous ovules w
long stalks.

Combretum virgatum

Nat. size

September 27. *Combretum platypetalum* subsp. *baumii* with dainty four-winged fruits and silvery leaves. Cwelei

Combretum gnidioides

C. arenarium Engl & Gilg

Cwelei
27 9 25

each with several shoots bearing their clusters of flowers. There was a good deal of south to south-easterly wind and red shoots, grass and all waved in the breeze. Mr Bailey told us the Kwando was a day's journey – the boys said no, impossible; they went on, my *tipoia* leading, till nearly sunset, much too late.

Then they landed us in the centre of a village, right up on the hill where I was surprised to find a tent and a white man – Mr Reinhart on his way to Kunjamba, where Mr Muir is ill with fever and its after effects. In about 20 minutes Miss Bleek arrived, but no sign of the porters. Mr Reinhart kindly invited us to supper with him, so with the remains of our two chickens (already used for dinner last night and lunch today!) we went along to supper – very glad to do so. Then, as the tent had not yet materialised but our bedding and my bed had, we put my bed and the two *tipoias* to act as three sides of a square, spread the hammocks on the ground and made our beds thereon and thankfully (so far as I was concerned after our late night and early rising) got to bed.

Sunday June 28th—Next morning we found that three loads, including the tent, had come in overnight but no food, so that Mr Reinhart's invitation to breakfast, by a note handed in over the up-turned bed, was most welcome. We made our *laager* just outside the village, only 100 yards or so from where Mr Reinhart was. Soon after, the boy we had sent to Muié arrived with the *guia*, a note from Mr Wilson, a little pot of strawberry jam – my marmalade pot refilled – a note for Mr Reinhart and his bread pot which Mrs Reinhart asked us to take on for him and which we were able to hand over at once as he had not yet left.

AT KANZANGOLLA VILLAGE
Then we bade him goodbye and packed up our things and shifted to our present camp above the village, quite a pleasant spot. Here we erected our tent and I dug, or rather had dug, a hole to sleep in – I'm going to sleep on the ground if possible in future, it's so much warmer than a stretcher – arranged my bed and *tipoia* to make a little shelter and got grass etc. put ready in the sun to dry. Then the rest of the porters arrived and we gave out a cup of salt and a 50c piece each to buy food, engaged extra men, and generally sorted out things. In this country one should never be in a hurry! Then I made a cake – whole meal but two eggs, milk, raisins and some fat. It baked beautifully – rose and has a lovely brown top – am just going to try it (3.45).

After lunch I retired with this book, cushions etc. to the shade of a nearby tree and Miss Bleek disappeared, I think down to the village to look for the little Bushman woman I saw there this morning – such a tiny little person with a baby a few days old. She was lighter of colour, more typically Bushman in features and size than any we have yet seen. Miss Bleek photographed her and questioned her, but she knew no Bushmen, had been with the villagers since she was a tiny child. Now Miss Bleek has gone off in her *tipoia* to see some Bushmen nearby and I have continued our trade in fowls, eggs, tomatoes, fowl food – mostly for salt, fowls for money. A pleasant Sabbath diversion for this village (Kanzangolla).

My cake is excellent, couldn't be better though I says it as shouldn't. The chickens are scratching around, thoroughly at home – those just bought are in the *camba*, but there are five others, such scraps too. Two are at present drinking the soapy water I have used for washing my hands.

Soon after we started yesterday I saw two large greyish, long-necked birds out in the valley. They were too far off to see very plainly but were, I think, secretary birds. They looked very affected with their grey colouring with black below, and looked huge – at first I thought they were ostriches!

7.45. We have just finished dinner and are now holding a reception – of young matrons and girls from the village behind us. A party of ten came up with broad smiles and seated themselves round our fire – evidently a visit of ceremony prompted by curiosity! They are decked with brass curtain rings as bangles and pendants, metal Maltese crosses and china discs slung round their necks, and their hair dressed with red clay.

Miss Bleek found several Bushmen – grandpapa, sons and daughters and grandchildren, some relations of Golli-ba, and had a very profitable afternoon. On the way back saw a Voodoo man in full rig and watched him dance. She got back just at dusk, a few minutes after 6. The evening was concluded by a service at which somewhat to our amusement Fuelu preached!

Monday June 29th—Trade in chickens and eggs continued up to the last moment and we left (8.20am) well provided for the journey, striking westward across the wooded heights. About 10 we came to a small river valley, with a few holes with water – the upper part of the Mueri. Across that we stopped for a short while, then went on for a couple of hours. With the exception of this river and a few open patches of grassland on the top, the whole trek was through woodland, sometimes quite thick. At the Mueri I found a second *Protea* – leaf much wider and slightly tomentose, but only found a single head and that not too good.

AT THE KWANDO RIVER

The afternoon trek was a long one and heavy for the not very experienced *tipoia* boys, as the path wound in and out of thick bush. Finally about 4 we could occasionally catch a glimpse of the distant blue of the Kwando valley. It was not till nearly 5 however that we could really get a good view of it. Then Fuelu pointed out the Kembu valley right ahead to the north-west and the Kwando on the right coming down from the north or north-east – a very lovely sight through the trees. The path wound on and on, and presently the silver ribbon of the Kwando appeared here and there through the branches. Finally about 5.30 we reached the river between two villages and camped just above its banks. Here it is right on the east side of its valley. The digging of my bed aroused great interest, a crowd of interested spectators gathering round to see it being prepared. The soil here is a red sand. Again we had a visit of women from the nearby village; they came while we were sitting at the fire after dinner. Then the boys came along and had a service – me with two towels, a vest and a pair of knickers hanging over my knees to dry!

CROSSING THE KWANDO

Tuesday June 30th—With some difficulty we got started about 9, thanks to me – the boys wanted to wait till it was warmer! A quarter of an hour's walk along the river bank brought us to an open green grassy patch where a family was already waiting to cross the river. Then followed a wait of some half an hour after which several *watas* arrived, much wider and steadier than any we have yet come across, some taking two boys and their loads in addition to the paddler. After Fuelu and some of the loads had gone, my *tipoia* and I followed. The river was lovely – palest green, clear, translucent, the sandy bottom clearly visible through the swift stream. In mid-stream the paddler stopped and he and the boy kneeling in front pointed out that the Kwando came in on the right looking up, the Kembu on the left.

A short channel through the reeds brought us to a landing where my *tipoia* was waiting and I got out – as a matter of fact the boat ought to have taken me much further on. The channel was deep for some way. I tried walking through the grass but it soon got too wet, and the *tipoia* boys had hard work wading up to their knees and sometimes even deeper, and feeling every step. In one particularly deep place, the front boy lost his footing and softly sat down sideways. A shout from the other boys made one wading in front turn round and he caught the pole just as it was dipping – not before the bulge in the hammock (which was me!) was resting well in the water. I was surprised at the speed with which that thick canvas got wet! However no harm was done and the boys were more careful afterwards.

Then we came to a raised, dry part, where a fire was burning; there I got out and walked, and the boys would like to have waited and warmed themselves, but I led them off at a good round pace – far more warming. Soon however swamp began again – deeper than ever – they had to tuck their loin cloths up in front, and even then got a bit wet. Finally after passing several swift little streams besides the actual swamp we came out on to firm dry ground, the path following a bend of the Kembu close to the bank of the clear swift stream. We followed it up for under a mile – most lovely, then struck up into the wood where we stopped for the rest to come up. Miss Bleek stayed to see all over, then came on. By 1pm all had arrived. From Mr Bailey's itinerary we had expected to camp there – he said the crossing would take all day. However from the boys' general attitude I could tell that they expected to go on, so did not unpack bedding etc., as had been my intention in crossing among the first.

THE KUTITI RIVER

After a light lunch we set off for the Kutiti where they told us we were to camp. I felt energetic so walked all the way. The path was very pretty, winding up through open woodland, where we saw spoor and holes dug for roots. Telosi said these were made by *ntengu* (large buck about the size of a donkey – Kudu?). Soon we caught a glimpse of blue through the trees, a sure sign of a river valley – this they said was the Kutiti. However, the path turned to the left instead of making for the river and we continued for some time through the wood, then finally came out to the edge of the river plain – the prettiest we have seen. We followed up this for some way until about 5 – the most strenuous exercise I've had up here – three hours' steady walking without a stop. I had certainly earned the thirst which consumed me! We camped on a little knoll projecting out into the plain. Suddenly I noticed Telosi sharpening a knife on a lump of stone! Enquiry brought the information that it came from the *donga,* so presently I went down to see – there really were rocks! On the bank above the stream (looked same type as on edge of Falls in the Rain Forest) and in the stream itself – a lovely little river some 8 ft wide and 4-6 deep it looked, running over boulders. It is a very young little river, making S-curves from side to side of its valley.

Just at sunset there was great excitement and Fuelu, pointing out a buck feeding on the far side of the valley, asked us to shoot it with the pistol! However, when we pointed out that it would be necessary to get within a dozen yards of it, the eagerness of the would-be hunters quickly evaporated.

TREKKING ALONG THE KUTITI

Wednesday July 1st—The morning's trek was lovely – all up the little gem of Angolan rivers, the Kutiti. Last night, for the first time since I have been sleeping on the ground, I felt it was cold – no wonder. This morning my hand happened to touch the pole of the *tipoia* at the head of my nest – it felt extraordinarily smooth and attracted my notice – the mist had frozen all over the pole!

By the way, I forgot a funny little incident yesterday. Soon after leaving the Kembu we crossed a piece of grassland, not swampy but moist, the path deep sunk between little banks and grass, very narrow so that one had to watch one's steps rather carefully. Walking along thus, I happened to glance up and across the field of my vision there moved three curious shapes. To my startled senses was conveyed the vivid impression of three caricatures of ostriches, and instantaneously I found myself wondering what animals these nightmare shapes could be! A second glance however resolved the puzzle – a procession of women carrying large calabashes of water on their heads and carrying baskets (carried by a strap round the forehead) with loads topped by babies on their backs. Several others followed with calabashes instead of babies topping their loads. The villages are often quite a long way from their water supply and it is one of the women's multifarious duties to carry up water.

To return to the Kutiti: we followed up the right bank for a mile or so and then sturdy little *Protea* bushes, covered with last season's heads, appeared. I photographed one with a boy (about 15) standing by it and just after that the path turned sharp to the right and crossed the stream – here rushing down over rock. The proteas were very numerous and must have looked lovely in the flowering season. Then we went on up the left bank, passing several outcrops of rock. In two or three places the whole of the river plain was covered with the proteas – leaf narrow like Kutsi one, but heads more open and less cup-shaped. I got down to have a look at them and found they were in very swampy ground with pools between, in which pale blue and white waterlilies and arrowhead etc. were growing. Then they suddenly disappeared and so did the rocky outcrops. Shortly after, I found another kind of *Protea* – a tree up to 20 ft high, slender, growing at the edge of the woodland. I thought I had seen a *Protea* tree a couple of days before, but could not get it; here they were fairly numerous and quite large. Unfortunately I could not photograph it as I had finished my roll and had not another.

About noon we crossed the river again – this time we had some difficulty – my *tipoia* less than Miss Bleek's, as two of my boys cut down young trees and used them as a bridge. Here the path leaves the stream, so we stopped for lunch – a perfect little paradise – the valley is narrow, very lovely and coming down at the right is a tiny tributary, about two feet wide and one deep, clear flowing water. It would be a lovely spot to make a home. However there was certainly a lion in that paradise – I saw his track down to my stream – and probably many serpents!

One of the porters (carrying my suitcases) has a sore foot. He was far behind here, so we sent one of my *tipoia* boys, Domingo, back for him and made him take his load. Giving the boys an easy time does not pay – the *tipoia* boys lag behind, and take every opportunity for lazing. We made a long trek through wood and across several of the curious burnt-looking grass and bush patches in the middle of high woodland to *donga* Wavenga – a flat open valley with a lot of red growth which we crossed. There was hardly any river, just swampy pools very red in colour. Some men were gathering red clay in the middle. On the far side I walked again – a very sandy track covered with spoor of guinea fowl. We saw a man constructing a deadfall for big game, several old deadfalls of the same type and also snares for guinea fowl, field rats (a favourite 'relish'), etc. Here too we saw two men with guns and cartridge belts, probably native soldiers or police. We camped about 5, half a day, they said, from Kunjamba, at a spot where there was an old shelter with a little table in front of it.

Thursday July 2—It was difficult to get the men started – it was past nine before we were under weigh. It was cold again –though I thought it was quite warm, I found ice in the bottom of the little basin. The trek was at first along the river then through woodland – a steep bit uphill through short thick bush. My *tipoia* lagged badly and then I found the top very much to pieces – made them stop and tie it up. Then as the bush was thick I walked on again; when next I looked round I found them carrying it edge up, and my press on top of my cushion on a man's head! Then I did get angry, and at the first opening in the bush, got in and stayed in, even though I would rather have walked until we reached the river beyond which Kunjamba lies.

KUNJAMBA MISSION

This is a wide open valley, the woodlands bordering it very thin and that on the south-east very low. We skirted it for some time and then crossed it – a fine swift little river some 12 ft broad. The ford was deep, up to the boys' thighs. A mile or two along the far side followed – quite pretty, then the path struck off away from the river, made for the far side of the plain and ended in a strada leading straight up the hill. There we were met by a mission boy – stopped for a time for lunch as it was past 12, then proceeded up the steep sandy hill to the desolate-looking mission station where Mr Reinhart and Mr Muir gave us a hearty welcome. We found they had waited dinner for us, so we had to eat again, and drink, lovely fresh milk by the cupful – such a treat!

The afternoon was busy – I had to bake bread and also made a cake, the latter with Mr Muir's flour, both good, baked by Telosi. Then I found Mr Muir had some lemons, and offered to make him lime juice and marmalade. I was still cutting up the peel and the cake was still baking when the bugle went for evening service, and at the same time our porters came up and began to make some request – for food apparently. The peel was hastily finished and put on to boil, the cake (or cakes rather) rescued and turned out beautifully baked, the boys told to wait till after service (incidentally they disappeared and we heard no more that evening) and we set off for the tiny little church. The boys sang remarkably well, harmonising beautifully.

After service the evening passed in talk, supper, fudge-making by Mr Muir (very good), marmalade by me, getting Portuguese and native phrases, reading and writing letters etc., and it was midnight when we finally got to bed. Much to my consternation the two men (Mr Muir is only just up and about after two weeks in bed with fever) turned out and slept in Mr Reinhart's tent. I had my bed on Mr Muir's very comfortable bedsprings. This house is an old one, propped up in places – his own house was burnt about Christmas time.

Friday July 3rd—Signalized by two events – our first lion, even though at a distance – and a strike on the part of the porters. We had intended to make an early start – Mr Muir told us Kunzumbia was one-and-a-half days' trek, not one day as Mr Bailey had told us. However, the porters in a body came and asked to be paid off, they wanted to go back. Fuelu adhered to his intention of leaving us here and received his pay, as also did the porter with the sore foot. The others Miss Bleek refused to listen to, but after consultation, as they said they wanted food, we decided to give them each a spoonful of salt. This, however, they refused and we asked Mr Muir to come out and speak to them, which he did, whereupon they left in a body, saying they would get their pay at Muié.

We took it calmly, took off our hats and sat about reading papers etc. By and by came a deputation, in chastened mood, and with Telosi and Mr Reinhart as our respective representatives, it by and by appeared that they had bought plenty of food with the salt given them, but had gorged themselves thereon and really had no more left. They could not get sufficient at Kunjamba which is a poor country. Finally Mr Muir said he would sell them some meal, although his reserve was small. There was plenty of money – the '*piastres*' we had given them had not been spent at all! Thus matters were arranged and we all retired for lunch.

Lion

Then just as Mr Reinhart was about to leave (his porters with his gun etc. had already gone) there was a great outcry from the valley and boys darted out with spears and rushed down the hill – a lion, said Mr Muir. We could see nothing and he said he thought it was in the bush across the river. After Mr Reinhart had left, a cowherd with his cattle and several little boys came along. The cows started feeding and the humans came and sat down in front of the house. Then followed a graphic account of how the cattle were feeding near the river. The herd stooped to drink, and suddenly saw a lion spring at a cow and seize her by the nose. The cow threw up her head, catching the lion (who must have been young and inexperienced!) on her horns and throwing him over her shoulders. He made off down the river – it was this side and he remained on this side. The cows and their calves surrounded the herd and all dashed for home, the man with hand and mouth raising a great outcry to give the alarm.

The action of the story teller was so expressive that a knowledge of the language was not necessary! We asked which cow it was and the heroine was pointed out to us, a small black and white cow with comparatively short straight horns. There was no sign about her of her recent adventure, except that while her sisters were already busy grazing, she was standing meditatively gazing straight ahead. Mr Muir could not find a scratch on her, so concluded the lion had sprung for her nose but had missed. 'It is the Lord helping us', he said; then in Mbundu, 'Let us pray to God', and he and the little group of herd and his audience knelt down on the sand and prayed, giving thanks to the Lord for his mercies.

About 1.30 our tardy porters came up and tied up their loads, and at 2pm we at last got started. The lion acted as a spur – there were no stragglers! The path led south-west down the hill at a slant, finally passing through a village and reaching the edge of the plain of the Kunjamba, up which we went for some way. Then we crossed and about 4pm made camp. The old difficulty of making the *tipoia* boys bring wood for our fire followed. They pitched the tent and made my sleeping place and brought a few small sticks, getting plenty of large ones for themselves. Finally I called out '*Waha tipoia, nehe vibuna vilatu vikama*' (*Tipoia* boys, bring three logs), and the 'old soldier' brought across his smallest at once, and eventually we got a fairly decent fire. We should really have gone much further up the river. It will make tomorrow's trek very long.

TREK TO KUNZUMBIA

Saturday July 4th—Managed to start at our old hour of 8.10, the porters all making a point of starting in a body. Poor old 'uncle' with the food box, always last to get his load, and slow at doing anything called '*Kanda! Kanda!*' (not yet), and I'm afraid Miss Bleek listened to him – much to my annoyance as it isn't right to let them call out commands to us like that. I went on and the bulk started; she waited to see the tail come on. The morning's trek was very beautiful. We crossed the river almost at once, not however before we had seen spoor of a large buck (*ntengu* = ?Kudu) mixed with that of a lion on the path not very far from our camp, then cut up through the bush over a steep spur of land and down again on the far side to a river which was also the Kunjamba. One, they told Miss Bleek, was the little, the other the big Kunjamba.

This was a narrow valley with steep wooded sides and a fine echo most of the way up. It wound in and out, and we crossed it four or five times. We had brought one boy, Lomeii, to take Fuelu's place in the kitchen and another one from the village near which we spent last night. The latter acted as guide and just before 10 when we made one of our numerous crossings he said that was the last water for a long time. We stopped for the men to drink and fill their calabashes, also for the porters to have a bit of a rest. Then we started, expecting to have seen the last of the river. The dry valley, however, continued for several miles, winding in and out, in many places floored with red clay with which several of the boys smeared their faces.

When we finally left the river valley we went up through wooded country on to much higher ground. Then we came out on to one of the grassy openings in the woodland, with scattered low bush, recently burnt, and over this we went, on and on. By and by I saw swirls of what I at first thought was smoke – then found to be tiny whirlwinds raising ash and burnt leaves like puffs of smoke, perhaps caused by the still warm ash.

It took us a good hour to cross and then about 12.30 we stopped at the edge of the unburnt bush and had a hasty lunch. Tea with one of Mr Bailey's limes instead of milk was very thirst quenching. We had seen lion spoor again early in the trek. The porters were eager to start, and after half an hour we went on again – on and on through bush, all small stuff, practically no large trees and very difficult for *tipoias*, so I walked most of the way – it was hot! I was glad when at last we got into larger stuff again and there was some shade. Here we met two men and a dog from Kunjamba, and soon after glimpses of distant blue began to appear through the trees – sure sign of a river valley (Query: is the intense blue a sign of the greater humidity of the air in the valleys?) which they said was Kunzumbia. A long way off, however, and our way wound on and on – a narrow path which sometimes one could see only for a yard in front of one, and at others could see only several yards ahead, while at one's feet it seemed to disappear.

Miss Bleek and about a dozen men were ahead of me, then I headed the rear-guard. It is wonderful how little trail a dozen or more men walking single file leave on these narrow sandy paths – each one partially obliterates the footprints of those in front. Often I had to pause and look ahead for the path, or where it made a sharp bend through bushes I would overshoot the mark and have to turn back. I was glad when we were through the close low bush and I felt I could get back into my *tipoia*.

At last we came out of the wood and on to a wide extensive grassy plain – no water yet however. We went on along the edge of this and then across it, and it was not till 5pm when we neared the river itself that we got water. There were a lot of *Protea* plants – the narrow-leaved one and a broad-leafed species very similar to, if not the same as, the Mueri one. At the first water I got out, as my boys wanted to drink, and walked on till it got too wet. I found a yellow-flowered floating *Utricularia* here – possibly a new species; the leaves looked rather different from anything I've seen. The river was pretty deep – not so clear as the other rivers, though very swift. The boys were in up to their thighs. Two natives in much patched shirts and trousers, evidently from their haircut from a Portuguese post, were at the stream and directed our crossing. Once across, twenty minutes or so brought us to a good camping spot a few hundred yards from the edge of the river plain, in a very pretty open wood, mostly *munyumba* – *musivi* and *mukusi* (bark tree) conspicuous by their absence.

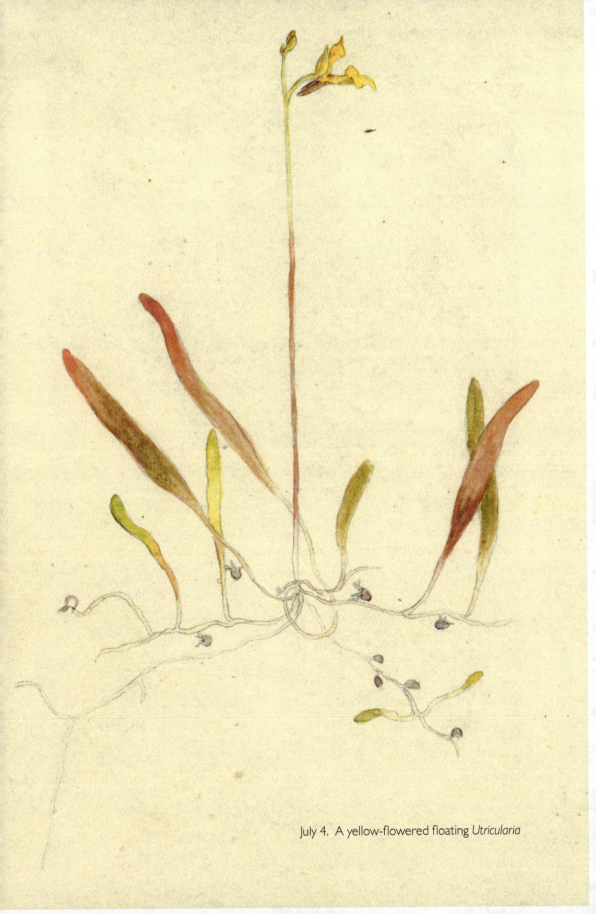

July 4. A yellow-flowered floating *Utricularia*

July 6. 'A large, rotate white and red flower' (*Napoleonaea gossweileri*), Kunzumbia

July 9. 'Crash! Splash! – an *nsonge* which had been lying in the grass dashed off and away!'

July 10. Lechwe in the short grass

A QUARREL

Here we were all glad to camp. The boys felled a large tree, we got the tent pitched and my cubby hole made, this time at the edge of the circle of fires. Telosi had been very good in bossing the boys and directing operations, but he is young and has not much authority. He told them to make a circle and they coolly ignored him, so he made a dash for the fire of one group who had settled themselves in the centre, picked it up and threw it in the direction he had indicated, whereupon the boy in charge said something. Telosi lost his temper, went for him with a stick, got slapped in the face and a free fight started. The other boys fortunately did not join in but separated the combatants who, however, at once started again and had again to be separated. This, though not pleasant, seemed to clear the air a bit. I longed for a few forcible words with which to express my feelings, and could only remember Telosi's name. He, of course, was not to be blamed for trying to do his duty even if he did lose his temper in a most un-Christian way!

KUNZUMBIA CAMP

Sunday July 5th—I was anxious to get our stockade built but we did not wish to employ these men further. So first thing, after counting out and arranging their pay (we have taken them on at so many different places that there were half a dozen different rates) Miss Bleek called their names, I handed out the respective pile of iron, or pile and paper voucher for pay at Muié, and Miss Bleek paid them off. They took it like lambs, so evidently it was more than ample (the full from Muié to Kunzumbia was $18.50 plus the salt they had for food on the way for *tipoia* boys, and $9.50 plus salt for porters), and there were no complaints but a good few came to buy salt.

We found they were not returning till tomorrow so that the need for haste in building was not so great. In the afternoon, as we started out for the village, we met the chief and a retainer (the Chief's name is Kamundonga) who laughed when we explained our intentions of building a kraal as protection against lions and said '*Ndomba wahi!*' (no lions) – on this side of the river.

STALKING A BUCK

In the morning word came that there were buck feeding nearby, at which Telosi again asked us to go out a-hunting. To satisfy him – and incidentally to please myself – I took the pistol and off we started, wading with great noise half a mile or so across the swampy plain, and sending two boys round to drive the game downwind towards us. No good – the buck turned out to be dead trees and though we saw two large black birds about the size of turkeys (I had seen two in the Kunjamba valley) – they are black with red heads and white tips to their wings – they were too far off to get a shot at. I did want a gun! So we started back – there was heaps of spoor and the forms in the deeper grass where the buck had been lying.

Just as we were leaving the plain I glanced round to the north (my right) and there was a grand chap – *nsonge,* Telosi called him – feeding quietly a couple of hundred yards off – a beautiful shot for a rifle but no good for a pocket pistol! We got into the wood and I tried to stalk him. He looked up and watched us but did not move off. I got near enough in the grass to have a good look – rufous brown back, white underneath, large rather humped hindquarters, white streak down his face and large slightly curved horns – a good deal larger and heavier than springbok. Then he moved off – not very far or fast, and we continued our stalk through the woodland, in vain. I could not get near enough and Telosi's effort to circle and drive him up finally put him to flight – past us not away from us – up the river and downwind, canny beast!

Telosi tells me the birds we saw were not the same as those on the Kunjamba – the latter were *mungombe* and had red heads, these two were *makande* and had not red heads. They were too far off for me to see that detail, whereas I was quite close to the Kunjamba ones. As we stalked the buck, I thought to myself what foolhardiness it was – stalking game in lion country, (armed with a pocket pistol!) and had visions of us being stalked by a lion. However the chief's contemptuous little laugh at the idea of lion on this side of the river does away with that picturesque little detail. I made bread today – prune yeast made nine days ago which had spent its life trekking in the sun. I put the sponge in the billy in the sun, made the bread midday and baked about 5 – result excellent except that the chief's visit rather distracted Telosi and he did the baking too quickly and burnt it slightly. We are sending one loaf to Mr Muir with letters by one of the porters tomorrow.

VISIT FROM KAMUNDONGA, THE CHIEF OF THE VILLAGE

Spent the evening writing to E.L.P., G.V. and Jeanie Causton. The boys bought a lot of buck meat, very cheaply, this morning. Telosi tried to get some too but was too late. Later in the day, however, the chief sent us a present of a leg and a promise of *Vasekele*, for which he was rewarded by a 'present' of salt. After meeting him we turned back to camp, with him following us, and they squatted down by the fire. Kamundonga is an oldish man, with rather a nice face – the features of the people round here are rather pleasing.

By and by, after the salt had been given and appreciated, Telosi came up to me with a request, something about writing at Kutsi. I could not make out what. Finally Miss Bleek interpreted it as a request to show the pictures and got out her book. That was nearly right but not quite – it turned out to be the paintings of Bushmen I had done at Kutsi, so they were produced and provided much interest and amusement, as did my hand mirror and camera. The old chief was rather pathetic – he gazed at himself, stroked the sides of his face, touched the hollows in his cheeks and the white patches in his beard, and evidently was rather surprised at the signs of advancing years. In the book, the pictures of women left them cold and uninterested. By the hunting pictures they were fairly amused, but the dog in the *tipoia* as usual brought down the house!

NSINGE BUILT

Monday July 6th—Miss Bleek, Telosi and I finally selected the spot for our new camp. Miss Bleek signed on 13 men and stayed to superintend operations – no sinecure as the men here don't understand our type of building as the Muié ones did. However by 3pm the *nsinge,* of the same type as the Kutsi one, but deeper, was complete and the workers exhausted by their long day's work (10 to 3 with an hour off at midday and frequent rests!). I stayed in the old camp, made bread etc; we decided not to move camp but to stay there another night, fear of lions being removed. In the morning I photographed a charming flower Miss Bleek found – one of the many geophilous plants, with a large, rotate white and red flower, very sweet scented, possibly Apocynaceae, but I've not yet attempted to run it down. Finally got to bed, strangely weary. In the afternoon of Sunday, by the way, I squeezed the ten small limes from Mr Bailey's tree, getting the honey bottle full of juice, cut up the skins, put them to soak, and today made excellent, fruity marmalade.

Tuesday July 7th—Miss Bleek again went down to the *nsinge* and I stayed – again to make bread! – and superintended the moving of our goods at the old camp. It was competently done by Telosi, Lomeii and two small boys, while Samolova got grass, and Miss Bleek started the men (only four with three boys) at building the shelter for the tent. They were very slow, and at 3 when they knocked off, had not started thatching the roof. However they have got the framework up, and tomorrow I hope to get that and the fence etc. finished. Samolova and Lomeii have just built a very nice table. The shelf is half finished and we have been visited by several women and babies. Two, one elderly, the other old, we think were the wife and probably aunt of the chief, or possibly two of his wives.

I saw a pretty, thrush-like bird – back dust-coloured, throat and breast spotted, yesterday. He has a clear whistle of two or three notes repeated twice. This afternoon saw a beautiful little blue butterfly with iridescent wings – a little gem.

Wednesday July 8th—Well, old Kamundonga is apparently 'queering our pitch'. After all the pleasant speeches and promises of Bushmen after the porters were gone (according to Telosi) he came into the camp alone this evening. Last evening, by the way, Miss Bleek sent a present of some sweets and a red kerchief by Telosi, who came back with thanks saying the chief would come himself next day. This accordingly he did, bringing a present of a basket (conical) of meal topped by a tit-bit of buck's meat which nearly smelt the camp out.

We at once turned the conversation to Bushmen and in return got a long, lyrical, mournful story of which at first we could not make head or tail. At last, so far as we could make out, the story emerged thus: The white people (Portuguese) had demanded tax from the Bushmen, and when it was not forthcoming held Kamundonga responsible, accused him of hiding the Bushmen, bound his arms behind his back and beat him. How much is true, how much we have misinterpreted, remains uncertain.

That there have been Bushmen here and lately is certain. Women and men however persist in saying *'Vasekele wahi!'* (not here) but that may be according to instructions – all delight in seeing the pictures of the Bushmen and Telosi says *'Kamundonga sweha'* (i.e. is hiding) the Bushmen. We wonder whether this is all a story and that this is the village

Hide and seek

A ten days march to the south-west brought us to the Kunzumbia River, a tributary of the Lomba where many Bushmen were said to live. Here we played hide and seek with [the Bushmen] for several weeks, finding many recently occupied encampments, but only twice getting speech with the people themselves. The Kangali chiefs were evidently keeping them away. I think the local officials had been trying to make the chiefs responsible for the Bushmen paying hut tax and had beaten the chiefs, when the payment was not forthcoming. Naturally they denied the presence of the little people, and tried to prevent any Europeans from seeing them[1].

which according to the story we heard at Kunjamba fell foul of the Bushmen and gave them poisoned beer to drink, whereat the Bushmen retaliated and shot them with bows and arrows!

WORK ON THE SHELTER FOR THE TENT AND THE KRAAL

Tomorrow Miss Bleek is going to visit some of the other villages. Work at the *helu na mbalaka* continued this morning; men were signed on, some for grass and some to start work on the *cepatonga* or kraal. The grass gatherers arrived and the three youngest boys started in on thatching the roof. I was just looking at it and thinking that if the timbers collapsed, not much of the tent would survive, when crack! – the centre cross beam broke in two right above the tent and there was a hurried scramble of the three boys for safety. Fortunately it did not quite part and the roof hung suspended. We quickly removed the tent and all therein, and then tried to get the men to put two strong poles to lever up the ends. No, that would not do. Much loud talking and gesticulating. Another attempt at advice on our part; still no good, more discussion etc. Then I managed to make one specially intelligent helper understand that if a temporary cross bar were put in to take the weight, the broken one could be replaced. '*Eh wah! wa ceti!*' a little more discussion and half went off to get *ceti* and the rest sat down to rest, smoke and look at us.

Then they came with a middle-aged tree trunk which was tied to one of the supports with a few twists of bark and held by the united embrace of two men while on to it was placed one end of the new cross beam (*munyumba* – the other was the wrong kind of tree and a rotten one at that). The other end was raised by half a dozen men while one swarmed up the upright and with head and hand tried to lever it into position – quite in vain, of course, as it meant raising the entire roof by himself, clinging to his pole with one arm and feet.

Then they got another pole which they shortened (too much) and placed under the end; then all who could get hold of it lifted, pushed and shouted at once. Still it was several inches too low, but nearer than before. Suddenly another brainy one (a little pock-marked man who has worked each day) had an inspiration, seized the logs of our fire –two large ones, all red hot and smouldering at one end – piled them up and explained his idea. Standing on these (one man actually stood on the burning end!) they tried again and very nearly succeeded. A little more arranging, putting one log on top of the other (Telosi meanwhile having poured a little water on the hot ends!), resting the lever on them etc., and success at last rewarded their efforts and one end of the new beam was in position, the other temporarily so. Followed, of course, a rest.

Then work started at the other end; after several efforts to lift it by main force, they desisted and one man (the old chief's henchman) taller and braver than the rest, assisted by the young boys, pulled down all the grass already placed and most of the cross timbers. Then the second end was got into place, the boys again climbed up, put back the poles, directed by the same old chap, tied them across the centre and put on the grass. By then, 1.30, they considered they had done enough and gathered for their *fueta* – one escudo each and a spoonful of salt.

Thursday July 9th—About 10 Miss Bleek with four *tipoia* boys and Telosi left for the next village up the river. As they neared it they met a man coming down, who said there were four Bushmen in the village. Just before reaching it, the boys stopped to have a drink and one ran on. When they reached the village he was sitting talking to the men in the central opening and the Bushmen were *wahi!* In fact the spokesman went so far as to say that he had not seen any since he was a tiny child! Thus completely giving the show away. They got back a little before 2.

I meanwhile had been giving audience to a crowd of women and children who stopped on their way to hoe their garden for a little amusement. They asked for the pictures of the *Vasekele,* which caused great delight. Then followed the Statham book, till finally I said *Kunahu.* Lomeii was at hand to interpret – he is quite quick at understanding one.

Then one explained that she wanted medicine (*cihembu*) – so Lomeii made her come forward. She had a bad sore on her ankle; she had clapped a leaf on it to keep out flies. I gave her some permanganate with instructions to put a few grains in hot water and bathe it thoroughly, morning and evening, and then put some ointment on it – also a couple of strips of old vest to tie it up with. Next an old dame (she of the beads) complained of sore eyes; she was given boracic with instructions to put it in water and bathe her eyes. Then, not to be outdone, the other old dame said she had headache etc., so to her I gave some fragments of quinine which I made her eat at once. I think she rather liked it!

About 4 Telosi came and asked me to go out after *vansonge* – I said I'd go later, and about 5 we went out. We had an exciting time stalking the splendid beasts – there was a herd of about 24. We got fairly close then tried to stalk them – Telosi circled and tried to drive them towards me, but they broke and made for the reeds towards the river. We followed, up muddy sides through long reeds, ankle deep and sometimes knee deep in water, most exciting. Then I tried to stalk one (the others had disappeared) which was standing looking towards us beyond the reeds. We followed him and then Telosi drove him past me, but too far off for the wretched little pistol and as the water was getting very deep we turned. Just then a duck flew over our heads. I called out 'Look there's a duck!' Very stupid, as Telosi couldn't understand, and not a dozen feet to my right – Crash! Splash! – an *nsonge* which had been lying in the grass dashed off and away! The only chance lost. Finally we turned back and got home after dark – very wet, very hungry but all the better for the exercise and the interest. I dried my muddy clothes, after rinsing them out, and planned to go again – Telosi eagerly assenting – early tomorrow.

Friday July 10th—At day break I got up and dressed and by 6am (light, but sun of course not up), had summoned a cold and protesting Telosi, very loath to leave the warm fire! I ate a couple of biscuits (I baked a cake and some flour and *wunga* biscuits yesterday) and started off down the river ostensibly for buck but also in hopes of finding traces of Bushmen at the first village. However the latter stands far back from the river on the hillside so we went on past it. We saw a couple of buck and then a solitary one but did not attempt to get near them. Further on, however, there was no sign of buck and we turned back up the river. The sun meanwhile had risen – a great disc of flame colour coming up behind the blue line of the distant woodland beyond the wide river plain – a glorious sight! – and at once the air felt warmer. The whole plain was golden.

A BUSHMAN ENCAMPMENT
We came back along the edge of the plain at a good pace – tried to stalk the solitary buck, quite impossible of course, as he was right in the open. Then Telosi stopped just as we were passing a path leading down from the wood to one of the little bays in the woodland – in these little bays of grassland, springs are thrown out and here are the water holes – and pointed out the fresh foot mark of a Bushman. We followed up the path into the woodland and came to a neat little Bushman encampment – three shelters made of leafy branches, several fires each with two neat little beds made of a little grass spread on the ground with a 3-inch pole at the head for a pillow – certainly a round dozen of people besides children. All was fresh and clean and there were signs of a hurried departure – pestle and mortar and one or two pots lying about, but the ashes were quite cold. Then we saw Bantu footmarks, some of a man running, leading to and from the camp, and the conclusion we both drew was: 'Kamundonga!' Evidently he had sent a messenger to warn them to take to the woods – who or what he thinks we are, goodness knows! This of course is proof positive that when he and his say 'Vasekele wahi!' they are lying, or at any rate if not actually lying, since it is probably true that, at the moment they are not here but in the woods (*mu musengi*), it is only because they have been sent there.

A HERD OF LECHWE
Leaving the Bushman encampment, we cut obliquely through the wood back to the river past the path leading to our *nsinge*, and up to where we were last night. There was the herd of *vansonge* – some 30 strong – and a few outlying couples or single animals. Two were feeding near one of the bush peninsulas and I sat behind a bush and watched them for some time. One went on feeding (?doe) the other, with a fine head, watched me, or where I was, snorting from time to time. When they run in the short grass they drop their noses, stretch out their necks, and the hind quarters are humped up high above the shoulders – they look almost ungainly. In the longer stuff, however, the powerful hind quarters come into play and they proceed by great leaps. They are much

larger and heavier than springbuck, rather a richer, redder colour, with long outward-curved horns and large ears. We think they are what Statham calls lechwe or *songwe* (Here *nsonge* singular, *vansonge* plural – Luchazi). After watching these two we went on through the wood.

Suddenly, a few yards in front of me, out stalked at a slow stately pace, a large bird – some kind of crane or stork body, tall slender legs which doubled up in the middle, glossy black body with a brilliant white patch behind the root of the neck, long neck, knob-like head and an enormous red bill. He had evidently been sleeping in the wood and was wandering out for his morning meal of frogs or fish. Later we saw them (two of them feeding out in the plain) and I also saw one rise and fly – a grand stately flight – and come to rest, stretching his long stilt-like legs down to the ground before he alighted (*likande,* plural *makande*). I longed for my camera – he was near enough to have made a good photo. I must go out again one morning with the camera.

Then we set to work to try and stalk our *vansonge*. We are both tyros at the game however, and as the land is all swamp it isn't comfortable country to do a proper crouching stalk! We did not get nearly so close as last night. I reached the river, which takes a bend westward just here – evidently a spot much affected by the *vansonge!* Incidentally, I collected two slimy leeches which, however, I discovered clinging to the woolly stockings I had on and dislodged them before they got through to my ankle. We tried several times but never got near the buck. Then on the way back Telosi saw something move and thought it was an *nsonge* feeding. So I gave him the pistol and let him have a try – he stalked it only to find it was a *likande,* which I, watching, had long since discovered as I saw its mate sail out of the wood and start feeding likewise.

I collected a few specimens and then, feeling decidedly hungry, we made for home. Samolova, fetching the calabash-full of water from the river (calabash bought last Sunday for one escudo – a fine big one) had joined us. No wonder I was hungry – it was 10 o'clock when I got back! I had a hot bath and then about 12 had early lunch – not an economy at all, as I certainly ate an enormous breakfast and an enormous lunch!

Miss Bleek meanwhile had been holding the usual séance, *Vasekele* and other pictures. By the way, she offered a piece of stuff to anyone who would bring her a Bushman, and yesterday evening one of the men we met coming from Kunjamba came in and with lowered voice and mysterious air promised to bring one in today. I had a good laze and sleep in the afternoon after my strenuous morning, and about 6 our friend arrived with bow and arrows – got up for the occasion! – but without the Bushman. He'd go out again on Monday – he had not been able to find one. He is a bit suspicious and his story or stories don't altogether correspond and we suspect he has been put up to it by Kamundonga to allay our impatience and prevent us looking for them ourselves.

A man from Muié came in today – says he met Mr Reinhart on the Kwando. He is going back on Sunday and offers to take in anything for us. Wish he'd brought the mail!

Saturday July 11th—Miss Bleek went out early with Telosi to see the encampment and photograph it. She left about 7 and got back a couple of hours later. Then I set the other two boys to complete the fence, or rather that part of it which was started, by filling in with leafy branches. The Muié man happening along, asked if he might help, went to get his axe and in a few minutes half a dozen helpers had arrived – we took on the first but refused the rest. A couple of hours saw the work finished. It is open to the east of the shelter but in the circumstances we don't mind that. I made sponge for bread early (half local meal) and it worked beautifully. At 11 mixed the bread and set it to rise, which it did not do very much. Finished my baking by 3.30 – fairly early.

Meat is getting short and we had to kill one of the 'precious seven' last night. We must eke them out as far as possible as we can't replace them here. As it happened, we could have postponed it, as about 3 o'clock a small boy – one of the most industrious workers – brought in a bird something like a quail, small but very plump, which we bought for 50c.

The boys were paid – Lomeii and Samolova both asked for a week's advance and bought *tanga*, the latter adding his belt for five escudos (the same Domingo wanted to sell us for 10!). Telosi asked for an increase in his wages! He already gets 12 escudos per week. However we decided to give him 14 and Miss Bleek gave him her old white dress as a present. He was pleased with the latter but evidently thought he ought to get more money! He is a good boy and does his best but does not carry much weight with the others.

We had the usual request for the pictures of the *Vasekele*. We are pretty certain that they know old Kavikisa or Kasindzela. They always ask for him, roar when they see his picture and today one of them, when I showed it, said to the other: '*Kasindzela, na Mueri*'. I said, '*Na Kutsi*' and he corrected himself '*Na Kutsi na Mueri.*'

The audience yesterday was chiefly composed of the men from the village Miss Bleek visited the day before. Today one old man wanted his eyes doctored – accordingly had them bathed with hot water and boracic. Tomorrow Miss Bleek is taking Telosi and Samolova, her blankets and some food and going down the river to try to get in touch with the Bushmen and possibly find a spot to which to shift camp. She may be back the same day or may stay away for the night – it depends on circumstances.

Re-started packing away and writing up specimens – only managed a couple of hours at it, what with superintending *cepatonga* etc., bread making and baking, and so on. Am enjoying '*Virgin Soil*' at intervals – have really got into it and am appreciating it. Our last book! I do hope M. will send the London Times or some other paper. We ought to get our post tomorrow or Monday, if they have sent it off promptly. It is getting cold and late (9.45) so I'm off to bed. Last night was the first cold night we have had here.

Miss Bleek goes off to reconnoitre

Sunday July 12th—This morning, after an early breakfast, Miss Bleek set out at 7.15, leaving Telosi and Samolova to follow a few minutes later with her things and their blankets. I was going to walk part of the way with her but, as she was in a hurry to be gone, she started on while I was putting in my film and putting on my shoes. She omitted to tell the boys which way she was going! So it was as well that I did stay behind.

I followed as quickly as possible for a quarter of an hour, but she must have gone fast as I could not even get a glimpse of her, though I could see her foot prints. I gave up the pursuit and turned to the river – waded through the swamp to the first of the bramble knolls which I noticed when I came over and spent half an hour examining the flora of the curious little islands in the grassland. I was bending down collecting specimens when I heard an indignant 'Hooff!' behind me, looked up, and there was a buck gazing at me, about 200 yards off. He resumed his feeding but was rather suspicious. Several buck tracks lead to and from this first island. I suppose it is a favourite grazing ground. It is only a few yards across.

There were a lot of small birds around – some seemed to be '*klappertjie leeuwerik*', anyway they clap their wings as they fly like those little Cape larks. Others in the swamp, seed-eaters I think, are rather like bullfinches in colouring and build, though rather smaller. Some of the tiny ones build beautiful little nests in the drier grass – domed like a little hut, with the entrance to one side just above the ground. The one I saw was evidently being used as a sleeping place – the little occupant flew out (this was the other evening when I was out tracking with Telosi) and I looked in – no eggs at all.

I got back about 8.30 having left here at 7.20. Put away specimens, much interrupted by Lomeii with his '*Nji xaka kulandu muivi*' (turned out to be a leather strap) – not for sale and certainly not for one escudo, which is his present worldly wealth! That settled, by and by it was '*Nji xaka kulandu lesu*', which after much difficulty I made out to be a red kerchief like the one sent to Kamundonga as a present – also not to be bought for one escudo! He and Telosi have evidently done a trade, as he is going about attired in the somewhat awful old white dress Miss Bleek bestowed upon Telosi yesterday. I wish he could buy something and be satisfied! The Muié man then came and again asked about the writing for Muié, borrowed a *guya* (needle) and thread (which by the way he has not yet returned) and then stayed and talked to Lomeii. Wish he'd stay altogether! Then perhaps I could have a little peace – the youth tries to be helpful but I do not like him any better.

BUSHMEN IN THE VILLAGE

About 2pm I got out my sketching things and tried to finish the two Kutsi ones. The Muié man was here and the boy with the feather came with food for Lomeii. Then he ran off and I heard the word *Vasekele;* Lomeii shortly came up and with lowered voice explained that there were two *Vasekele* in the village – I must go with him. I tried to get him to go and bring them here, without success. Finally I went with him – was reluctant to leave the *nsinge,* but thought the Muié man was staying. However, he went too, and I risked it. There has been drumming going on all day.

We went right up into the poor little scattered village and there in the centre was an open hut beyond which were five or six men astride their curious drums, beating them with their hands – three or four of them men who worked for us here. Several men were in the shelter, two of whom came out and greeted me. Another was working the bellows of a small forge, while two or three more forged steel blades for knives. Among them sat an old Bushman, evidently quite at home; by and by he came out and pranced about – very much Kavikisa's type.

Then in front of the drummers, who had their backs to me, danced out a figure – a man either painted over arms and body or in skin-tight garments – the former I think – with skins flapping around his waist and some kind of headdress on. He came down the slope from the bushes, dancing and kicking up the dust, while the drummers beat hard. This continued for some time – every now and then he'd retreat into the bushes, or someone would run out from a hut with an offering of a mealie cob – these were collected by a youth from among the drummers. Now one of the latter would get tired and another would take his place, and every now and then several women would come out of one or other of the huts and chant a chorus. I should have liked to stay and watch for some time but did not like to leave our belongings for too long, though I believe they are safe enough. Explained to Lomeii and the toothy man with the string round his head that I couldn't talk to the Bushman, and that if they came to the *nsinge* I'd give them salt.

Everything was as I left it – I repeated the above to Lomeii who began talking some nonsense about cutting throats if they came to the shelter – perhaps romancing that they were frightened or trying to frighten me, I don't know which. However, we had not been back long when my toothy friend arrived with two Bushmen, as calm and collected as possible, no more scared than I was and quite refusing to do any play acting!

The old man, nearly toothless, looked very skinny and empty, poor old thing – very like Kavikisa in build and colour, the other big and broad of face, very dark, also without some of his teeth, a very mixed specimen. When asked for Bushman names they said their names were !Kõ and Golli! When simply asked their names they gave Sungensu (!Kõ) and Kantema (Golli). I'd already shown them my Bushmen pictures which they accepted quite calmly.

Then I tried to take their photos, and that also left them cold. They took up positions quite readily and I'm sure they had either been photographed before or had heard what we had been doing at Kutsi. Unfortunately my shutter stuck just like the Vest Pocket Kodak, and I could not take them. (So much for having my cameras overhauled by Kodak's!) I gave them each a spoonful of salt, and Lomeii insisted that they wanted meal as well, so I gave them a cup each while the toothy one promptly picked up the old broken calabash to put it in, and they departed, saying they'd come tomorrow.

Lomeii, very lonely, has made several journeys to the village, came back once saying there were now four Bushmen there and now (5pm) has just brought the Muié man here once again. The drumming only stopped half an hour ago, and is now succeeded by the sound of pounding meal. Now I must start letters to send to Muié.

Hope Miss Bleek will come back tonight, but as the Bushmen have come in, she probably won't, and I shall not know what to do except give them more salt and meal! Lomeii has again come to try and buy his *lesu* – I'm sorry but he has already had a week's advance.

The 'pea-shooter', a bush about five feet high, has been shooting all round all afternoon – crick! crack! One on the far side of the fence – about seven yards off – has just shot a seed right on to my bed. It is an explosive mechanism with a vengeance. The seeds, though flat, are quite large; usually one valve, as well as the seeds, is shot off.

Monday July 13th—Lomeii was very restless and at dusk he came, with his everlasting '*Nji xaka...*' – this time to go to the village to sleep. Poor wretch, it was too lonely for him by the kitchen fire. First I told him he could sleep by our fire, then decided to send him for Parata (the man from Muié) and ask him to sleep here, which he did, going off armed with the axe and coming back contented with Parata. I sat writing letters till late – my watch said 9pm but it seemed much later – possibly it is ten minutes late or so. Wrote reams to the usual source, and also to Miss Saltmarsh and Mrs Bailey. It got cold and Lomeii had not made a good fire, so I was glad to get into bed.

This morning I made Parata happy with the promised *piastre* and a spoonful of salt, but he wanted a letter for Kunjamba too, so I scribbled a note to Mr Muir. A little later our pock-marked friend came to know if I was going on with the building of the *cepatonga* – I said no, but eventually decided to do so and sent the bright specimen to engage the said would-be worker and one other. Of course he wanted to get more. He came back with two, quite different ones, and I set them on to the back wall of the tent shelter. Others, of course, came up and in the end I engaged five and got the wall done and the fence completed. The fifth was an old man from the upper village – I did not want him at all but finally let him earn his escudo and salt. He has a piece of figured blue stuff as shirt and a coat, so is evidently a man of property!

Two girls of 15 or 16 came and after sitting watching me work for some time – occasionally chanting a rhyme which I suspect to be either a rhyming begging or a description of us and our queer doings – probably the latter – they finally asked to see the *Vasekele*. I showed the paintings and promised the book later as such shows invariably stop all work. They finished about 12.30, and then followed the longed-for show. They always ask specially for *Vasekele,* then *Gengi* and finally *wata leuwanika*. New spectators were the two girls, the old man from the further village and the ancient nearly blind old boy who came to have his eyes doctored two days ago. The rest sit around acting as *cicerones* – pointing out the special features of each.

BUSHMEN APPEAR AT THE CAMP

Before it was finished, just as I, very hungry, was about to dismiss them and have lunch, behold the Bushmen! – three this time. I again tried to get their names, not very successfully, photographed them all separately and old Golli twice – once side face, close up – it's a nice old face, much more attractive than the usual run of Luchazi face – then told them to wait. Lomeii took them out into the kitchen premises and entertained them there. I'd had a peaceful morning by sending him to the river to do washing!

The birds have been whistling a lot this morning – a couple answering one another, I suppose somewhat of the Bokmakierie type, though the note is quite different, and the thrush with his two shrill, sweet metallic notes, at an interval of a fourth or thereabouts between them, as well as the doves. The fowls are absurd – they now come, not only

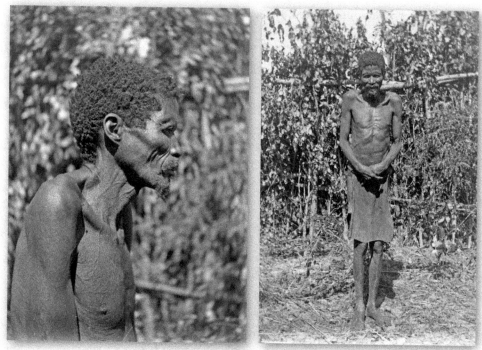

July 13. Old Golli, waiting patiently for his salt, Kunzumbia

when I call them, but almost whenever I move they stand round and gaze fixedly and longingly at me. They even gathered round my bed this morning and gazed expectantly through the mosquito nets. And yet people say hens are stupid!

Before I had finished lunch, all the rest of the village had re-assembled, men, women and children. They waited in the kitchen till I'd finished and then flocked in. The old boy from the up-river village had a sore toe which he wanted doctored, preparing it for me by scraping it hard with a piece of wood and washing it with water for which he asked Lomeii! I bathed it with hot water and potassium permanganate (gave the fowls the remains of the latter for the good of their health!) then made him rub on a little zinc ointment, and before the breathless admiring crowd cut a strip from an old vest with my big scissors (they'd evidently never seen such things before!) and tied it up. Then with a flourish, I cut the ends of the tie with the scissors and a perfect roar of laughter went up!

Boy-with-the-feather then piped up and said the *munakasi* (a buxom young matron with red-clayed hair) hadn't seen the *Vasekele,* so the old show had to be gone through again, somewhat abbreviated. Then, seeing me put away my camera, eager interest was aroused and a peep show followed – old Golli, sitting waiting patiently for his salt, acting as a model. At last (3.45 by my watch) the three *Vasekele* having been given their salt, the crowd has gone and peace has descended – broken only by Lomeii chopping (the cloven hoof appeared – he spanned in the youngest *Vasekele to* fetch water from the river!) and the pea-shooters cracking first on one side and then on the other. I ought really to have baked today but as Miss Bleek has the 'baking tin' billy I did not. I am just making some scones instead.

Tuesday July 14th—I waited till after dark then decided to have dinner and was just finishing when (7.15) I heard voices and Miss Bleek and the two boys returned, very weary after the two days walking down to the Lomba, up the Kusezi, back across the hill to the Kunzumbia and so home. No sign of Bushmen, but they had seen a smelting oven for iron, a dancing place (men only) and various other items of interest – taken fish out of a basket trap etc. The night was spent in the far village where a newly built grain-storage hut was placed at Miss Bleek's disposal for the night. We waited all day for the Bushmen – not a sign of them.

In the morning, about 9, I took my camera etc. and went up the river to the big bay just below where the *vansonge* usually are. It was a curious, overcast, leaden kind of day – not a sign of life on the plain, and even the birds in the trees were comparatively quiet. It was delightful – apart from the occasional distant sound of a drum and the footprints on the paths, not a sign or sound of man – not a hut to be seen – grass rustling softly in the breeze, such a treat after the last two days of constant visitors to the *cepatonga*. I thoroughly enjoyed it, though I collected next to nothing and it was too windy (wind was not strong but the steady breeze made far too much movement in the plants I wanted to photograph). Further, a pleasant feeling of laziness made it sufficient to stroll along, enjoying the colour and breeze, and the distant cooing of turtle doves. It was getting hot so I came back in about an hour and a half. Still no sign of post. We have decided to get a man or some men to go in to Muié and fetch the next post. Miss Bleek engaged four men to build a fence round the kitchen for the boys.

Wednesday July 15th—We got up extra early, had a hasty breakfast and by 8 o'clock were off for a visit to the Bushman encampment. On the way, we came across another smaller one. Miss Bleek cut down to the river while I kept straight on and as I neared the encampment a long-tailed bird flew out with his shrill chattering cry – an ominous sign of the absence of man in the neighbourhood! Then came another and another – five or six in all – one or two, I think, females with shorter tails, but of that I am not sure. There is a good deal of colour about them – golden red bills, white band in the wing, but I want to have another good look at them.

The Bushman camp, as I feared, was still deserted, the salt I left untouched. We made our way up (I waited 15-20 minutes and Miss Bleek at last appeared, after I had called several times) to the higher part of the hill and came back by way of the cultivated land, getting on to the high level path by which Miss Bleek returned on Monday. It winds along, now just rising, then dropping again and so on every ten feet or so, most monotonous. At first there was a gorgeous view over the plain, most colourful in the unusual light – sky cloudy, sun shining through the clouds. Later this was hidden by trees lower down.

Just as we approached the village we heard voices, of which I thought some sounded *Vasekele*-like. We turned aside to investigate and there sure enough were four *Vasekele* – my three friends and a younger boy – having a warm discussion with three *Vankangola*, our big old friend (Chief's henchman) among them. Three of them had their bows and arrows; after talking to them for a time we brought them down to the *nsinge* (which we were surprised to find actually in sight) and Miss Bleek spent the rest of the morning talking to them. What the discussion had been we were not sure – I think the Bushmen wanted to go and hunt, and the others wanted them to come to us, but it may have been the opposite! Miss Bleek bought a quiver and some arrows from Golli for two yards of blue stuff which delighted his heart, but he refused to sell the bow. They said they could not come tomorrow as they were hungry and wanted to go and hunt food, but promised (apparently) to send their women next day. Miss Bleek said she wanted the two younger boys in the afternoon, but they did not come.

We sent Samolova to the village to get a man to go for the post. He returned with two, one the old man with bad eyes. We offered $9.00 for the return journey. It seemed to satisfy them, but they won't go until Friday – we wanted them to start tomorrow. It's unfortunate but can't be helped. About 4.45 we went out up the river. The birds were most interesting. A black chap about the size of a thrush, large head, sitting rather straight with slender tail almost vertical, was calling out – several different notes, very varied, and being answered call for call from another tree. In the same tree with him were a number of little grey-green birds with yellowish breasts, dropping from branch to branch like falling leaves, then making circling darts, returning to their branch (flycatchers of some sort apparently). Then we startled half a dozen wood pigeons feeding on the ground, which got up with a scattering noise. We saw no sign of *vansonge* though we went up past where the path to the upper village comes out to the river. We came back through the wood by it, as Telosi and I had done after our evening stalk – rather pleased that I found my way back straight away.

Thursday July 16th—No Bushmen again – not a sign of them! I have been painting a few flower specimens, with more or less success. There are however not many to do and those chiefly small – it is close work and tiring, – very difficult to make out details of structure even with the lens. The production of the lens brought Lomeii and Samolova out eager to see. The latter *could not* see through it for a long time; finally managed to do so and was duly impressed.

Yesterday, by the way, the whole community was excited – Bushmen included! – because Madame, who had been very restless since early morning, dashing out into the bush, climbing on the roof, the hedge, up and down – wanted to lay! Samolova fixed the *camba* in a tree, put a little grass in, caught panting Madame, and put her in. Followed a great scratching and turning round – not grass enough to cover the hard ridged bottom. She was given more grass, offered water, and finally settled down to the business in hand, while everyone waited expectantly! By and by Samolova opened her door and out she came, the deed accomplished and her tiny egg laid! It will be interesting to see how long she will continue laying before getting broody again. The others are growing – the croupy one is better and turning out a very pretty little hen.

Salikunda, with business instinct, turned up with a very small knife which he said he himself had made, and duly received the coveted one in exchange. Hearing Miss Bleek endeavouring to buy the Bushman bow, he turned up with a rather inferior one which he bartered for a yard of stuff, and a tall folding chair – rather nice for a sketching stool, though not very secure. It was rather late so we told him to wait for the morning, and eventually we bought that too, for a yard of material. Our next cavalcade will be even more peculiar than before.

This evening we took our walk up past where we found the Bushmen, past and round the back of the village and then some way on the hill behind. Here a large bird got up from the sandy ground with a great clatter – looked rather like a sand grouse, similar to, but larger than, the one we bought the other day. The path began to turn south so we left it and cut across very burry country into a most inviting autumnal tinted valley to the west and north of the village. Up this we wandered then turned and made our way down it, past a large enclosure fenced round like our *cepatonga* where drumming and dancing was going on. A man perched up on the wall spied us, jumped down and the chanting stopped, but soon began again when they saw that we were not coming up to it, but merely passing. We had to cut across from path to path to avoid the enclosure, winding in and out of the bushes. There were a dozen or so paths, all apparently leading to it. After passing it, we cut through one corner of the village and spoke to one or two of the women. One young thing, nice and clean, had a particularly nice clean baby whom she showed off with pride.

Just as we were having our dinner the *nsinge* was hailed and two men arrived, one the Portuguese individual who took my note and the marmalade to Mr Muir the Monday after we arrived. It was the long expected post – a very meagre one so far as I was concerned, only two short letters and no newspapers. With the post was a long strip of smoked pork from Mr Muir – it looks and smells for all the world like a rectangular strip of tarred rope!

The evening was spent in reading and writing letters, the latter arousing much interest, particularly in the boy who accompanied the mail carrier, and who propped himself against the pole at the head of the table and gazed!

Friday July 17th—Again no Bushmen and we wanted to go out and look for them but had to wait for our mail carriers who were very tardy. After sending two messengers for them, the old man turned up about 9 o'clock, was given the parcel (enclosing letters, tin for fat, 10/- for Mrs Bailey to buy fat, potatoes, rice and a basket if she can, and a parcel from Telosi for his wife – the dress Lomeii wore last Sunday!) By the way, we expected ructions, as the matter was reported to Telosi who came to me for confirmation. Instead he merely wrapped it up and asked if he might hide it (*kusweha*) in the *nsinge*, and in a most Christian spirit helped Lomeii to make himself a pair of trousers! We told the messenger that if he got back by tomorrow week we would give him extra pay, but as he started so late I don't think he will. Then I painted the slender red composite, watched again by a breathless, admiring group of sightseers – after having set sponge for bread (three-quarters local meal this time, to try).

KAMUNDONGA'S VISIT

About 11.30, just as I was starting to make up the bread, in walked Kamundonga, accompanied by wife, child and dog. We explained what I was doing and I gave him a slice of bread to taste. Then, as conversation was flagging, I produced the painting I had just done. It really was very funny – his face lit up, his mouth opened and he shook a roguish finger at me – why I don't quite know – and called to his wife who had betaken herself to the kitchen, to come and look. The other drawings interested them, but being less spectacular did not arouse quite so much admiration. Then I showed them the paintbox. Later all but the chief departed. We were getting very hungry and wished he would go too. Then in dashed Samolova, seized the bow and one of the arrows and to our astonishment went off with them. A minute later Telosi came and ran off with another arrow. We three all followed to see what was happening. Samolova had shot both arrows through a slender light-coloured snake, (our first up here by the way – except one yellowish one whose tail I saw disappear down his hole) which he has finished off with a stick. All very well, but the bow and arrows were not bought for that purpose!

Finally old Kamundonga 'got it off his chest' – he wanted a piece of sacking like that round the tent. We eventually gave him the small empty salt bag and it appeared that he wanted it for separating honey from the wax. There is apparently a big trade done by the Portuguese in the latter commodity – old Golli-ba was a wax collector among other things. He proceeded to unpick the stitching. I couldn't wait any longer for lunch and called to Telosi to know if there was *mema a tu puku*, whereupon Kamundonga took the hint and retired to the boys' premises and we at last got lunch.

Had a good laze with one of the two-months-old *Weekend Argus* and found quite a lot of interest in it. The flies were very tiresome. The swatter alas shows signs of a speedy demise. It has been such a boon. We have used it too hard and the wire is breaking in places. About 4.30 we went out, made our way across and up the far side of the valley we came down last night. It is pretty – reds and browns of autumnal tints contrasting with dark green of the evergreens. We saw a couple of groups of wood-gatherers and passed the dancing enclosure on our left. As we passed, a drum was beaten several times – it made me think that perhaps it was some mysterious signal giving warning of our coming – in all probability nothing so romantic! We went right up the far side. Near the top we put up a dark bird with white stripe in the wing and soft noiseless flight. It flew a short way and then came to earth again. I followed and caught another glimpse as it made another short flight. It was, I think, a nightjar, but I could not get close enough to make sure.

Near the top of the rise are several fairly large *musivis* (*Sterculia*), scarce around here owing, I suppose, to continued cutting. The bush round them had been cleared, so they stood out well and in the light of the setting sun the trunks, shining red, and the dark green glossy foliage looked very beautiful. We turned down past them and returned along the floor of the valley, the sun behind us bringing out the rich reds, browns and golds of the old leaves – most lovely. This time we passed above the dance enclosure – quite quiet now, and as we did so Miss Bleek remarked 'that is a curious place' and I found to my utter astonishment that she had not seen it! We were walking towards and past it for nearly half an hour last night with drumming and chanting coming from it, and passed close to it again on our way out this evening!

FETISH POLES

As we entered the village they were burning rubbish and had several fine bonfires. The women came and talked to us, wanted to know what I was going to do with the flowers and leaves I had gathered, and explained what the fires were. Then we noticed the fetish poles – several small ones, painted, a kind of little cradle with sticks on it, a group of spiked sticks with heads of *bambi* (duiker) and one of *nsonge* (and horns). Laid in front of the sticks were several leaves of *Sansevieria,* from which, they explained, they make fibre for the string to set the traps or deadfalls to catch the buck – very interesting – evidently laid there to get the benison of whatever deity the sticks honour – anyway to bring success to the hunters! Evidently the Bushmen have adopted that custom since in their little encampment were similar sticks which Telosi explained to me were used by them to spread offerings to *Njamba* to ensure success to their hunting.

Many of the women are of a surprisingly pleasing type – rather finely formed features, not nearly such thick lips and generally Negroid features as the Mbundu and Luchazi. We had noticed a different type in some of the men, e.g. Kamundonga himself, but it is even more pronounced in the women. His wife was distinctly Semitic in feature – nose aquiline – and in some the colour is bronze rather than black. Probably there is a slight admixture of some Semitic strain, probably Arab.

This afternoon the birds near the *nsinge* were charming – a couple with rich clear notes were calling in the tree just opposite my bed – about the size of a thrush with the quality of the blackbird in their call – and others were trilling like mountain canaries. There certainly is plenty of bird life about here.

Madame laid again today. She cannot get to the *camba* herself – Miss Bleek put her in this morning – so after she had finished I arranged a little platform of interlaced sticks and am curious to see if she will make use of it for going in and out. Blackie attacked a tiny mouse which was running across the *nsinge* – it escaped and disappeared. 'I do hope it

> ### Fetish sticks
>
> *In every little Bushman encampment I saw, there were forked sticks planted upright in the ground near the huts or sleeping places. On these all implements for the hunt or chase are hung or laid, and the sticks smeared with blood from any animal killed 'to bring luck'. The whole thing is copied from neighbouring villages, where fetish sticks of every sort abound[2].*

July 17. Fetish sticks, Kunzumbia

hasn't gone into my boot,' I exclaimed, picking up the said boot and shaking it. Out tumbled the poor little mouse, to take refuge next under my bed, which had next to be removed and finally the mouse escaped through the wall. It was a tiny grey thing with a long slender nose, possibly a shrew mouse, a pretty little creature but not welcome in either boot or bed!

Vansonge

Saturday July 18th—After an early breakfast we both went out – Miss Bleek up the hill in search of *Vasekele,* I through the wood and up the river to try and get snaps of animals and perhaps sketch. It was lovely out – a strong north-easterly breeze blowing over the plain. No sign of my lovely red-billed storks, but far up the river I came on the *vansonge,* 50 or 60 strong, and for some way I stalked them, or rather tried to, but there was no cover and they were shy so I did not get very close. I took several snaps, though I am afraid they were too far off to be any good with my camera. The creatures looked lovely outlined against the sun. They went off at a long loping run, changing as they got into the swampy ground to great leaps. They did not go far and when I got into swampy ground they all stood still and watched me hopping from tussock to tussock with the greatest of interest! There was a curious slightly raised patch with a few deciduous trees and many proteas for which I was aiming – evidently a favourite grazing ground. The herd – or herds rather, for there seemed to be really two – made for the east of this. I cut across to it and had a grand view of them all standing at attention with one fine old buck with magnificent head in front. I tried to get nearer, but off they went northwards along the dry patch, round it and me and back down its west side.

Then, wet of foot and growing weary, I left the swamp and made for the wood through which I returned, sometimes along the edge, then further in where I found the broad track up to the northern village. Looked for tracks of the little people and followed one or two a little way up into the wood, only to find they petered out. Back at 'Nsonge Point' I found the best view I could with the red sprays in the foreground and spent an hour making a sketch of the river valley. It is unfortunate that my yellow ochre has disappeared – one needs it for everything here. No other yellow takes its place.

A variety of birds

By the time I had finished it was past noon and I had a hot walk back, first through the water of the upper path to try and wash off some of the red mud, then through the wood past the old village clearing. Near the *nsinge* several birds flew out of a tree and swooped down into a further one – four of them were hoopoes – rufous backs, and crest erected as they alighted, and alternating light and dark bands on the body – pretty things. I could not get very near as they continually flew to a further tree, but as they alighted there was an impression of the rufous colouring down the back. On the 'redheads' as I came to them, by the way, was a long-billed blue-black bird gathering honey – apparently they are bird pollinated. From the hoopoes' tree also flew out some dark birds with forked swallow tails and, as I was going up through the grasses, various kinds of birds rose. Some tiny ones clinging to the long grass stalks were singing their little trilling song, then letting go had to fly with all their might against the wind to maintain their position. A larger dark yellow-green bird flew up – probably was feeding or drinking in the swampy ground – much too wet for a nest – and a speckled lark-like bird also rose. A grey heron (it appeared, was too far off to be decisive) was feeding in the distance and, as I was sketching, two hawks hung and soared above the next woody peninsula.

On my return about 12.30 I found Miss Bleek back, without having had any success, though she had found plenty of Bushman tracks. She had just been asked by the boys for their *fueta* – had paid Telosi who wanted stuff. He refused the white, refused to pay the price for the blue, even though Miss Bleek explained that though she really wanted it for the Bushmen she would let him have it as he has been a good boy and even made a reduction. No, he wanted it for half the price, wherein we traced the influence of the specious 'Christian' boy who brought the post, who refused the stuff on the ground that it was too dear, that Mr Muir let them have it for $5.00 – at least so I understood.

A 'STRIKE'

Then the other two asked for their wages, which both had had in advance the previous week, Samolova apparently genuinely muddle-headed as to the amount of his pay, although it had been explained to him to begin with. He asked what his pay would be if he stayed next week, was told $7.00 and said then he would ask for his *guia* and return to Muié tomorrow. Upon this Lomeii chipped in with his usual graceful manner and said he would have his too. Miss Bleek said 'very well', and asked Telosi to remove his money as she wanted to work at the table.

There matters rested till I had got back, when all three came in a body and said they wished to return tomorrow. So that's that. We rather took the wind out of their sails by consenting immediately without protest. They – at least Lomeii – have come grumbling from time to time on one ground or another, asking for a present, forsooth! Much to his annoyance he was sent off to the *donga* to wash. Miss Bleek gave him some things but he hung about till after I got back, so got a further instalment. Thereupon he evidently thought he'd better go before his pile of wash got bigger, and went.

It is unfortunate that we sent post to Mr Bailey yesterday (as a matter of fact they did not go till this morning), otherwise as the Bushmen don't seem to be coming we'd get porters as soon as possible and trek on. As it is, since Telosi won't carry letters to Mr Bailey – at least not quickly. I think it is as well Telosi should go back soon – he seems to be getting corrupted. He spent last night out altogether, was not even here to do the breakfast. He is now comforting himself by singing hymns – very much out of tune! He'd better get back to his regular services and routine of mission life. As for Lomeii, I shall be thankful to see the last of him. He is an objectionable, ill-mannered lout, though he can work quite well. Samolova is a nice boy I think, but indolent and without much backbone – much older, of course, than either of the others.

Later. The strike is ended! Telosi and Samolova came up, stood one on each side of the table and Telosi started to orate – as usual talking so fast that we were left miles behind. Eventually we gathered the gist of it: 'The pay is good, very good. We don't like to leave you alone here in the wood', etc. etc. We took it calmly, said if he wanted to stay he could stay, if he preferred to go, well and good. Samolova tried to put in a word or two, without much success. Finally Telosi returned the $1.00 given him to buy food for the journey and Samolova followed suit. Both were accepted, and we returned to our respective occupations. By and by, Lomeii comes up and proffers his $1.00, which Miss Bleek refused, saying he was always wanting, wanting – which brought a huge chuckle of delight from the audience – and Lomeii retired. I think the other two will be better without him, particularly as Samolova will now get more pay.

Sunday July 19th—We told Lomeii he could stay till tomorrow if he liked, and he elected to do so. He is now attired in trousers – product of his and Telosi's joint efforts, and a Joseph's coat of many colours made out of the two selvedge ends of my old Cambridge hammock (now my *machila* cover).

EXCURSION DOWN-RIVER

We got up early and before 8 o'clock started down river – cutting through the wood and coming out on the river at the first large bay. Half a mile or so below that, the river makes a huge bend. The tree-covered right bank rises high above it (some 30 ft or so) – so steep that it is rather wonderful it is not constantly sliding in. Here the river winds and doubles back on itself in a series of S-curves – it is deep, olive green with long feathery grasses edging it, white waterlilies and arrowhead growing on its surface. We left the river edge and followed the little path up over the hillside whence are lovely glimpses of the river below. Beyond, the path comes down again, only to ascend once more over the next peninsula.

About here two brown birds with white streaks in the wings flew out with soft noiseless flight from the bush over towards the river – they were about the size and colour of hamerkops and at first I thought that was what they were, but the heads did not seem the right shape. We passed the nest – a large affair of sticks, but open basket-shaped, not spherical with a side entrance, as the hamerkops' usually is. I intended to photograph it on the way back, but missed it by going a different way. Beyond these two bends the river passes further out again. On the far side scattered trees come close to it so that the valley, or rather plain, is very narrow. Between the right bank and the river here were numbers of proteas, a clump of which I photographed. They must have been a sight last season, as they were covered with fruiting heads, evidently last year's.

There are great bends in the edge of the plain below this, the woodland forming heights above it. We went round one or two of these, then up the next point where there was a lovely view of river plain and woodlands with a blue valley (probably only a large bend in the Kunzumbia) in the distance, so I sat down and did a hasty sketch (about an hour) while Miss Bleek returned over the hills.

I came back along the river edge under the cliff and there found growing both on the edge and actually in the river, trees of the plant with gum-like flowers, of which there

July 16. *Brackenridgea arenaria*, Kunzumbia

July 19. Kunzumbia River

July 19. Kunzumbia River meander

July 19. Brown bird with white bands in the wing

a Crane soaring

band

July 19. Tree with gum-like flowers
(= *Syzygium guineense*), Kunzumbia

Muhemele
Eugenia Owarensis

263
403
809

are numerous small bushes in the Kutsi. Here it is a tree with a 6-8″ trunk, some 15-20 ft high. Having come past the bend, I had to go back up the hill to get my photo – wished I had taken it earlier, as it was a much prettier light. Then I returned to the river and collected some specimens – one a new one – solitary shoots coming from the ground on the swampy edge of the river, with green, oak-like leaves and red and green round 'fruits'. It proved to be a fig! It is rather a nice thing.

Soon after I had got this, a large bird passed overhead soaring upwind – a lovely sight – wings with front edge absolutely straight and showing dark with a white band as seen from underneath – feet and tail outstretched, rudder-like behind. I hastily felt for pencil and book, but before I could find them he was out of sight over the wood behind and I had to try and remember his lines – very difficult I find it! I should think he must have been a heron or something similar.

It was getting late and hot – the last bit through the wood was very hot and long! As I neared the *nsinge*, there was a chattering alarm cry and two birds flew off with a flash of iridescent turquoise on back and wing. I could not see more than that. There certainly is a wealth of bird life about here. I got back hot and hungry about 1 o'clock to find Miss Bleek who had been back some time, busy indexing.

Before we set out Telosi had asked for the blue stuff – the cause of the strike! We had told him to wait, and put his $20.00 away for him. When Miss Bleek got back he said he would take $20.00 worth of the white stuff instead (I was glad we'd made him wait and think things over!), had been given it and was at work on another garment. He had been to the river to wash and had seen an *nsonge,* and was anxious to go after it with the pistol. I had it, so Miss Bleek could not let him go, but after I came in and we'd had lunch, we let him go, though I really wanted him to bake bread. However, he and Lomeii went off and spent a happy afternoon chasing the buck. He fired once and says he hit. I hope he did not!

I painted two or three more flowers, to the great interest of various spectators who evidently think it a foolish and childish waste of time to paint all these *miti*, but enjoy watching.

Monday July 20th—Lomeii was given his $1.00 (wages for yesterday), a packet of sweets and letter for Mr Muir, for carrying which he was given $2.00, whereupon he asked for salt! Then, after eating his food, he departed – seen off by Samolova. I fancy he is not much regretted in the kitchen! We neither of us felt like exploring this morning and stayed in and worked, Miss Bleek as usual indexing, I painting specimens, making yeast etc. Madame laid again, her fifth egg. She appreciates the platform I made for her and goes to her eyrie quite happily. The others are growing well, and they certainly add to the interest of life. When they begin to feel hungry, they collect round my feet and gaze up expectantly. Madame bosses them all, and when she goes to lay, they lie about under her tree.

PHOTOGRAPHY SUBJECTS

In the afternoon two girls, one with a baby, came in to have a look at us and see the *Vasekele sonneke*. The other girl had an elaborate pattern tattooed or rather cut on her body – a series of small cuts leaving raised scars. I don't know how they do it, but it's quite becoming! I photographed the two of them, giving them each a spoonful of salt as a reward.

Then we went up into the village where we saw the dancers all togged up, looked at the forge which was not working and took several photos – one of a woman with a hoe standing beside the fetish sticks outside her hut, a woman pounding meal, Kamundonga and a crowd of children, and a general view of several huts, people etc., with the fetish tree and sticks fenced round, at one entrance to the village.

Then we went down to the river escorted by our little pock-marked friend, to see we did not go near the dance enclosure. We think this must be the boys' initiation, which would account for the daily drumming and dancing which has been going on since we came here.

A TRAP-DOOR SPIDER

On the way back we investigated the curious 4-cusped marks in the sand. They are made by a dense web, forming a sheath over a hole in the sand – dry on top but quite wet below – wherein dwells a little dark-coloured spider. I lifted one and the occupant put out his feelers. I attracted him with a piece of grass, and he then decided it was *san gaucho* and closed his door by pulling down a piece of the dense web which lines his hole, since I had removed his proper door. Having closed it completely he retreated – the hole is about one-and-a-half to two inches deep. I collected three specimens with their doors, but doubt if they will be of any use. There is nearly always a little corpse just underneath the door web. It looks like the brown, empty skin of a spider, so perhaps my little friend is a lady and this the skin of her dead, devoured mate – I do not know.

Saw one of the black, fish-tailed birds – he has a very sweet song or rather whistle, not unlike the blackbird at home. I think he was the one who was visiting the red flower, but am not certain. Now (8.15pm) there is a curious little cooing note coming from the wood, which I think is a nightjar. I believe the drumming from the dance place the other evening was to announce our proximity, because this evening the same thing occurred. Our 'escort' gave a whistle, whereupon it at once ceased. Should like to try again!

COTTON SPINNING

Tuesday July 21st—Soon after breakfast we took Samolova to carry my sketching things, chair etc. and went up into the village to see the iron working which, however, was not going on. Near the hut, however, sat our old friend Sangevi (Chief's henchman) with a spindle, twisting cotton thread – the empty cotton pods lying on a heap nearby. Two pairs of bellows, one old outside, the newer one inside the shelter, without the skins, several tools, a tiny anvil and some smelting cones lay about. Outside were various stools and stumps.

July 20. Young woman on right with elaborate 'tattoos' on her stomach. Both women have clay and oil styled hair and filed front teeth, Kunzumbia

July 20. Chief Kamundonga with children, Kunzumbia

Bushmen tattoos

Women are often tattooed at the time of their initiation. Either their father or a medicine man makes a number of parallel cuts in the face or upper arms or thighs, rubs in charcoal and lets the places heal. I am certain that they are partly for ornament, though their being done at the time of initiation points to some religious significance as well[3]. It is clear from Mary's diaries that both Bushmen and Bantu speaking people 'tattooed' (cicatrised) using the sap from the young fruit of rothmannia and gardenia plants. The sap is rubbed into superficial cuts in the skin causing permanent raised patterns after the wound has healed.

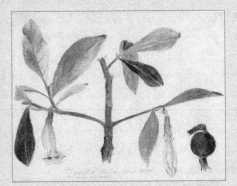

Rothmannia engleriana. (vern: *muwanguloangu*). 'Sap from young fruit used for staining cuts, tattoo marks'. Menongue to southern Bihe

July 20. Fetish poles and woman with hoe, Kunzumbia

I started a sketch and soon had half the village round me, blocking the view. Miss Bleek announced that we should want porters next week and wrote five names for *tipoia* boys, then she and Samolova returned and I went on with my sketching. Fortunately pencil is not so interesting as paint, and after half an hour or so all dispersed to their several avocations – fetching water, clearing ground etc. – except Sangevi and a couple of young men lazing in the foreground. A yellow dog circled about in the background barking lugubriously at me. By the way, yesterday we saw six fascinating pups just able to crawl, after one of which, mongrel though they are, I have a bit of a hankering!

I was left in peace for some time, then a couple of men came and sat near and watched, absorbed, particularly one. Soon someone recognized Sangevi twisting *wanda* and the news went around and in a few minutes a number of spectators had gathered! He had gone away some time before, but now returned to have a look, much amused. It was getting very hot so I put things away and got one of the boys standing by to carry my things back to the *nsinge*. Sangevi came along too, and I pointed out the stool he had been sitting on, a particularly nicely carved one, and explained that I'd like to buy it. He asked what for, and we said '*tanga*'. '*Shono tanga*', he replied, whereupon Miss Bleek brought out the white stuff and offered him a yard. His eyes sparkled and measuring off two yards, edging his fingers ever a little further along, he said he'd sell the stool and two fowls for that! I bargained for his snuff box, a very nicely carved one, for another half yard (probably far too much!).

We marked the piece agreed upon with pencil, and he went off, returning in a little while with the stool and several little boys, one of whom was carrying two quite nice little hens. We tore off his bit of stuff – nearly three yards, the old beggar! I now have the

July 21. Sangevi spinning cotton, Kunzumbia

cituamo (which is nearly half a porter's load in weight) and the snuff box, and our family is increased by two. One hen is very light in colour, slightly speckled, the other dark with white splodges, a tuft, and a few feathers down one leg! He emptied the snuff from his box – I gave him a piece of newspaper from which he cut off a tiny square, emptied the snuff into it, then offered me the empty box to smell! I thanked him and refused his kind offer of snuff. It's very strong stuff – I caught a whiff from the empty box.

All afternoon there has been a succession of visits to see 'Sangevi' in the picture. One group – a couple of women, several of the usual little boys and some small children – was particularly interesting and I was sorry the light was too bad to photograph them (it was nearly five). The two women leant forward with '*Eh wah!*'s and handclappings at each new discovery (there are several huts, three men, a baby, a girl with a basket, fowls, dog, fetish sticks etc.!) and in front of them a small child of about four or five, bangles round its ankles, leant forward, hands on its knees, also exclaiming '*Eh wah!*' and coming in at appropriate intervals with tardy handclappings. It was so comical that I burst out laughing. The child caught my eye, gave a howl and fled behind its mother for protection. She gave it a hand and in a minute it was on her back astride of her left hip and continuing its observations over her shoulder!

I spent the afternoon working at my sketch (which being of the caricature stamp appeals to the childish mind!) and finishing drawings of two specimens. About three I put things away and we were going to photograph the Bushman camp again when a couple of boys came, introduced by Samolova, to buy stuff. By the time that was finished we had decided not to go till tomorrow as I want to take the photographs with plates and had a roll of film in.

TWO YOUTHS IN COSTUME

A little later there was a rustle of skins and in pranced two youths dressed up in white, red and black tights from head to foot, the garb being made of string netted with a kind of loose twist. While I was sketching this morning one of them, wearing a red mask as well, had pranced up holding out his hand. As I wanted to photograph one, I made him pose and then gave him a *piastre*. These two, with feathered headdresses, one with two red discs over his eyes and gauntlets on his hands, the other with bow and arrow, I took out into the sun and we made them dance while I photographed them. Then we gave each some salt. I had a good look at their armour – it is made in bands of various colours – natural, red and black. The headdress is chiefly of cocks' feathers. Both had the usual skins round their waists.

Miss Bleek went out but I wanted to finish my drawings which took longer than usual, so did not go. I caught Madame standing on her platform after laying, having a preliminary cackle before descending, and snapped her – the light was not too good. The nightjar was calling again just after dusk.

AN ABUNDANCE OF BIRDS

Wednesday July 22nd—We had an early breakfast and by 7.55 had started up the river for the first village. It was lovely – fresh breeze and such beautiful colours over the grass. Birds everywhere – as we passed the grain field a flock of sparrow-sized grey-brown birds arose; and nearer, from the edge of the field, with a whirr of tiny wings, a flock of small brown birds, startled by the former, rose like a dark cloud and made off. A canary-like song was coming from trees near the river, answered from further off by a similar one. The black fish-tail bird was in evidence – his, I think, was the sweet melodious whistle, three or four notes in a descending scale sounding over and over. Further on, there was a sudden noise and up rose three or four of the big black turkey-like birds and made off with heavy flight towards the river. Nearby, three of the stork kind were winging their way with slow powerful strokes across the plain.

Rite of passage

The masked youths photographed by Mary were *likishi* (initiates) dressed for the *makishi* initiation ceremony dance. Their remarkable armour and gauntlets were woven from twine made from the bark of the miombo tree (species of *Brachystegia*) and took several months to prepare. The various colours would have been achieved using natural plant dyes. The costume is an important accessory in masquerades and story-telling. Despite cultural and religious changes and the aftermath of the Angolan war *makishi* initiation ceremonies still take place in eastern Angola and north-western Zambia[4].

String net garments, Kunzumbia

DUIKER

Vansonge as usual were feeding out in the grass. Some way further we came out of a strip of wood down to the edge of the river plain and then through a bit of very open scattered small trees with grass. Miss Bleek, who was ahead, suddenly stopped and pointed – I could not see at what. Then out of the bushes across our path about 30 yards away came a small light-coloured buck – *bambi* or duiker I believe – wandering down towards the river. It did not see us and I pulled out my camera hoping to get a snapshot. The slight noise I made attracted its attention; up went the beautiful little head, snuffing the breeze and gazing round. It caught sight of us and darted off a few yards. I approached, still on photography bent, and then it went off with great leaps through the grass. It is the first duiker I've seen, though I believe there are plenty about.

A BUSHMAN SLEEPING PLACE

After about an hour and three quarters of steady walking we found ourselves almost in the village, so retreated a few yards and made a wide circuit through the wood above the village. There were various tracks going off – some obviously only chance ones, others wide and well trodden. We passed an open grassy patch, evidently cultivated at some time, and then on the far side of the village we came upon several little tracks, one of which we followed. It led us to a Bushman sleeping place! There were five little joss sticks, one with a small skull (about the size of a rabbit) and five fires with several beds, a small mortar much cracked round the edge, a battered pestle and heaps of husks. The ashes however were quite cold. Most of the beds had a fire on each side, and in some at any rate there was a log down one side next to one of the fires instead of at the head for a pillow, as I think was the case in the big south encampment.

Then we wandered on a bit, still circling, got nearly to the river, then turned up a path towards the village. In the distance we saw a line of three women and children with calabashes, fish baskets etc. evidently going from the village to some fishing pool on the far side of the next woody peninsula, as they disappeared into it. Further up, two paths branched off to the right, and following the second of these we found another sleeping place – three beds and five fires (fire on each side of the bed). The ashes of one fire were warm, but I think not from last night's fire. We looked about a bit but found no more traces so as it was getting very hot we went into the village, which we found absolutely deserted except for the old chief Ndumbe, his wife and two small boys, apparently theirs.

The chief moved two log stools into the shade for us, sat himself down on a tree trunk and we made conversation. He is a courteous old man – elderly rather – somewhat like Kamundonga in appearance and manner. We told him we wanted porters next week. He said that there were only two or three in the village, they were at present away but on their return he would send then down to sign on. It is quite a small village: a circle of huts – not very well built ones – round a patch of partially cleared ground. We noticed a bird in a cage and he sent the little boy to bring it. There were two tiny green and yellow ones rather like Cape canaries, one merely a fledgling, in very neatly made little oval cages, with water or seed in calabash necks. Near several of the huts were little grass traps – round with a narrow passage leading in – evidently for catching the birds which they sell to the Portuguese.

We had a long hot walk home – left the village 11am and arrived 1pm – explored one side track fruitlessly and had two short rests but walked all the rest of the way steadily. *Vansonge* had all disappeared! I think they go to the wet reedy ground for the hot hours.

A welcome rest after lunch refreshed us, then I made a baking powder loaf (ought really to have baked today) and a raisin loaf, in the middle of which Kamundonga arrived with a retinue of small boys and girls, three fowls and $7.50 - to buy *tanga*. (He wouldn't sell his snuff box!). Then I showed him various drawings – the retinue crowding round to be shooed off periodically, only to return – and the map. Salikumbi, who was here before eight to buy safety pins, only to be told by me to come later, then arrived. I told him he could have two large and two small gilt ones, and $1.00 if he brought me a tool for hollowing wood. As I thought, the 'I, myself, made it' as regards the knife was, let us say, a figure of speech, as 'I myself' did not seem at all keen on making this little tool and tried hard to persuade me to accept *wunga* instead!

Then Telosi wanted his 'shirt' or blouse, so I had to look through my things and finally found a navy blue silk blouse and a new vest. He did not fancy himself in the former, so took the vest and later appeared attired in it and his new trousers which he has just finished – and marked with his name in blue thread – a vision in white!

The fowls are too funny – they will not mix with the low speckled company which has invaded their domain. When the latter went to roost in their sleeping place, they left it and have gone to roost in an entirely new part of the fence! Poor Madame was much perturbed as we ate both her eggs and did not leave her one! She had to be shut in before she would lay. One little cock started crowing yesterday – choking himself each time by his efforts, so today it was decreed he must die, poor little beggar! And very good he proved. About 5 our hunter (the man with the large dark skin – sable antelope?), arrived with two legs of buck, one of which we bought, leaving the other for the boys.

The new moon appeared this evening, and our Telosi in all the glory of his new white vest and trousers has gone off to the celebrations in her honour. There is quite a different drumming going on and so far (8pm) no chanting – a single drum struck with a double stroke at intervals.

End of Volume III.

THE DIARIES
Volume Four

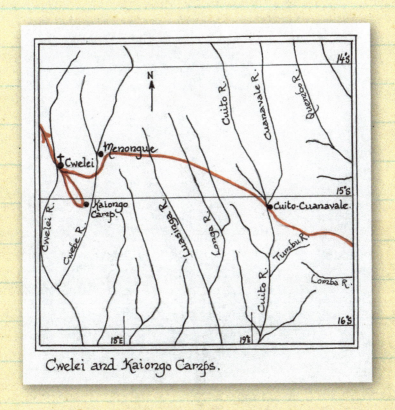

Cwelei and Kaiongo Camps.

Kunzumbia to Cwelei Mission Station
(July 23 to September 9, 1925)

'Sometimes the dancers danced up to first one then another of the ring of men clapping, shaking and shivering.'

July 24. *Eulophia saxicola*, Kunzumbia

IN CAMP ON THE KUNZUMBIA NEAR KAMUNDONGA'S VILLAGE

Thursday July 23rd—This, the fourth volume of my Angola diary, is inaugurated by a 'dull, domestic day'. It started with making the sponge for bread which was mixed and buried in the warm ashes by 8am, not however before three or four villagers accompanied by one or two little boys bearing a fowl, a sifting basket (old and damaged) and a mat (also old and much slept upon) – the two latter needless to say belonging to our bargain-hunting friend Salikumbi – had arrived on shopping bent. All three, it is hardly necessary to state, were refused, (with 10 fowls and a leg of buck already, we need no more of the former here!) whereupon the owner of the fowl produced $10.50 and asked for *tanga*. He was shown the amount of cloth that would purchase and asked if we would accept a basket of meal in addition. We said certainly, but he must first produce it – if he brought it and the money tomorrow well and good. No, he'd take his money's worth at once thank you, no risks to be run!

Then I resumed the work of packing away and writing up specimens till 11 o'clock when the bread had to be made and put to rise. That was safely buried, and I went on to make biscuits – mainly flour, sugar and water, with a modicum of the small remaining amount of fat; as only four to six will go into the pan at once, baking is a slow process. Baking, rolling out etc., combined with putting away specimens took till nearly 3 o'clock. Then Telosi baked the bread, very successfully – although two thirds local meal, it rose quite well.

In the morning I had engaged two small boys, Lopali and Kakupa, to get grass from the river, and in the afternoon Samolova recarpeted the *nsinge*, which now looks delightfully fresh and clean. It was as well that I moved my specimens this morning, as the ants were getting to them. They have eaten holes in the bottom sheet of drying paper and had started on my holdall and Miss Bleek's basket.

About 5 we went for a short walk to the river plain – the lights were lovely. As we came out of the wood on to the plain we saw one of the black and white, red-billed storks. It saw us a long way off, spread its great wings – upper part black, lower quills white – turned round for another look and then with slow majestic flight moved up and off towards the river. I think it was probably one of those and not a purple heron that passed above me the other day.

Friday July 24th—We were up at 6, had a quick breakfast and at 6.30 just as the sun rose we left the *nsinge* and made our way up the hill at the back, keeping well to the left of the village. After making our way through the open bush for some way almost due west, we struck a path leading south-west and turned into it. The nights have turned cold with the advent of the new moon and last night was one of the coldest we have had here. My hands were frozen and it took quite a good while of hard walking to warm me up – as for Miss Bleek she wore her woolly till about nine o'clock!

A VISIT TO THE VILLAGES

Having gone out prepared for sketching and not for collecting, I proceeded to find some nice specimens – a sweet-scented, pretty white-flowered creeper, asclepiad I think, flower and fruit of a shrub which I have not found blossoming before, various odds and ends, another two blooms of the small *Eulophia* and on the way back a fine *Stapelia*, unfortunately not flowering.

Our little path led up and up till we reached the top of the hill, here rather bare so that we had a glorious view over the surrounding woodlands looking back north, with the line of the river to the right; a little further and we were looking over the river plain which here makes a great bend westward. On the big bend there is a series of small bays of grassland with wooded peninsulas between. Far to the south were one or two blue valleys and another over to the south-west marked the line of the Kusezi. Our path led down to the river plain, then along its margin, every now and then crossing a bit of a grassy bay, or leaving the riverlands to cut off a peninsula. About 8 o'clock we reached the path leading up to the first village; leaving this we went on to the point of the peninsula and across to a kind of island of woodland which lies half across the neck of the big grassy bay, on the hill at the head of which stands the second village.

As it was still early and a cool breeze was blowing, we decided to go on to the iron furnace and on our return to visit the villages and try to recruit porters for Cuito. Shortly before reaching this second village the path left the river for rather a long time, cutting off quite a big bit. As we were in the middle of the wood there was the sound of wings and two soft brown birds – rather a light cinnamon brown – about the size of hamerkops, flew up and perched in the top of a bare tree, so that I was able to get a good look at them. They had long curved yellowish beaks and long tails. In a tree nearby was a dilapidated-looking nest of sticks perched apparently precariously in the fork between two branches, possibly their nest – perhaps last year's nest to which they were contemplating returning. They showed up beautifully against the sky and though rather far off I thought they would show up sufficiently and got out my camera. Unfortunately it was not ready. They waited till I had got ready to make the exposure and then flew away! When we got back to the river, the path cut right across a bay through a piece of water, through which we waded ankle deep.

THE SMELTING FURNACE

To return to the wooded island: it was very pretty – open wood and the path led right through its length and out at the far end on to the grassland across which it ran, leaving the second village far off on the right on the hill at the head of the bay. Some way on we passed the third village, and then the path wound on and on. We followed a side track up the hill, hoping it would lead to Bushmen, but it went on and on and we finally left it and cut down through the wood – very thick here – back to the river path. One or two more little peninsulas, and just as we were beginning to get rather weary of the chase for the elusive furnace, round a corner it appeared – a queer humpy thing of yellowish-grey clay with rusty specks in it. In the grassland beyond were feeding two of my red-billed storks which I longed to photograph. However, they were far off and I did not feel inclined to stalk them.

July 24. Miss Bleek standing next to the smelting furnace, Kunzumbia

The smelting furnace was quite interesting – the front facing the wood was flattened, the back round. At the foot of the front was the hole for the fire and this led out of a shallow space dug about a foot deep in the sand and some five or six feet long. The front is considerably higher than the back and above the furnace door are two large bosses – presumably forming the entrances to the smelting chamber. The fire place led right up into a big space opening out at the top near the back, while a through draught is ensured by a small opening level with the ground at the back. I took a photograph and then after a short rest we started back. We reached the furnace soon after 9.45, so that, allowing for one or two stops and a detour, it took us a good three hours.

The first village
At the first village (Cinkanda), after greeting one or two women, we made our way to the centre of the village, and several men and boys at once came forward. We explained our errand and booked five or six names. We asked the way to the next village, thinking there might be a way along the hill. However they directed us down to the river.

Making a bark blanket
Nearby the place where we interviewed them, a man was making a bark blanket. He had done the hammering and was working it with his hands, thumping it on the tree trunk and then pulling it out. As they apparently strip off a complete circle of the bark some three or four feet long, it is not surprising that in some parts – this for instance – the

July 24. Beating and pegging bark cloth, Kunzumbia

bark trees have almost entirely disappeared. We told him we would like to buy one, but that did not seem to interest him at all! Then one of the men pointed to my camera case, and asked if it was the gun, and could they see it. I took it out, pointed it at them and they immediately scattered in all directions! Only, however, to gather round laughing and eager to look when I explained what it was – so far as I could! One of the women standing round had a tiny baby – only a couple of days old by the look of it – with tiny pink feet and long fluffy black hair.

The second village

To get to the second village (Tukovota) we went through woodland, up and down and then round the head of the grassy bay and up a very steep path – in its upper part adorned by many fetish poles of various kinds. A damsel going down to fetch water saw us coming and returned hurriedly to the village! Near the entrance an old dame was working a black oily liquid in a calabash – probably castor oil and clay to adorn somebody's locks. At the men's meeting hut an elderly man came forward, placed his stool for Miss Bleek and made a little boy bring a stump for me. Then we went to work: the men of the village were apparently all away except one or two, but the old chief promised to send us some and we signed on one fine youth and a couple of boys. Meanwhile the usual audience had gathered and the group was so picturesque that as we left, led by our future porter to show the way, I turned back, took out my camera and turned it on them – whereupon some of the women fled in terror, so I had to stop and let them look at it.

The men's dancing place

I forgot to say that as we approached the first village up a long slope from the south, we came upon a grass enclosure with many pointed poles – the men's dancing place. In the enclosure was a series of little rooms leading into one another, and each with a dead fire. Beyond the enclosure was a long bare strip of ground with old fires at intervals – the dancing ground – and at the far end an arch made by a bent sapling from the top of which hung two intertwined grass rings. On the tree at the top right hand corner a parasitic plant (?*Loranthus*) was growing. The village was just beyond the far end of this

July 24. Initiation dancing floor, Kunzumbia

dancing floor. Many of the huts in both villages had most fascinating collections of totem poles, in addition to the usual group of sharp fetish sticks with skulls of animals on top.

We asked our old friend the way back to Kamundonga and were told that we had to go back to the river and past the next village before we could get to the hill path, so we had to do all three. Coming down the hill from the village were an old man and a girl, the former carrying a bark receptacle of charcoal, which he said was for use in making the steel for axes of which he carried two. I think they were bound for our forge. Here again a steep path led up to the village and as we crossed to the centre, a boy with *tanga* on came forward and handed us a letter which we were surprised to find was addressed to us! It was from Mr Muir recommending the bearer as an excellent boy, very helpful, who wanted to enter our service. We told him to come to the *nsinge* later on.

At this village (Mbumba) the usual courtesy was wanting – perhaps it was a bad time – 1 o'clock nearly, and the half dozen men, all elderly, who were lying round the shelter were taking their siesta. However it was, not one of them stirred except one who was sitting on a long drum, and even he did not offer us a seat! However, he greeted us, and by and by one or two more turned up and said *Bon dia!* We explained our errand once more, and again all the younger men were away. Then we asked about the path and the drum sitter, quite an old man with a few very long teeth, got up and took us along a path up the slope beyond and so into another path leading through the woods. When Miss Bleek gave him the last pinch of salt, his old face lighted up with pleasure and gratitude!

A walk of a little more than an hour on a path running more or less parallel with the outgoing one but further west, brought us to the village. I cut down to the right, skirting the village and making a bee line (more or less!) for the *nsinge* which gleamed corn-yellow in the woods below us. It is growing horribly conspicuous as the grass dries – yellow grass roof, brown leaf walls, sometimes rather an advantage as one can see it so far off! It was two o'clock when we reached it, very thirsty but not nearly so hot and tired as after our up-river walk two days ago. I was very stiff however, so had a delightful hot bath as soon as we had had lunch. After an hour's scrubbing I really felt clean!

The Kunjamba boy, Muyeye, arrived about 4.30. He is older, stronger and altogether more promising-looking than Lomeii. We decided to send him over with our letters to Kunjamba with a request to Mr Muir to send them on at the first opportunity. Miss Bleek wrote to Mr Bailey asking him to send our post on to Cwelei.

In the evening I heard Telosi reading something in which one phrase came again and again with a lot of semi-familiar names. We guessed it must be the Gospel according to St Matthew, called him and asked and found it was – a copy brought over by Muyeye, so evidently the printing is complete. We are asking Mr Bailey to send us copies. The evening of course was taken up with letter writing, and I did not get to bed till nearly 10. I drew the *Eulophia* and the pretty blue scroph. which I have at last succeeded in getting home with at least one flower on it – but could not finish the painting before the light went.

Saturday July 25th—The 'merchant', who by the way brought the hollowing tool I wanted last night and received in exchange his longed-for safety pins, $2.00 and some salt, arrived before 8 o'clock again. We wonder whether he has been set to watch our movements! Muyeye left early (about 8.30) with the letters. I have spent the day first finishing the two specimens I started last night, then working at two sketches, till about 3pm, after which I put away specimens till 5.30. Two small people arrived with a tiny basket of fowls' food and asked triumphantly for *mongwe* in exchange, very proudly taking it.

ATTACK ON MADAME

We had just finished dinner when there was a great squawking, cries from the boys and a rush with the lantern – after searching round, Telosi found poor Madame in the hedge, very frightened, very indignant but fortunately not much hurt I think, though she has a nasty bite near her tail. The boys say it was a *muswe*, a grey cat-like animal. A man brought a skin to sell the other day but refused the $2.50 we offered for it. It had evidently jumped up and caught Madame, who fortunately was sleeping head in towards the *nsinge,* and dragged her down. We proceeded to put her in the coop and catch four of the others. They made a most fearful noise, evidently thinking their last hour had come, and before we had finished, Salikumbi and the pock-marked little man had arrived to see what was wrong. Silly fowls – they've left their old sleeping place which was the hedge between our *cepatonga* and the boys' and therefore fairly safe, and gone to sleep elsewhere and this is the result.

I have just (9.15) heard a noise outside the *cepatonga* – had a look round but saw nothing, except a large bird flying from one tree to another – probably an owl or a nightjar. A bird which the boys call *liku* has been calling quite nearby – I think it is a nightjar but am not sure.

Now to my nice warm bed. Miss Bleek has followed my example and today had her bed taken out and a grass one made on the floor of the tent.

By the way, yesterday just before we finally left the river we came to a small bay – Miss Bleek went round, I cut across through swampy ground and found the large *Lycopodium* again. Two long-legged birds were feeding – they differed from the red-billed chaps in the shape of the tail which was long and drooping. They looked dark grey but were rather far off and I could not really see details. When I first saw them they looked extraordinary – both had their wings outstretched and heads tucked in under the left one, evidently preening it. Looking for specimens in the swamp I forgot to watch them, so missed seeing their flight. As I left the swamp I looked round and they were gone.

A DISTURBED NIGHT

Sunday July 26th—Such a night! Hardly was I settled in bed and just getting to sleep when 'squawk!' from the fowls, cries from the boys and general pandemonium. I heard something in the *nsinge* knock against the sticks of Miss Bleek's bed and after calling to the boys repeatedly, (they did not understand 'It's here!') I reluctantly got out, lit the lantern, the boys came and Telosi produced another fowl, a speckled one this time, which had either been dragged into or had taken shelter in the back fence of the *nsinge*! They then collected the rest of the five and put them in the fence near them, and we hurriedly betook ourselves to the warmth of our beds – it was horribly cold! Soon I heard a sound like something jumping up against the *camba* followed by 'squawk! squawk!' I shouted, clapped my hands etc. and Telosi rushed out. This time he said it was *kasila* i.e. a bird. It might have been flapping against the *camba* and thus have made the sound I took for jumping.

Then for a time, peace. By and by, just as I was once more asleep, it was there again – again I shouted and clapped. This went on at intervals all night it seemed, though it probably only happened three or four times! I was finally roused about 5.30. It was neither dark nor light – what is called 'false dawn', I suppose, because dawn actually begins about 6. What the intruder was I don't know. They say it wasn't an owl, though what night-flying bird would attack fowls I don't know, if not an owl. Anyway, it was most persistent.

It was a very cold morning again. After breakfast I took sketching things and camera and we went beyond the first big bay up the river, where after some search I found the red flowers I had noticed the other day. I sat down and started to sketch. Miss Bleek came back. I went on for about an hour, when it grew too hot and I packed up to return. On the way, however, I stopped and did another sketch just at the edge of the bay, finding a patch of shade to sit in. Unfortunately, my stool is always too low for the view I want. It is coming to grief too – the canvas top, after repeated mendings, is going in a new place. I got back soon after noon, rather hot and extremely hungry. I had intended packing away specimens in the afternoon but instead started working at the half-finished sketch, the edge of the wood, and spent the afternoon at it.

Madame was very sweet this morning. She was caressing her sore tail and I stooped to have a look at it, whereupon she fluffed out her feathers well away from it, just as if she wanted me to have a good look. In spite of her adventure, she laid early in the afternoon. Samolova has built them a sleeping place of bark. I hope the enemy is not a rat, as it is on the ground round a small sapling and a rat could easily dig his way under the bark wall. The new members of the flock are getting much tamer and one or two are beginning to know what it means when we call them, and to come. They are specially fond of coming into the *nsinge,* which I do not encourage.

Another Kunjamba boy, quite a lad of 13 or so, has arrived with a request to enter our service – he wants to go to Cwelei! As we will very likely be short of porters, we are taking him as far as Cuito at any rate.

MUSIC AND DANCING IN THE VILLAGE

The last few days, since new moon, the character of the dancing in the village has quite changed. There has been chanting, apparently the women, for about half an hour after dark, and again for another half hour in the morning from dawn till about 6.30 when the sun rises. This evening for the first time since new moon there has been prolonged drumming and chanting of the usual type, first in the village and then further off. Presumably they had moved to the men's dancing place where they are now. First a single voice holds forth, then the chorus joins in – the same over and over. Occasionally without drums, then with a clapping – how made I haven't yet learnt. It sounds like two wooden instruments being knocked together. I have seen only long drums here. At the third village we visited on Friday, besides the long drum on which our old friend was sitting, there was also one of the wedge-shaped ones. I should have liked to examine it. I think one of the long sides is covered with hide but am not sure. They give a very pure sonorous note. (My curiosity was satisfied later on when trekking to Cuito Cuanavale.)

Monday July 27th—Breakfast as usual at 7.30 – porridge made of local meal, quite good. We opened our last tin of jam (white fig), save one of marmalade. My delicious lemon and lime marmalade is all finished, alas! I am *hoping* Mrs Bailey will send us out a few more limes with our letters, potatoes etc. Other stores, too, are getting low –I am eking out flour with local meal, making the bread two thirds meal, one third flour – quite good. Even the sugar bag is getting very light, and we have less than half the last bag of salt from Kunjamba. Our dried fruit, too, is getting low, and we have as vegetables only about two cups of rice, less than two of beans and some bean-like seeds which Miss Bleek bought yesterday.

The Bushman sweets we are eating lavishly – the last chocolate and crystallised fruit we are saving for Cwelei. However, we have fared very well, if simply, and here we have

July 23. Red-billed stork (?) near Kamundonga's village

July 24. Two cinnamon-brown birds with long curved yellowish beaks and long tails

July 24. The elusive smelting furnace

July 26. The fowls' bedroom

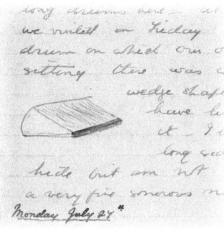

July 26. Wedge-shaped wooden drum

July 26. 'I found a seat, a bent tree on to which I could jump…'

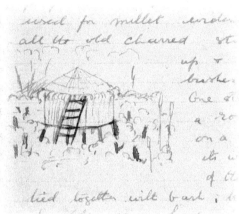

July 26. A storage hut with walls formed of stalks of millet tied together

July 26. Black forked-tailed birds

July 30. 'They looked rather like black geese – big bills with a raised reddish part near the root'

a very good, though irregular, supply of venison. Whenever I kill a fowl, a boy is certain to arrive with a couple of legs of *bambi* for sale! ($1.50 apiece, and that is 50c too much! But having given that in the first instance by mistake, it has fixed the price for us.)

Well, to return from this culinary digression: – After putting the sponge to work, I collected my sketching things and went off for the clearing on the hillside. It was not so beautiful as on the cloudy days, but still quite sufficiently so. I found a seat, a bent tree (*munyumba*) on to which I could jump, just inside the wood so as to get a little shade, whence, with a little manipulation – breaking away of dead branches, bushes etc. and arranging a couple of dead boughs as a footrest etc., I got a view of the scene I wished to sketch – the clearing with a couple of storage huts, woodland, plain and hills beyond. In the foreground a newly 'cleared' bit of land – cleared only by cutting the bush growing on it, the branches, chiefly *munyumba* left as they fell to dry, so that the leaves were all shades of purplish brown – the older clearing, used for millet evidently, still with all the old charred stumps standing up and coppice-shoot bushes all about.

A STORAGE HUT

One storage hut was a round affair built on a 3 ft platform. Its walls were formed of the stalks of millet tied together with bark, the conical roof of golden grass had a fringe of loose pale yellow grass, while in front was a rough ladder arrangement. As I was sketching, a woman came along, and suddenly I was surprised to find that the position of the shadows on the roof had altered suddenly and completely. This took me by surprise – for an instant I wondered if the sun had taken a sudden jump upwards! To get out the day's (or week's whichever it was) supply of millet heads, she had tilted up the roof, and then proceeded to ascend the ladder (much fewer stages than I have shown) and dive into the interior! About 11.45 I stopped and came back at a good pace, putting up two partridges on the way. Why does a rising partridge always send cold shivers down one's back? I suppose it is the sudden clattering, whirring noise coming suddenly out of silent grass right under one's nose.

I was back by noon, made the bread and spent half an hour at the sketch. Then we had lunch and after a short rest, reading bits out of an ancient *Weekly Times,* five or six months old, I spent the rest of the day putting away specimens. I have now put away all except those in the press, thank goodness! I must now go through those and put away all that are dry – it is much too full. I have plenty of herbarium paper free now, but have nearly come to the end of my supply of newspaper, and it does not seem likely that I shall get any more.

Late in the afternoon I was attracted by the sound of birds and there were two of the black, forked-tailed birds I have noticed several times, whistling and bowing to one another in a most amusing way. They have a very pretty note and seem to vary their song a great deal. This duet and dance I hadn't heard or seen before. They are pretty little chaps with a rather large head, a yellow bill I think, and the long forked tail which they move about a lot – often it is almost straight down at a wide angle to the wings which project behind. The sleek black plumage has a sheen of blue in some lights.

Miss Bleek went out for the usual 'evening constitutional' about 4pm. I did not go but stayed to see the baking of the bread accomplished and to finish putting away my specimens. One of the men from the village came to tell us that the *Chefe* of the post in whose district this is was coming to collect taxes. As a matter of fact he arrived about 5 o'clock this evening and has camped between us and the village. Perhaps when he has gone we shall get the Bushmen! About the same time Muyeye returned from Kunjamba bringing a note from Mr Muir. He has done well – exactly according to instructions. No sign of the two men from Muié as yet – I don't suppose *they* are making much speed!

Tuesday July 28th—This morning I went down to the point to the right of the water holes, sat in the shade of a tree and did a sketch of the holes with the far peninsula and some women getting water. Bits of it are not bad, but all the sketches are growing very much alike. Miss Bleek meanwhile held the fort, i.e. went on with her indexing.

ENTERTAINING THE *CHEFE*

I returned about 11 and, as I approached, noticed blue smoke rising from the kitchen. I guessed what that meant and, sure enough, I was greeted by the news that the *Chefe* was coming to dinner. He had sent a boy down to ask us to go up and see him. Miss Bleek went, found him a very decent young fellow (not young at all – he had been in the country 30 years I think he said!) and after some conversation invited him to dinner. He accepted the invitation with pleasure so on her return she set Telosi to roast the leg of buck and to make soup of the remains of the chicken. Soon after, arrived a gift of a piece of buck and ten eggs! I set to and made a pudding etc. and we waited till 1 o'clock, the hour Miss Bleek had named – no *Chefe* and the boys as well as the chief Kamundonga said he had gone away. It appeared he had departed by *tipoia* for Mbumba (the village up river) and would return here about 3 or 4 they said, so evidently he had not understood that he was invited for *midday* dinner! I hope the dinner won't be spoilt in consequence. The soup at any rate will be all the better for more cooking.

Kamundonga brought a couple of men to see the *miti* (not the *Vasekele!*) and they were most impressed by the drawings and sketches. The Mbundu name for a Bushman is *Kasekele,* plural *Vasekele* i.e. the chased people. The old chief was quite hilarious – I suppose because the ordeal of tax collecting was safely over! Three small people also came and added the remarks of experienced sightseers to those of the men to whom the pictures were quite new.

I worked at the sketch a bit, changed some of the papers of the specimens in the press and now (5pm) having got the dinner safely on, I'm waiting while Miss Bleek has her bath. The return of the *Chefe's tipoia* has just been announced.

7.30pm. Well, he has been, and gone. About 5.30 Miss Bleek wrote a note announcing that dinner was ready and Senhor Vieira came forthwith – a very decent elderly man, small, spare and I should say very abstemious, with a sense of humour. He rather reminded me of Mr Jalland! The conversation was a bit difficult – fortunately he talked quite a lot, with a few remarks from Miss Bleek, while I listened or talked

Mbundu! Simply could not remember any Portuguese. He offers help with porters etc. if we have any difficulty. He does not give the people round here too good a name, and as for the 'click, click' people! They apparently according to him are most fierce! We held our peace.

Wednesday July 29th—We had just finished breakfast when we heard a voice outside the door of the *cepatonga* which some zealous soul among our staff (Telosi I think) will persist in shutting. I called to Samolova to open it and in walked our friend the *Chefe*, Senhor Vieira, very kindly come to *visé* our passports and generally fix us up. He wrote us a *guia* for our porters to Menongue, leaving us to fill in the names, so that little difficulty is settled. He was just about to leave for his five days' journey westward, to the source of the Lomba, we gathered. He showed a laudable lack of curiosity as to our doings. Perhaps he already knows all about them.

That finished, I went down to the river and tried another sketch of the water hole (the far one) from the far side —not a success. It is patronized by very few of the women. I want to try the other one again and do pencil studies of the various women who come. Back at 12.30. At lunch Miss Bleek turned faint and I gave her a dose of brandy. She had a good long rest and sleep and seems all right again. (Anyway, the brandy has justified its existence!) I lazed for some time too, then worked at my sketch till 4 o'clock when we went for a walk.

As we set out this morning, I happened to glance back, and there, coming full pelt after us as fast as their little legs would go, was our family of nine, Featherlegs in the van with Madame a close second! They followed us some way from the *cepatonga*. Poor Madame – she has lost a lot of feathers – the whole of the left side of her tail, so that the lining shows on that side! And a good handful of the soft feathers near it. Yesterday she showed distinct signs of going broody again, but nevertheless laid today, her tenth egg. She is a fierce little body and is very much lord (or lady!) of the flock. They all went to bed in the new hut of their own accord tonight for the first time. The larger brown hen had to be gently persuaded but eventually went in. The new ones adopted it at once. In the wood this evening I put up a small tan-coloured animal about the size of a dog, which made off full tilt through the wood so that I only saw a rather broad head with two upstanding, short ears. I suppose it was somewhat of the jackal tribe, though the gait was a rather stiff one, rather more like the bounding of a buck than a smooth running, perhaps owing to the bushes etc. which encumbered his path. Post not in yet (7.10pm).

Thursday July 30th—An uneventful day. The weather cycle is round again. The very cold nights are past. Today was windy with large white clouds and last night was very mild. I put the press etc. in the sun, finished changing the papers and restarted packing away, but soon stopped as the wind was too strong. Lazed, reading old odd sheets of the *Weekly Times, Argus* etc. Then Sangevi brought two women to see the paintings and that took some time. In the afternoon, was driven to solving the *Argus* 'H.R.H. The Prince of Wales' crossword puzzle. Then packed away for some time. That helpful merchant,

Salikumbi, brought the same youth to get the other $5.00 worth of *tanga* – he had a tiny bird in his hands, which he wanted to sell us. We said we did not want it and he kept on teasing the tiny creature, and finally wrung its neck and threw it in the ashes. I think even he realized that that was not the kind of thing we liked!

Went for the usual 'stroll' between 4 and 5 – found a new specimen, a small shrub with green fruits, evidently closely related to the brown velvet fruit tree. As we got near the river, four big black birds got up, beauties, jet black with white tipped wings. They looked rather like black geese – big bills with a raised reddish part near the root. They flew out into the river where they alighted in the short grass. I have seen them before in the grasslands, but never so close as this. I wish one could get near enough to these creatures to get photographs, or that I had a telephoto lens. There is certainly plenty to do here (in a scientific way I mean), but one has to look for it. On our way back, in sight almost of the *nsinge* we came upon a Bushman sleeping place, three fires and three beds, just to the south-east of us. We must have passed on both sides of it again and again. It may even be of quite recent origin.

Two puppies

The party was increased by two today. Telosi arrived from the village with a small, pathetic browny-yellow puppy 'as ought to have been home with its Ma'. Soon after, Samolova appeared with a huge fat one, double the size of the first (who is a lady), while the second could not be mistaken for anything but a 'bouncing boy' dog for an instant. Telosi christened his Kunzumbia – hearing them try to teach her her very uncallable name, I told Samolova he'd better call his Cwelei. He proceeded to do so at once and the puppy nearly scrambled over the log in his efforts to respond.

Friday July 31st—We have decided to start on Monday, post or no post. Stores are getting so low that they must be replenished at an early date – even the sugar bag is nearly empty.

After setting my sponge, I went down to the river to collect. I did not get much but found two new specimens – a mauve flowered aquatic composite plant and another aquatic with tiny blue speedwell-like flowers, and also a very fine delicate *Najas*, apparently rooted, in quite shallow water, an eight-inch pool on the path. Could find no sign of it elsewhere.

On my return made the bread and a few scones to go on with, and after lunch set to work putting my specimens away, a tedious job and much interrupted by the fowls – who, missing their recent filling meals of *wunga* (there was very little left for them today), gave me no peace but persisted in walking over and sampling my specimens – and the flies, which were peculiarly annoying and sticky. We sent Telosi up to the next village, Swamondumba, to announce our departure on Monday and sign on further porters. One, Maliti, sent by Sr Vieira, came yesterday to sign on. (He proved to be an excellent boy, willing and intelligent). Telosi returned about 5 – porters were away, but the chief would send them tomorrow.

THE MAIL ARRIVES

In the midst of my work, just as I had set Samolova to bake the bread, Muyeye arrived with great élan, ushering in a very much clad native – jacket, trousers, cycling sock tops as gaiters – bearing a large parcel – our post from Mr Muir, and still no sign of our carriers! My share was much more satisfying this time – two from E.L.P., one L.G.P., one with enclosures from G.V.P., one from Uncle John with four *Courants*. G.V, mentions poor Auntie Carrie's death on the 15th June, 'supposes I had heard about it' – how? I wonder what Auntie May will do. We wrote hasty letters and sent them back to Kunjamba by the messenger who may return, possibly, as a porter. After finishing the specimens, baking etc., I settled down for a good read, such a treat! The last three *Courants* and the *Times* from Munnie remain for tomorrow.

Saturday August 1st—A day of baking – all morning made biscuits, baking in the pan and pot lid, then made raisin cake – nearly finished the flour – there is only enough left for one more baking of bread. In the afternoon had a good long rest.

Sunday August 2nd—Another busy day – made bread but couldn't get any *wunga* till late afternoon, so had to make the sponge of flour only. Finally the boys came to the rescue and lent us three cups of meal so that I was able to mix nearly the usual quantity, leaving a couple of spoons of white and one of whole meal for gravy etc. in case of need. The intervals in baking etc. were occupied in packing away specimens. I had already decided to transfer them from the lidless box I got from Mr McGill to the now empty store box. Halfway through, I found the capacity of the latter was much less than I thought – most annoying, but I couldn't re-pack again – so have had to put some in a paper parcel outside. About 3.30, went with Miss Bleek to photograph the Bushman camp – took two views of it, one of the river and a fourth of one of the little geophilous plants. Got back about 5 and again went on with packing etc.

READY FOR DEPARTURE

Monday August 3rd—9.50am Here we sit among our goods, all ready packed for the road, and only waiting for our porters who are conspicuous by their absence! This, although Telosi and Muyeye respectively were sent to the upper and three lower villages on Friday and Saturday to notify them that we would leave first thing on Monday. Apparently these people cannot bring themselves to start until a day or two after the time arranged. When we had just finished packing, the purchaser of cloth (Salikumbi's protegé) arrived, and was horrified to hear we intended starting today! Next, evidently having been told by him, Kamundonga arrived and orated loud and fast. Then finally he departed and now all five porters (including Kakupa) have just arrived in a deputation to say they must fill their insides (this by signs!) today, tomorrow they would go! Much orating and excited replying on the part of the deputation and Miss Bleek. I doubt if we shall get off today in spite of all our protestations, for unfortunately the deciding factor is their dilatory selves.

Departure delayed

Tuesday August 4th—9am. We didn't, and as yet there are no signs that we shall do so today. Everything is ready, save the tent which will not come down till some at least of the porters have arrived. Telosi left an hour or more ago to go to the lower villages, Muyeye for Swamondumba – the latter has just, as I wrote the words, returned with one solitary porter, Maliti – probably met him on the way. We spent yesterday restfully after the speechifying etc. of the morning was finished. Miss Bleek as usual doing writing in connection with the Bushman work, and I painting. I painted the portrait of old Golli which I started weeks ago – rather spoilt it to my mind. The first shading looked rather well. I needed my model so badly! Then I started a sketch of what I could see of our 'yard' and kitchen premises – an awful production but, like the village one, it served its turn. They were so pleased to see a painting of Samolova with his pipe (not the least like him) and Muyeye's back and the fowls' hock that it was certainly worth the hour's work.

I must say, apart from the annoyance of the delay, the restful lazy afternoon was rather welcome. But how annoying to have to unpack all one's bedding etc. and to know it must be re-packed in the morning and not a yard of our way covered. Today I don't suppose we shall do more than a half day's trek, and that probably not a very full one. Miss Bleek sent a message for distribution in the village by Telosi last night that if the porters didn't turn up early this morning she would send to the post to complain of them and ask for porters. It does not seem to have had any effect, except to make Telosi suggest going to the lower villages.

9.10. The tall man from Kunjamba (Ulombo) has just re-appeared, rather to my surprise. So the number is slowly mounting up. If we can get five good hefty men for the cases, we could put much of the smaller stuff (or rather softer packages) in the two *tipoias* and walk this first day, but angular packages mustn't go in, or the hammocks would be ruined.

Bushmen arrive

About noon Ulombo with a couple of men came up to say that there were Bushmen at their village, the first village down river (Umpongo) – they had been away and returned the day before. Miss Bleek said if one of them went down and fetched some, she would give him a spoonful of salt for each one he brought. By the way eyes and teeth flashed and sparkled we judged he'd bring *some* by hook or by crook. Then followed a long wait and finally about 3pm voices were heard and Telosi, porters and four Bushmen, two men and two women, arrived. Miss Bleek started in at once with the latter while I signed up porters and allotted loads. All but three, who were still getting *wunga*, had now presented themselves and I would have started on, leaving Miss Bleek to follow, but Kamundonga, who had come for another lot of boracic, and to see the porters I think, with Sangevi, said we could not reach water that night. I consulted Ulombo, the much-dressed Kunjamba man, and he bore out the chief, so there was nothing for it but to unpack once more, re-erect the tent and spend yet another night at the Kunzumbia camp.

DESCRIPTION OF BUSHMEN VISITORS

The Bushmen were a delight. First came in two tall lanky youths, dressed in the usual skin apron, carrying bows and arrows, wearing fez-shaped caps of skin, one with the fur, rather a short shape, the other cleared of fur, and decked out with beads and bracelets. The front of the hair was worked into manes adorned with beads, skin danglets etc.

Then Telosi said something about *munakasi*, and out from behind the *tipoias* came two visions – tiny little women somewhat of the !Kõ type but rather darker, one young and really a pretty little thing, the other elderly, but very like her – mother and daughter. The daughter, especially, was decked out wonderfully – bead chains, charms – half a dozen of them – fastened to her hair, in her nose yellow flowers fastened to sticks. The younger boy, too, had his nose pierced and tried hard to insert in the hole a tiny brass safety pin I gave him, with the result that he pricked his nose and handed the safety pin to the old lady!

Round her neck the daughter had a whole curio shop of charms – two calabash necks, a rather nicely made brass bell about two inches long, four cartridge cases tied together, a large brass ring, the top of some other plant etc. A leather strap was worn tightly round her body above the tiny dark-coloured breasts. The charms were attached by strings to several strands of string wound tightly round her neck. Her ear lobes were pierced and through the holes she wore two pieces of 'fire stick' about 12″ long. Her safety pin went at once into her ear, where she already wore a three-cornered piece of ostrich egg. Both she and the elder woman had pretty little hands, feet and ears. Her mother also had various attachments but not nearly so many as the daughter. Their skin clothes were much more ample than the Kutsi people.

Of the men, the elder was broad-faced, with big feet, rather dark (had he been washed he would have been *much* lighter!) and altogether much heavier in build than the other. The younger, though very tall and at first sight ungainly, on closer examination

August 4. Young Bushmen with bows, wearing fez-shaped caps of skin, Kunzumbia

August 4. Bushman mother with her daughter who is wearing beads, charms and flowers in her nose, Kunzumbia

proved to have a really pretty face – straight profile, thinner lips (even the women had markedly thick lips) and pretty little even white teeth. All four had quite good teeth – not filed. By the way, the Kangali apparently don't file their front teeth – at any rate not the people of Kamundonga. All four had various dark marks about their faces and one youth (the younger) the cut between the eyebrows.

The boys kept on saying they were hungry and must go and eat, but after having been given salt, *wunga* (our last, save half a cup!) etc. they stayed on quite happily, tied up their precious salt, beads, tobacco, had a smoke and went on talking for quite a long time. The women fastened their beads etc. in corners of their skin garments. The two boys gave the older woman their *wunga* to carry (one apparently was her stepson, the other her nephew, or rather husband's nephew). The men tried to sell some of their arrows for *tanga*; the elder woman, on being promised salt and *tanga*, said she would go with us to Cuito, but I don't suppose she will.

One of the men produced a poisoned arrow and was telling Miss Bleek what plant they used – a tuberous one – so I drew a rough sketch of the *Euphorbia* I found at Kutsi, received by boys and elder woman with acclamation. They squeezed and pounded the tuber and got white juice from it and dipped their arrows in it. Altogether a profitable afternoon and worth the two days' delay. The porters promise to be ready early tomorrow, soon after sunrise! Time will show.

DEPARTURE FROM KUNZUMBIA CAMP
Wednesday August 5th— 7.40pm. In camp near a water hole in the plain of the Kusezi. Round us are the porters, in three groups – our lot round one fire, Kamundonga lot to the left, down-river lot beyond the kitchen party. A quiet amicable tone pervades all, very different from the worried feeling of the Kutsi–Kunzumbia trek – long may it last! The moon, just past the full, rose half an hour ago – a great disc of gold. We are in open flat country – the flattest camp since we left the Zambesi – very sparse woodland, a tongue of trees widely spaced just where we are, to the west the river plain, burnt black.

Talk about Alice and her croquet party! That was nothing to collecting porters. On Monday we had our *tipoia* boys at hand, though tardy, porters simply not there. On Tuesday with great effort we managed to collect the porters. This morning the porters were there and all the *tipoia* boys began to cry off. When routed out, one had a sore toe done up in bark, another had vanished, a third had a bad cough, a fourth, an old man whose heart had evidently failed him at the last, brought a young boy to take his place. The first was written off, the sick one was given quinine and the chair to carry today and told that he'd be better tomorrow. Various other expedients finally raised two men to carry my *tipoia* with tent, kit bag etc. and five for Miss Bleek's. By 8am all loads were given out but it was past 9 when I, with the first instalment, finally left the *nsinge* and set off via the village. There, to my annoyance, they stopped to eat! I expressed myself so forcibly that Nyundu (old pock marks), who was sitting on his pack, started off forthwith at a good round pace while the others stood round simply throwing lumps of porridge into their mouths!

Cold night, hot day seems to be the rule here. Our route lay up the side of the valley and soon led into open grassy land – a kind of wide grassy passage through bushland, and very hot and unsheltered. The breeze was behind us and, after an hour's good walking, I was very glad when we reached a fine, large tree and our leader, Nyundu, stopped for a rest in the shade thereof. We waited there till the rearguard came up, then we trudged on and on through the sandy track, and still no Kusezi appeared. 12 o'clock came and passed, no sign of a river. My hunger and thirst increased alarmingly. Finally, at 1 o'clock I stopped and, with Telosi and Muyeye who were in front with me, rested delightfully in the shade of a tree. About 15 minutes later Miss Bleek and the rest came up, and we had lunch. Then on again – sometimes through thin short bush, other times through dry open woodland with some fine trees scattered about.

THE PLAIN OF THE KUSEZI

At last came a break in the trees. An open grassy plain appeared and Nyundu announced the Kusezi. No sign of water however and on up the valley (or plain rather), not across, went the path. By and by, three black forms appeared in the distance. 'Mpulu', said Telosi, 'Vimpulu vitati'. They were in the tawny grass but, on seeing us, dashed off to a black, burnt patch and waited till we were near – tails switching in the air – then, as one turned to look at us, I could see the humped shoulders, evidently wildebeest.

Then we passed a shelter made of branches and soon after, about 3pm, Nyundu left the path, put down his pack and made for a patch of long green grass in the middle of the burnt plain. They said it was the only water until we reach the Vimpulu, so as we had all had enough, we chose the least burnt patch which had a little shade and made camp. Everything went quickly and smoothly, the boys hopped about, were sent for water etc. Ulombo, Muyeye, Samolova and Kakupa put up the tent and the two first *without being told* started to dig our beds and sent my two *tipoia* boys to fetch grass etc. – a most calm and peaceful camp!

I intended to be in bed by seven, but have been writing up this diary and now it's past eight. 6am to 8pm is quite a long enough day. A beastly boil started just at the point of my left jaw below the ear three days ago. Fortunately, I have a tube of Stannoxyl which I at once started taking, and today (this morning) it opened. I've kept boracic lint and basilicon on it and this evening, it opened well – I think even the core came away – lucky if it is so. I suppose I'm missing the *makovi*.

CLOSE ENCOUNTER WITH A SNAKE

About midday I was trudging along, with Telosi a few yards ahead, when suddenly he gave a shout and jumped wildly sideways. I just saw a light tawny yellow band undulating through the air towards me, also gave a shout and straddling wide jumped in the other direction while the snake (*dinoke*) passed up the path exactly between where my feet would have been had I waited a second later! Our second snake in Angola. As we sat in camp we could see three or four black and white storks feeding beyond the water holes.

Thursday August 6th—I was wakened by the sound of birds singing all about us, rather like small larks and canaries. The sky was cloudy and the moon still shining in the west, time 5.45. Half an hour later, as the sun rose, all the birds were silent. Sunrise was lovely – a great red, fiery disc gradually came up above the low bush hiding the eastern horizon. We were dressed and packed by 6.45, and finished breakfast soon after, but in spite of that we did not get started until a quarter to eight, and even then quite a lot of the porters stayed behind, just sitting. In expectation of having to wade across a river bed I wore my old canvas shoes, but nary a river, not even a damp spot did we see. Some way above the camp we crossed a depression floored with yellow-red soil, very like the upper Kunjamba on a small scale, evidently the upper Kusezi in the wet season.

For over an hour the way was across open grassy country. It was a full hour before we reached a shade tree where we stopped for five minutes, chiefly to let Miss Bleek's other three *machila* boys come up – her two were puffing and coughing after 15 minutes' carry. For another hour the country was still fairly open but with more trees, – some very fine *musivis*. One fine big tree was covered with fruit. It looked very beautiful – the seed-cases had opened and, in each, the seed with its bright orange-red aril shone against the rich glossy green foliage. Then came a strip of more thickly wooded country. All this part, Telosi told me, was formerly a great centre for rubber collecting, and the *kambungo* was practically exterminated – taken in for sale to Cuito and further east to Muié. We passed several old shelters.

A LONG TREK

The trek proved a very long one. At 10.30 we stopped for 15 minutes, then on again. At 11.30 we reached an open sandy stretch again and had another rest for a few minutes. Then we reached a stretch of lovely woodland, trees close, moss here and there, and several flowering shrubs – the large-flowered 'azalea' legume, quite absent from Kunzumbia, among them. Just before this, crossing a bare sandy patch, recently burnt, I found a new specimen – low growing, tufted with yellow heads, at first sight like a composite but actually one of the Thymelaceae, with a delicious faint scent, the scent of cowslips. I enjoyed it for the rest of the trek. Ulombo and other boys told me that the Bushmen obtain poison for their arrows from its roots, which I find hard to believe. Several corroborated this. Soon after, we crossed a dry *donga* – head of the Vimpulu.

We finally reached the village (Litwe) just about 1pm, and that is a good ten minutes or more from the water! Here the hope of porters induced us to make camp, although we had again done only five hours' trekking. It was not a very nice place for a camp but we picked the best spot we could find and it proved not so bad, but the peaceful, amicable atmosphere of last night was lacking. The *Mueni*, Litwe, and a lot of men came to the camp in the afternoon and were asked about porters and Bushmen. On the first they were non-committal, on the second point – '*Vasekele wahi*'. They were much impressed by the two paintings – old Golli with his pipe, and the scrawl of the *nsinge* kitchen premises, all I had unpacked. Women came bringing sweet potatoes etc. for sale. Meal apparently was not too plentiful, though the boys bought a fair amount. Miss Bleek bought a little for our breakfasts, but we failed to get potatoes.

We had had dinner and were sitting over the fire when we heard greetings, 'Sachingumba', and the words 'wangamba mikanda', and there – almost too good to be true – we saw our dilatory messengers to Muié. Tomorrow it will be three weeks since we gave them the letters and, as we thought, dispatched them. As a matter of fact they did not leave till Saturday, and got to Muié the following Tuesday. Mrs Bailey gave them the things – rice (a huge bag), sweet potatoes, fat, salted pork and joy! a bag of lemons – and they promised to start the next day – Thursday, a fortnight ago! I suppose they stayed for a week. Anyway, they got back to Kamundonga last night, received our message that they were to come on to Cuito and we would leave their wages at the post, and immediately came on at full speed, catching us up in one day. We were so glad to see the wretches that besides pointing out the time they had taken and getting a good hearty laugh at their expense from the listening crowd of porters, we paid them their $9.00 and $1.00 for 'crossing the Kwando' for which they asked, and gave them a cup of rice into the bargain, because they said they were hungry!

All other efforts to raise porters having failed, at Telosi's suggestion Miss Bleek went and interviewed the chief and said if he did not supply us with at least three, we would leave some packages and send back for them from the post. This, apparently, had the desired effect. By the way, Telosi's dog, Kunzumbia, remained behind. It was exchanged, I believe, for a small black chicken which the 'kitchen', now consisting of Telosi, Samolova, Ulombo, Muyeye and Kaliye, ate. Samolova made a neat little cage, door and all complete, for his, which is a fine pup. He, Cwelei, rides on top of Miss Bleek's kit bag, on his master's shoulder or head.

Friday August 7th—No porters appeared. Ulombo, sent to summon them, came back without. As it was a large village and there were dozens of men about, we knew it was all nonsense. So Miss Bleek went down with one package which was dumped in the centre of the village. Then the old chief said the men would come as soon as they had eaten. We waited ten or fifteen minutes, then started for the village, where we found a number of men sitting round the fire. Nearby was a three-cornered drum at which I was glad to have a good look. It is a hollow section or sector rather of a tree trunk, open along the narrow edge. On the sides are two or three bosses of dark greyish stuff – clay, probably worked up with something. Striking different parts gives different notes, and striking the bosses yet others, and very resonant notes too.

THE VALLEY OF THE CAMBINGA

The chief moved to one side and started measuring out meal from a bark *wata* a small boy brought him. Then I started on with most of the porters. Miss Bleek, the *tipoias*, two loads and one or two porters waited in the village. Eventually they caught us up on the Cambinga about noon. In the end only two porters were forthcoming. However, as I walked all day except 20 minutes near evening, and Miss Bleek also walked a good part of the time, we did very well.

The trek was a pretty one – first through clearings, then woodland, up and down, some very beautiful, till we reached the valley of the Cambinga, a most beautiful one, full of colour – rich reds, yellows, browns and here and there patches of green. For some way it was dry and open, then we reached spots where the darker colour, and storks (black and white) feeding, indicated the presence of water. Then came a water hole just beyond which we stopped under some trees, and I had one of Mrs Bailey's excellent lemons.

Here Miss Bleek and the rest of the porters caught us up. We had stopped at one of the clearings where two men and a boy were pegging out the skin of a newly killed buck, and our boys bought a piece of buck, which sat on Miss Bleek's basket, much to the detriment of the hessian wrapped round it. We went on down the river for another half hour. Just below came a small tributary stream, Duva or some such name, then the Cambinga itself made a curve. It is a narrow (8-10 ft) rust-coloured, swift, deep stream with long, feathery reeds on its far bank. A rude bridge of tree trunks, very unsafe-looking, led across. We stopped for lunch, then after an hour and a half started. The porters stopped to cook food or we would have started much earlier. Finally Muyeye started and I followed. (I'd tried to get the party started half an hour before, in vain). I took off my shoes, but got across dry shod, rather to my surprise. Fortunately, I had on my old canvas shoes as it was pretty muddy in spots further on.

The valley was very pretty – our path led down the centre for some way, then across to the far side and up the wooded slope beyond. Then up and down through the woods, past several small tributaries, the names of which the boys told me, in and out and round about – the way seemed unending. Then we crossed an open burnt part and here I finally gave my *tipoia* boys 20 minutes' work. They carry very well, steadily and quickly, but are puffed in no time. Another patch of woodland followed, in which I found another sweet-scented geophilous shrub – flowers small – in full bloom. Soon after, we came in sight of the Tumpu with the Cuito and Cuanavale in the distance. Here was a wide stretch of burnt land which took a long time to cross. Ahead on the hillside was a village; passing the path to this, I led my *tipoia* boys (we were right ahead of the line) to the nearest non-burnt patch of woodland just above the pretty little river. It was sundown and we were all only too glad to make camp.

CUITO-CUANAVALE

Saturday August 8th—We started at 8.30, cut across the tongue of burnt land to the pretty, clear little river over which I jumped, then through a piece of rather boggy land and in a few minutes were on the *strada* and in sight of Cuito-Cuanavale, perched on the hillside above the blue valley. Such a sandy, bare strip! The woodland paths are much more interesting and pleasant to walk on. The Portuguese seem to have the faculty of bringing ugliness in their wake – the *stradas* and the posts are bare, ugly scars on the face of a most beautiful country. By 10.30 we had reached the Cuito – a wide swift river, curving and doubling on its course. Above to our right lay the winding valley of the Cuanavale, beyond that of the Cuito. To the left from behind us came the Tumpu, and on the far side the little Tengu, from which comes the native name, the village having been here long before the post. Near the river we passed a large camp with still smoking fires. The boys said *'Cindele'* with porters going eastwards.

CROSSING THE CUITO

A boggy bit brought us to the actual bank, firm and grassy. I had got my feet muddy crossing the swamp, so had a good excuse for walking into the edge of the river and paddling while waiting for the dugouts to come across. There were three of these – one a huge one; the two *machilas* and four boys crossed in it. Then we followed, deck chair and small chair, sitting at our ease, with two of the boys in front. Two smaller boats also helped and everyone got across very quickly. Then we crossed a bit of swamp, climbed half up the hillside, then turned northward into the woodland above the river, and made camp a couple of hundred yards from the bank, under some largish trees. The river here makes a big bend and this side is a high bank of yellow sand, fringed below in places by a narrow band of grass and dotted above with trees – most beautiful. Above, the river curves right around on itself; at the end of the high bank is the landing stage for the post.

CAMP NEAR CUITO-CUANAVALE

We wanted to pay off the porters at once, but found, much to our annoyance, that Samolova, Muyeye and Kaliye had stayed behind at the village to buy food. The last named had the despatch case with most of the money in it, so everything had to wait for an hour or more till they arrived. We had a small clearing made, tent put up and got camp more or less in order. At last the boys arrived, we counted out the money and gave it out with a dessertspoon of salt which we could just manage. Thereupon, all up with their blankets and off – no grumbles or arguments or anything, and I've no doubt they slept at Litwe this night.

Then we had our lunch and about 2 went up to the post with three of the boys, saw the *Sous-Chefe* and *Chefe*, the latter a rather weary, elderly man (with the nails of the little finger on each hand an inch long!) who to our relief spoke French. He examined our passports very minutely, several times over by the length of time he spent over them, and then disappeared into his subordinate's office next door where they carried on a low-voiced conversation for some time. When he returned, Miss Bleek asked if there were any Bushmen in the neighbourhood, to which he replied that he thought not but had only been at the post since July. We then asked about porters and explained what we needed. Thereupon *Sous-Chefe* was called in and a voluble conversation in Portuguese followed. Finally the required porters were promised for *'le lendemain'* if possible. I rather stupidly took it to mean 'tomorrow' – probably meant that the porters would be sent for and told to come 'the next day'.

That finished, we proceeded to the store, where the storekeeper, a sad-looking individual, gave us chairs and we bought salt, sugar and flour. As soon as we got back to camp, I squeezed the juice of ten or twelve of Mrs Bailey's lemons, cut up the skins and put them to boil for marmalade. In the midst, the boys presented themselves for their week's wages and were much disappointed to learn that all the porters must wait for Menongue to get theirs – but Miss Bleek paid them all up to Wednesday when we started. Then they wanted to buy things and went on and on till I finally put my foot down. The sun had set and it was getting dark – dinner not cooked, water for baths not ready, fires for the night not made etc. Miss Bleek sold them bits of an old sheet to make shirts etc. with. She had already given them each a small piece to make bags for their salt and all including the *tipoia* boys had been busy sewing. Bed was very welcome, but my hole not very comfortable. The views from the fort are fine, only marred by the bare foreground.

Sunday August 9th—The sewing party continued – Samolova and Muyeye making shirts with Telosi's assistance and Ulombo, I think, trousers. They wanted to go on buying but as they have not much money, wanted to have an advance, so the decree went forth – no more selling till we get to Cwelei. That may bring us a little peace. I went down towards the post landing stage and started sketching a view of the river below us. There is a deep bend, the river coming close in to this bank which rises high above it – a steep bank of ochre-coloured sand.

By and by, two of the *tipoia* boys came past and, seeing me, turned out of the path to watch. One presently departed to buy *wunga,* the other stayed and later was joined by Ulombo and Kaliye; then two men from the post arrived and were added to the spectators. They announced with bated breath that I was 'writing' (or drawing, *sonneke* is used for both apparently) the Cuito and the trees, and were much pleased when they successfully identified the particular bush I was drawing at the moment.

In the afternoon I went in the other direction quite near our camp and did a sketch looking up. It was rather a hotch-potch as I started it mid-afternoon and finished just about sunset, changing the whole light and colouring of the sky and river. The two men from the post had also visited the camp and wanted to buy something from Miss Bleek. They told Ulombo that the *Chefe* had sent a messenger up the Cuanavale to get our porters.

Monday August 10th—I quite forgot to say that yesterday before sketching I put the bread to rise, finished the first boiling of half the marmalade and started the second. I got Ulombo to cut two crotched sticks across which I put a strong green stick on which I could hang my pots. Apple rings, and prunes for the yeast, went on to boil and then I left the three pots in charge of Miss Bleek and Ulombo, and went out. At noon I came back and made up the loaves. The baking I left to Telosi. This flour rises well; although the time was cut down rather, the loaves rose better than the brown has done for a long time. The marmalade was finished in the afternoon – two-and-a-half tins, quite good but not as good as the Kutsi and Kunzumbia lots.

Today I made another batch of bread which rose beautifully – right out of the tins! Telosi did not bake it quite enough, unfortunately. The sketch today was down near the lower end of the bend with a bough of the pretty pinkish-white leguminous tree in the foreground. I started it in the morning and went back again for an hour in the afternoon

August 6. 'The seed with its bright orange-red aril shone against the rich glossy green foliage'
(*Xylopia* sp.), Kusezi

August 10. A beautiful rose-pink 'skew pea' (*Vigna nuda*), Cuito Cuanavale

August 11. Bright yellow
labiate with edible tubers
(*Plectranthus esculentus*),
Cuito River

– really much too difficult for me. I wish I had some technique!

About 4, after having given directions for dinner, we walked out through the two small hamlets (of three or four huts each) getting magnificent views of the winding Cuito. One curve is like an Ω then away to the south-east it winds and curves. Its curves and those of the Cuanavale are quite impossible to sort out, by us at least. I ought to walk to the top of the hill again and have a good look to see where they really join. I spent some time photographing first the beautiful rose-pink ground pea (leafless, or nearly so, at present) with its twisted keel with a projection on one side to raise up the wing on that side, and then the Cuito and its valley beyond the village. It took a long time to find the best view point for the latter. Then I found another leafless pea – much smaller, and a darker red with white and yellow markings, a beautiful little thing.

Beyond the curve of the hill, above the valley of the Tengu, we came on a simply magnificent patch of the red clustered-flowered shrub – several plants in full blaze – better specimens than any we have yet seen. The azalea-legume, too, is in bloom here – beautiful rose-pink flowers, as well as white. Altogether, this seems a very fertile spot. We walked up the Tengu valley to just below the post and then came back. In the morning Miss Bleek went up to the post to see if all was well, and saw the *Chefe* who said the porters were being fetched and he hoped to 'present them tomorrow'. I hope he will, though there is plenty here for me to do.

Tuesday August 11th—No bread today, thank goodness! In the morning I went down to the spot beyond the village above the omega curve which I had marked yesterday. It was difficult to get the view I wanted and also to get shade to sit in. Finally a fairly satisfactory spot was found and I set to work. I treated myself to a piece of the thick drawing paper, really a treat, but oh for some yellow ochre! I want it for the sky, the trees, the ground, above all the grass, the people, specially the Bushmen, tree trunks, everything! Fortunately, I have plenty of blues, but my browns and reds, too, are not too good.

Hunger drove me back at noon (we had breakfast at 7). After lunch I worked at the sketch a bit, cleared up things generally, made my bed, sewed up my sleeping bag, and then we walked up the river – past the post landing stage and some way along the side of the river flats. I got a small white-flowered *Thesium* (?) in the swamp, 1-3 ft high, the mauve *Utricularia*, *Genlisea* (not in flower), a *Drosera* with captured butterfly, and a very nice Hydrocharitaceae with white flowers (male and female I think) and curious prickly leaves (three-angled in section) and stems. I hope it will press well – I've put it in the cardboard press to try slighter pressure.

We went a little way up into the wood, trying to get a view, and on the return journey I found a spray of the bright yellow labiate I found at Litwe. Thereby hangs a tale: Yesterday, as we started out, I saw a plant of the species near the path and marked it down to pick on our return. We searched for it high and low – not a sign of it. Then when I was putting away my specimens, two of the *tipoia* boys came and watched with great interest, telling me names. When they saw this yellow-flowered spray, the interest increased – 'Kulia!' they said. I asked if the part they ate was underground. 'Eh wah!' and one of them fetched a cluster of long potato-like tubers and showed them to me. The mysterious disappearance of my specimen was explained! Today I tried to get out the tubers but had only my knife and a stick and unfortunately, just as I was getting to them, the stem broke off, so though I got some small tubers, my specimen is not complete.

The boys are chanting and the *tipoia* boys dancing just behind me. There was a great battle of words an hour or so ago –evidently a battle of the clans, as 'Mbundu', 'Vangali', 'Luchazi' came again and again from the various speakers. The words to the chant are evidently a kind of petition to *Njamba* – '*Nji saha tanga*' came again and again the other day. At present one is making a rhythmic 'Tut–ti–ti, ti, ti' while another is chanting. On Sunday, just as we were going to have dinner, Ulombo came and asked Miss Bleek how long we were going to stay here. He then went on to say that the store-keeper had bought up all the grain from the country round and the people had consequently none to sell. Some had actually to buy from him, so they (the boys) could not buy *wunga* as they couldn't buy for salt in the shop. We gave them some rice for that evening, though probably it was not really necessary, and next morning gave them $2.00 each all round, to buy food while they are here. They have bought a bag of grain and I think are getting the women to grind it for them.

THE BIRD LIFE OF THE AREA

The bird life round here is very rich – on the river below are numbers of pied kingfishers. Yesterday, as I was sketching, I was startled by a 'thud! thud!' just near me. A pied kingfisher had caught a rather broad fish and was killing it against a log in the bank just below me. In the distance I've seen several black and white storks feeding, and yesterday four flew over where I was sitting, necks stretched far out in front, feet behind, lovely things. The woods are full of doves, night jars are creaking now, there are guinea fowl and heaps of small birds, and the riverlands are full of water fowl. Three ducks flew out from under the river bank the other morning, and there are brilliantly coloured birds which must, I think, be a kind of parakeet or love bird – electric blue-green backs and some red. The tiny birds are fascinating but elusive. This evening a black bird was sitting at the tip top of a tree calling to a friend who was answering from a little distance off.

I was putting my specimens away this evening when suddenly I saw Telosi climbing a tree a short way off; with the axe he cut steps for himself and finally reached the fork. One large branch at the fork was dead and he proceeded to chop it off, and then cut it up for firewood – a most elaborate business.

No sign of the porters as yet – we hope they will appear tomorrow, but won't be surprised if they don't! There is plenty to do for me – nice specimens and heaps of sketching. It's an artist's paradise – talk of the Serpentine at the Wilderness! Of course the sea is missing here, but there is plenty else – canoes on the river, trees of varied forms and beautiful views of the river galore. The fowls evidently find plenty of good things to eat. We hardly see them and they never come and ask for food. My bed is on the slope looking almost due east – the sun comes up like a red disc behind the trees. Every now and again the river comes with a little spurt and one hears the sound of the swift waters, then it is silent again. Or perhaps a breeze stirs the leaves in the trees above us and there is a rustling like falling rain on the leaves. Usually the boys' chattering or chanting nearby masks other sounds, then comes a lull in their converse and the sounds of the countryside come in. At present a bird is honking far out up the river – some kind of waterfowl, I suppose.

KINGFISHERS

Wednesday August 12th—This morning I went down to the landing stage below the post to sketch. Miss Bleek went with me and walked on along the track up which lies our way. It was a cloudy morning – I managed to catch the sun this morning at last – saw it come up a ball of fire, to plunge behind the clouds almost at once. The whole sky was hidden behind a pearly veil of cloud except for a peep of blue here and there seen through cracks in the veil. First I tried the bay with its reflections. As I sat just above the swirling, dark yet translucent green water, little gems flitted to and fro: a glorious little kingfisher – a deep rich blue, almost sapphire blue in depth, yet with an electric sheen over it, a long, powerful orange-red bill, and some red colour underneath, tail short and stumpy, the whole not much larger than a robin, if as large. They kept on flying past below me with their shrill chirping cry and disappearing under the overhanging bank just to the right of the landing stage and a few yards to my left. I looked afterwards to see if their nests were there but could see no holes. If they are there, the lip of the bank hides them.

WATER-CARRIERS FOR THE POST

After about an hour and a half, the light changed completely, so I left the sketch and packed up, intending to return, but the view up the river was so attractive that I turned my stool (which gets horribly hard!) through a right angle and tried another. I had just started this when Miss Bleek returned. I was interested in the water-carriers for the post – two men with a large drum, about the size of our hot water tank, slung on a beam which they carried on their shoulders. They made three journeys while I was waiting. I came back soon after noon, as the sun had finally come out and it was too hot to go on. I worked at the two sketches for a bit, particularly the second.

Then about 3 we started out, as I wanted to try another of the serpentine curves to the south. We had not gone far, however, when I found a beautiful spray of the yellow labiate with edible tubers, so while Miss Bleek mounted guard, I came back for camera and digger. It ought to make a very nice photo. It looked charming and made me long for colour plates. Then having with Ulombo's help (he and Kaliye happened to pass on their way to the village) dug up the plant with its two long young tubers and one old one, we returned to camp as I wanted to sketch it. Before we went out, we had set Ulombo and Muyeye to make a table – why we did not do it as soon as we got here I don't know – so for this afternoon at any rate we have had the comfort of a table.

PORTERS ARRIVE

Soon after 5, when we had resigned ourselves to several days longer here, Muyeye and one of the *tipoia* boys announced *'Vangamba li na isa'* (Porters are coming). I went down to the river bank and Muyeye pointed out the crowd of porters coming down the river path. So perhaps we shall get off tomorrow! We shall be very glad, though personally I have plenty to do. We did a little clearing up and general preparation, and a few minutes ago, heard voices in the distance. Apparently the porters are camping down by the river below the post. I don't suppose we shall be able to make a very early start, but still I'm going to bed now, though it is only just after 7.30. A few minutes since, a breeze sprang up and for a short while it sounded as though it were raining. The leaves rustle just as if rain were falling on them. So strong was the impression that I even imagined I felt a few drops and looked up, only to see a clear starry sky, without a cloud. The Southern Cross is right in front of me as I write.

DEPARTURE FROM CUITO

Thursday August 13th—Such a restless night – the boys talked and talked and I could not get to sleep, and woke several times during the night. I suppose the excitement of the arrival of the porters at long last was too much for me! Up before six and went to the river to see the sun rise – in vain – a thick pall of pearly cloud hid the sky and effectually veiled his majesty. By seven we were packed, by eight had breakfasted and generally cleared up, and soon after, we set out for the post. *Sous-Chefe* said the porters had not arrived! However, the interpreter told him they had, and soon the whole procession, 41 plus several boys to carry food, arrived chanting up the hill.

Our 20 were picked out, and by and by the *Chefe* himself arrived. Their names were taken and we returned to camp, as Miss Bleek hadn't taken up enough money. The porters were sent down to do up their packs, with instructions to return to the post. While I gave out loads and salt, Miss Bleek took the money up to the post. As soon as the packs were ready, I sent the 20 up to the Fort – stupidly I did not enquire first as to the route, and told them to leave their packs. Soon after, Miss Bleek came back, saying that we had to go past the post, not along the river as I had thought. So we, with our four *tipoia* boys, went up, leaving Telosi to see everything taken by the porters.

TREK TO THE CUMPULWE

We got our *guia* for Menongue, not Cwelei unfortunately, and set off, leaving the porters to follow. A sandy bit at the back of the post brought us to the *strada* which stretched away straight in front of us. We followed it for less than a mile, then turned off to the right into the bush, and for an hour followed a winding path which gradually slanted down to the river. Every inch was beautiful and I longed to paint again and again. The day was still cloudy, with lovely pearly lights alternating with bright sunshine, and the distance was lovely. Each turn in the path brought a fresh view, each lovely. Once we cut up over a high bank, the river curving in below us – kingfishers were darting below us and part of the bank was full of the holes leading to their nests.

Then the path went along the edge of the river. Later, in a deep bend, there was a beautiful little vlei with lily pads flecking its surface and a lovely reflection of the woodland beyond. On it were three small duck – rufous brown bodies, white heads, and black and white streak across the wing when flying. On the water there were three – startled by our arrival they rose – and behold there were six! – the reflection was so perfect. They settled a little further on and I had a good look at them as I passed. Later I saw two other lots of three each, in flight. They seem to hunt in threes. Then a larger black waterfowl, about the size of a coot but all black with a greyish beak, flew up. Once we put up a fine fat grouse feeding near the edge of the wood, and there were feathers and tracks of guinea fowl all about.

Of flowers there were plenty – the red clusters were magnificent; *Eulophia* – the tall yellow one – formed a big clump at one spot near the path. Once I was attracted by a patch of violet in the grassland – a lovely pale purple *Moraea*. The two ground peas, both beautiful, were fairly abundant, and a pretty golden-flowered leguminous shrub was fairly abundant in the riverlands. A lovely trek. Quite short – I took three and three-quarter hours, Miss Bleek four, actual trekking. We reached this camp, half an hour up the Cumpulwe at 5.15.

TELOSI LEAVES

Last evening Telosi announced that he had been sick in his inside since Sunday, and wanted to go back to Muié. Miss Bleek thought he was asking for medicine and doctored him. Then Samolova announced he wanted to return and was refused – his name is on the Menongue *guia*. Today Telosi looked really seedy, and at midday said he was feeling very bad and couldn't go on. However, after a rest, tea and medicine he felt better and decided to come on with us for the afternoon. However, he has decided to stay behind and has just had his pay, and will leave us tomorrow. He is disappointed not to be allowed to buy more stuff – also wants the blue kerchief but it's packed away in a case which I am not going to open. He has had plenty – more than is good for him and will get his present in the morning. Samolova had another try a few minutes ago, and was again told he must go as far as Menongue. We have written letters for Telosi to take to Muié with him, though I doubt if he will get there much before the next post from Cwelei. Now, at last (9pm), to bed.

TROUBLE WITH PORTERS

Friday August 14th—In spite of some delay due to hunting for Telosi's stuff and his 'present', we left camp at 7.30am, the earliest start we have yet managed, and looked forward to a comfortably early camp. But we reckoned without our porters. I was pleased and somewhat surprised at the alacrity with which they started in the morning. This was explained by the fact that, at the first village we came to, they announced that they wished to buy food. So we gave them 40 minutes and then forced a reluctant start. At 11am we reached another village – again a stop, and again right in the centre of the village – this time to cook their food. We moved down into the bush, and as by that time it was 11.30, we proceeded to have our lunch and then gave the porters a good hour to cook their food. Then, if you please, they announced they wanted to spend the night there and have their meal ground!

AN ATTEMPTED STRIKE

There followed half an hour of discussion, till finally I said, 'Kunahu! Ta ende' very firmly and, after seeing the fowls (which had been out for a run) caught and put in the camba, we started off, calling upon the tipoia boys. Then followed an attempt at a strike. One or two announced that they were going to stay – one of Miss Bleek's tipoia boys being the worst – and our friend Muyeye adding his note to the chorus. We trudged on up the strada, two lonely specks in the long stretch, casting anxious glances behind and discussing what we should do if our firm stand failed to take effect. As all have already received about two thirds of their pay, they are rather independent. However, after about a quarter of a mile, we saw three specks issue from the bush and start following us down the road.

After some time we stopped in the shade of a tree, for the specks to catch us up, when they proved to be Ulombo, Muyeye and Samolova. The first we sent back to tell the porters that if they did not come we should report them to the government. Whether he delivered the message or not, I can't say. I doubt if he got so far, as soon we saw other figures start out. He returned, saying that the tipoias were coming, and some at any rate of the porters. If some came, probably the rest would follow, so we proceeded on our way. At 2.20 the tipoias caught us up and I gave mine an hour's solid work. We arrived at our camp site at 4.35 and by 5.30 the rest of the porters had come in. Altogether they have done under five-and-a-half hours' trekking.

We have announced that tomorrow we sleep at Makonga, food or no food, so hope we shall not have any more trouble. As a matter of fact they are quite a good, experienced lot of porters (my four new tipoia boys are all really good) and the loads are all fair, and with two or three exceptions quite light, well under 60, and most, I think, well under 50 lbs. But, we being two females, more or less ignorant of the language and, presumably, of the route, they are simply trying it on. From what Mr Bailey told us, that is only what we might expect. I hope now things will go smoothly.

The trek today has not been nearly so beautiful as yesterday. At first we were in the river valley coming up the Cumpuluwe, and it was quite beautiful. Then we left it and got on to the strada and the beauty was gone. At the second village we saw a small boy, evidently about half Bushman – the first trace we have seen of them.

The bird life has been less interesting – a number of the shrill-voiced, long-tailed birds are in the woods, and we put up two large birds, with a fine spread of wing, possibly hawks. By the way, as we were waiting at the post yesterday, a fine large brown hawk came swooping over the clearing, probably on the look-out for chickens. His flight was a delight to watch. Late in the afternoon just before we left the strada, it did become beautiful – it dipped down, was tree-edged and covered with yellow (very burry) grass. The soft yellow road bordered by trees and leading to a glimpse of blue distance really was very beautiful.

Now it is 8.30 – very late, thanks to those wretches of porters – and one or two mosquitoes are singing, so for bed.

TREK TO MAKONGA

Saturday August 15th—7.45pm. In camp on the hill above the source of the Makonga River, 4800 ft in altitude. On arrival, I read barometer and thermometer – latter inside my case was 90°F! Dropped to 80°F (4.30pm).

We were up at dawn – 5.30 – sun, unfortunately, again invisible, so that I was again unable to correct my watch. This evening, hoping to check it at sunset, again was disappointed – at 5.30, just before it set, it sank into a cloud bank. I think the watch is about five minutes slow, ten at the outside. By 7 we were ready to start, but the porters were eating – actually left camp at 7.40. We cut up obliquely through the wood to the *tapalo (strada)* and continued along it. It was really beautiful – chiefly because it has evidently not received much attention lately. It is overgrown with grass and coppice shoots from the trees and bushes which have been cut down in its making, and in this part there are some really fine trees near it, so that the vistas from time to time formed pictures of real beauty.

THE SMITHY AT SAKAMBILA

After about one-and-a-half hours my *tipoia* boys wanted to turn up a side path, but I was suspicious and objected. However, soon after, we did turn up to go to a large village, Sakambila, where everyone started buying food at a great rate. It is a very large, evidently very prosperous village, with pigs, goats, fowls and even some cows. We sat in the village square while the buying went on – was most interesting. In the centre was a large well-built shelter – the smithy – fenced round, presumably to keep out little people, where several men were busy forging axes, and making handles for them. They proceeded steadily with their work and it was most interesting to watch. Two men were busy blowing the bellows. Of these there are two pairs, the nozzles in the broad openings of two hollow clay cones, the pointed ends of which point to a small pile of charcoal. The openings of the hollows are covered with two pieces of hide and on these rest two sticks which are worked up and down alternately by a man sitting beside it. A third man was forging the axe head – heating it and then beating it out on a tiny round mushroom-shaped anvil. A fourth was preparing the haft – doing pokerwork ornamentations when we were there. Outside, others were cutting shafts from pieces of tree, some sharpening knives on a slab of fine sandstone, etc.

A PHOTOGRAPH IN THE VILLAGE

Then a youth passed with a long forked pole and we enquired what it was for, but his answer left us as wise as we were before. However, later on I saw him at work and the problem was solved: – He was busy thatching a half-built hut or store-house and this was his ladder! The whole scene was so interesting that I decided to try a photo in spite of the dull cloudy light. I knew what the effect of producing my camera would be, and sure enough, as soon as I started expanding the legs of the stand, a large crowd lined up – several small boys, of course, well in the foreground. I screwed on the camera, stood up and, looking my sternest, pointed it at them, pressed the spring and shot it open – whereupon there was a wild stampede of small boys, and laughter from the men. Fortunately, though a lot of the men lined up, most went on with their work and the photo may be all right. Having finished, I called to some small girls, watching with great interest, to come and look, whereupon they hastily retreated! Ulombo, however, explained to the men and was only too glad to set the example and have a peep himself – the consequence was he was almost trampled underfoot in their eagerness to see – women and little girls hovering regretfully on the outskirts of the crowd. When nearly everyone had had at least one peep, the show was declared finished.

The women were interesting – various styles of head-dressing caught our eye, one a new one – a regular hat with narrow brim and a kind of crest running from back to front. Most wore many bangles and one or two were extraordinarily tall – one, I think, partly European. After giving the men 40 minutes, we started on along a winding path through the wood, which after some 20 minutes took us back to the *tapalo.* Along that we continued till noon. It was very interesting, however, as every now and then the woodland gave place to open sandy country, recently burnt, where were several very interesting geophilous plants – the cowslip-scented Thymelaceous plant, – fully out it smells more like *Buddleia* – and a very pretty yellow-flowered shrub with a delicate scent of spring, something between primroses and daffodils, and rather similar to St. John's Wort in flower, besides several others, three or four of them new.

AT THE HEAD OF THE MAKONGA

There were a few large birds – two flew across – a wide spread of wing, brownish-white with a black border, long neck and knob-like head, and a rather harsh cry – 'Kwa-quonk! Kwa-quonk!' They alighted not far off. They looked rather as though they might be a kind of large plover, but I could not see well enough. Later I saw one or two others in similar sandy country.

We stopped for lunch in the shade of a grand *musivi,* and then left the *strada* and cut across country, north-westwards instead of westwards, finally – before we expected to – coming out at the head of the Makonga. The view was rather lovely as we did so – a gap in the wood framed a vista of which the middle distance was a rather bare patch of rounded hillside sloping to a winding valley, and in the far distance the blue of a distant river valley – either a different river (Cweheleli I think), or a lower part of the Makonga. Here we found a shelter, presumably Mr Bailey's 'rest houses' high on the hillside, where the boys insisted we must camp, although they have to go nearly ten minutes down the hill to fetch water. However, here we have camped, in comfort, eaten, and now there is a cheerful sound of satisfied black humanity. Camp is always much pleasanter when we are not near a village. Last night the noise of talking, dancing and singing went on and on till I thought it would never cease.

We had two fires today – Ulombo, who made our midday tea, let the grass catch and there was a fine flare, and now they have set the grass alight to light them down to the river – wasteful wretches! Everyone bought so lavishly this morning that their salt is finished, and there have been several requests – but our porters' salt is finished too and I can't start on our small private store.

The kitchen as usual presented itself for its pay – Telosi's work is being divided up among them, Samolova doing most of the cooking. He has got tired of carrying Cwelei, and the poor little wretch had to walk all this afternoon – was quite exhausted, poor mite!

Madame didn't lay today – the last two days she (or one of the others) has laid in the *camba* en route! So this morning we had boiled eggs for breakfast.

JOURNEY TO THE LONGA

Sunday August 16th—8pm. In camp on the Longa, in the woods above the river and below the post. Alt. 4600', Temp. 63° F. We left camp at 7.25 this morning and, just after leaving it, we suddenly came in sight of Mr Bailey's 'Large Pool' – an extensive sheet of water lying in the basin between the hills on which we were and those on the far side. From the lower end, a stream issued and wound its way down the narrow winding valley. The valley bottom was deep green, in strong contrast to the burnt slopes on both sides. The whole thing was both unexpected and impressive. Our path wound up and up, first over open ground, where I found a rather nice perennial gentian, still in bud, and one or two other specimens, then into rather dense wood, still going up till we must have been well over 5000 feet.

I was walking at the head with only one boy in front when suddenly he stopped and pointed down into a vale just below us. There was a beautiful little *bambi* (duiker) who started off and then stopped to have a look. The boys whistled to attract his attention and twice he turned back before finally disappearing. Several times I noticed long tubular red flowers lying on the path, and searched round in vain. They looked like a species of *Loranthus,* and by and by I found a small green shoot, parasitic on a leguminous tree. Finally, where there were several flowers, the boys pointed out the plant, pendent from one of the topmost boughs of a high tree. Later on I saw a lot more of it, also high up in a large tree. I walked for an hour and a half, then rode a bit, but it was very uncomfortable in the bush and I was glad to get out again.

Later we left the wood and got on to a bare burnt part above the Cweheleli which wound in and out below us. It took a long time to get past this treeless expanse. Finally, about 11 o'clock, we got back on to the *strada*, which turned down to the river – such a pretty bit – a vlei to the right, green and brown river grass below it, forming a vivid contrast to the burnt slopes above and full of birds – one I noticed looked like a large lark. Then a picturesque bridge of poles carried the *strada* across the pretty, swift but rather shallow river, beautifully clear, like a Scottish trout stream or the Ness. A short piece more brought us to the trees, the shadiest of which I chose for lunch. There we stayed from 11.30 to 1.10, for a much needed rest (for the porters, that is!) as we had trekked steadily through with hardly a rest. Then we started on again up the *strada* which climbed steadily uphill for two-and-a-half miles (50 minutes steady walking). Then I thought my boys had better do some work, and rode for half an hour.

We trekked, with a short rest now and again, till 4pm, when we reached the Longa post. We went to present ourselves to the *Chefe*, but he was away, so we went on to the shop, kept by Mr Bailey's friend Sr Dominguez, where we bought salt, eagerly watched by my old screw-eyed *tipoia* boy, and as many more as could crowd into the tiny shop, where the trader and his little half-caste girl (of six or seven) were. This valley, too, is fine, but is all burnt out. On the far side are high hills up which our route to the Luasinga tomorrow lies. Now for bed, hooray! The boys have each had two large spoons of salt; our personal boys have been given extra pay for the Cuito stay, so all are happy. Muyeye and Kaliye have been asking for medicine and have been really seedy. I think it is the awful stale fish they have been eating. We half expected Muyeye to say he was too ill to go on, but I don't think he'll want to stop here! The fort was rather interesting – mud walls and a deep ditch to protect it, crossed by a narrow bridge. A few poor little banana trees, each fenced round with sticks, struggle for existence in front of it, and someone, either the trader or someone at the fort, has a mule. The first of the equine tribe we have seen in Angola!

A VARIETY OF BIRDS

Monday August 17th—In camp near the *tapalo* above the Luasinga River, 7.45pm. Alt 4800'. Temp. 62°. We left camp at 7.25 this morning, leaving our woody surroundings which had proved both beautiful, specially by fire-light, and convenient, and went across the slope below to the *tapalo* which continued straight down the hill and across the river plain. The river itself was crossed by a low bridge beyond which the *strada* was under repair apparently, and on this part numbers of birds were feeding – four glossy black crows (?) about the size of jackdaws, without the grey head. When we approached, they flew off. With them were flocks of smaller birds of two sizes. The larger were greyish, tips of wings darker, while at the root of the wing was a vivid orange-red splash of colour. The smaller were soft-looking grey-brown things. As we crossed the riverland beyond, there were numbers of birds on all sides. A tiny grey thing clung on to a grass stalk and watched us pass, another black one perched on the top of a small tree facing the sun and singing his morning greeting.

From several directions, far and near, came a sweet, peculiar song – trills on several notes, ending with a very deep falling cadence. Later on I saw the little songsters, something of the lark variety I suppose. They rise from the grass trilling continuously, fly up twenty feet or so, sweep round and drop down on a slanting line, the song falling to the deeper note as they drop. As we entered the woodland on the far slope, the number and variety of notes from every direction testified to the crowd of unseen denizens of the tree tops. We heard the ubiquitous turtle dove, more often than not ending its monotonous song at the beginning of the phrase, some singing like Cape canaries, the harsh chattering cry of the long-tailed jay bird, who has a lot of blue about him when he flies, besides the dark and light bands in the wing. There were some of the shrike tribe with a clear whistle, answering one another like the Bokmakierie, another with a clear bell-like note. Once we put up a fat grouse (?) which the boys call *ntento*.

It took us 35 minutes to cross the riverland, and then there was a long slope taking us up to 4800 feet. I walked for two hours, then rode and walked by turns, usually walking up the slopes, of which there were several, and riding down. After three hours' steady going, we took ten minutes rest – i.e. we and the *tipoia* boys. Muyeye was the only one of the porters with us. Then, after a level bit, came a fairly deep drop, followed by a long slope. Halfway up this, at 11.50, I found a fairly shady patch and we stopped for lunch. There I found a single shoot of a yellow *Crotalaria*-like pea. We started on again at 1.10, although very few of the porters had come up. It is a long trek without water.

On the long slope just before this we had met a curious procession. First came two boys carrying loads on their heads, after them came a riding ox which looked somewhat peculiar from the front. '*Vindele!*', said my boys, and as we came abreast the load on the ox's back resolved itself into a Portuguese man with a fringe of beard, who greeted me with '*Bon tarde!*' as I passed, and perched behind him, holding on with her arms round his waist, a native woman clad in clean bright cotton, pink round her body, blue on her head. At the foot of the slope was a fire where they had evidently stopped for a meal.

CAMP AT THE LUASINGA

After lunch our slope took us up to 5100', the highest point we reached. After that we gradually descended; then at 3.10, after two hours' trekking, one of which I walked, the blue of a valley appeared ahead, and my boys announced '*donga* Luasinga'. It looked very close, but the *tapalo* turned to the right and descended the slope obliquely, so that it was 4 o'clock before we reached the swift, clear stream with a deep, translucent green pool above the bridge. There we stopped for a few minutes, then climbed the slope beyond, leaving the *tapalo* and taking a path which led to the 'rest house' perched on the hillside and just above that, at the edge of the treeland, we made camp, or rather, sat and waited a couple of hours till our porters straggled in, and we could get camp made.

The valley looked very beautiful in the distance, but like most it is marred by having been burnt out. This country, like other parts of Africa, is being spoilt by repeated burning, not only of the grassland but of the woodland too. Today we have passed such a lot of burnt land, and shortly before getting here we passed a large fire in the woodland. Just below it was a burnt grassy patch and, as we passed, a small whirlwind swept over it, carrying off burnt ash and dust.

By the way, soon after we left Longa we passed a man and a boy sitting at the edge of the *tapalo,* and our boys stopped to buy skins from them (I was ahead and did not see this). Old One-eye, I think, bought two *bambi* skins and at our next stop there was a dress parade, each of the younger men trying on the new skins in turn. At lunch time Samolova appeared without Cwelei, and on enquiry we learnt that he had sold him for $12.50 to the skin merchant! He came with great pride to ask Miss Bleek to *kusweha* $12.00 for him. We are very glad the little dog has been left behind, as Samolova, grown tired of carrying him, had thrown away his *camba* and was making the poor puppy walk most of the time. Yesterday I stopped in the wood for my boys to tie up the *tipoia* covering, and Samolova passed me, followed by poor Cwelei, who on seeing the *tipoia*, walked up to it whimpering, from which I gathered he had been having rides in it! A little later we passed them, Samolova with Cwelei tucked under his arm, and the longing glances both cast at the *tipoia* as I passed were very ludicrous. It was hard to resist the pleading of the little dog, poor little beast. At night he struggled into camp absolutely exhausted.

Tomorrow we have a long trek, longer than today I think, though with water in the middle, so I must get to bed. The *tipoia* boys are fresh enough, but the porters quite weary and poor Kaliye really done up – he is still seedy. Muyeye seems all right or nearly so.

Tuesday August 18th—In camp near the source of the Kambumbi 'mwana Kweve', 8.15pm. Alt. 4950′. Temp. 64°. I roused this morning soon after 5am (my watch had lost about a quarter of an hour) before light, to see Orion and his dogs processing across the northern sky. Soon a faint light began to spread upwards and in the distance a bird started singing to greet the dawn. Our hillside faced eastwards, so we got the light early – earlier and faster than any camp since Cuito.

The porters were rather slow in getting a move on. After waiting some while to see the tent packed, we started at 7.25 up a narrow path which soon led to the *tapalo*. Most of the porters went up to the village to buy food, and were late in consequence. It was a lovely morning with a cool breeze, and the *tapalo* quite attractive. At first it was, as usual, straight and wide. Later in the day we came to a part which had evidently been lately cleared and from here onward it was half the width (some 12 ft. wide I should say) and wound in and out like an English country lane – apparently just the original track widened and cleared. Trees grew right to the edge, so that it was most attractive, yet with the usual contrariness of man, I missed the long distant vistas of the straight, wide *strada*!

Walking up the first long hill, Miss Bleek saw a *bambi* cross the road and bound into the wood on the far side. The birds again were numerous, calling to one another and singing, and several times I heard a woodpecker. After two-and-a-quarter hours' steady going (of which I walked one-and-a-half) we stopped for half an hour to let the tail catch up a bit, then went on for another two hours and more, till we reached the slope above the Kwatiri where we stopped under the shadiest tree we could find, for our lunch and noontide rest, which we purposely made very long. The trek was a good four-and-a-half hours and we wanted to give the porters a rest, but some did not come in before we left. Kaliye was ill again, and he and Samolova did not arrive till past 2, partly of course because Samolova had been buying food, which means dawdling and talking as well.

BUTTERFLIES

I had been thinking that I have not yet said anything about the butterflies, of which there were quite a lot in the dry woodland – the spotted browny-grey one which always alights to rest on the sand, and the small scarlet one with narrow pointed wings (really a moth I think and not a butterfly at all) which is always making me think it is a red flower, both of which were common at Kutsi. In addition, a large turquoise blue and black one with swallow tails, an occasional *Meneris tulbaghia,* a smaller black and white one, a vivid saffron yellow, a small blue and others. Often their shadow passes across the path as we walk along. At one point on the Cuito, as we were crossing a small stream, there were dozens of the turquoise blue and the small vivid yellow. I had been remembering them all and planning to write about them this evening. Then at lunch time I found a beautiful soft grey and brown moth resting on the bark of the trunk of a *mukuvi,* its colouring blending with that of its surroundings. After our rest we walked on down the river, leaving the men to follow, and there was a wonderful sight. We noticed from a distance crowds of butterflies, and as we approached, we saw that patches of damp soil on both sides of the river were literally covered by hundreds of butterflies, while thousands more hovered above, chief among them the large turquoise blue, but also numbers of the white and black, small blue (like the 'blue chalk' at home), a few yellow, and numbers of tiny ones, some probably moths. The ground was blue with them, getting a drink I suppose.

We did not know how long the trek to the Kambumbi was, though we had been very firm about it to the porters, who wanted to sleep at Kwatiri. Fortunately we had talked about it the day before, and since the first attempted mutiny when they discovered to their surprise that we knew a good deal about the route, they have been as mild as lambs. They may grumble among themselves but no one has argued against our decision as to the night's camp. We left midday camp at 2.20, and reached here soon after 5, having stopped half an hour or so on the way. We left Ulombo to follow with Kaliye and help him with his little load if necessary.

THE VALLEY OF THE KAMBUMBI
We reached the Kambumbi valley above the source of the stream and followed it down some way – a pretty winding valley, very narrow up here. The *tapalo* crossed the head and here runs along above the right bank of the stream, more or less following the bends of the valley. The porters straggled in after us, the last coming in about an hour later. Muyeye came in with me, Ulombo and Kaliye about half an hour later. Fortunately the tent arrived early, but the bath last of all! Meanwhile as the flour and food box had arrived I was able to make a few scones. We both feel we want something in the way of sweet bread stuffs and thoroughly enjoy the scones.

I've just discovered a letter from Mr Bailey which I did not know about. It must have come with those two carriers and I suppose I slipped it into this book in the hurry, and overlooked it. I must try to send an answer back with Samolova. Now for bed. Noon tomorrow ought to see us at Serpa Pinto (Menongue).

APPROACHING MENONGUE
Wednesday August 19th—Near the camp I found bracken for the first time – all dead and brown at present, of course. The boys call it *chiselu* and say it is *vihembu* (medicine) – the

rootstock, I suppose. Stupidly pulling at it, it cut my finger – a deep cut right in the bend of the knuckle. The fibres must be strong as steel.

PROTEAS

The country began to change from now on. The sand began to give place to rock and the road wound in and out along the side of the valley. Suddenly proteas abounded, large and small, and joy! – at last some were flowering. This was a large kind, a small tree, very often up to 25 ft high, other times bushy rather like *Leucospermum conocarpum*. The buds were blunt and covered with silvery hairs, and the open heads rather flat, large and lovely delicate shades of pink and silver. I collected several heads in various stages and took them along to photograph when we stopped at Menongue. Near the proteas was a lovely speckled red *Gladiolus*, flower hood bent down, most beautiful. I failed to get any corms out, unfortunately. The soil was very hard, rather clayey, and the corms deep. The flowers on this stretch were more varied and interesting than earlier. Just before we reached Menongue there was a vivid scarlet malvaceous flower growing in the middle of the *tapalo*.

As we neared Menongue the country was very bare – all bush absolutely cleared, then the ground sloped down to the river, which we could hear and see, rushing over rocks. The road passed a store, some barracks – bare and destitute of vegetation – then dipped to the bridge beyond which was a cluster of quite civilized-looking houses. The river at the bridge is lovely; looking up, it is a torrent rushing over rocks and divided into two streams by a large island, where some attempt has been made at forming an ornamental park – a very feeble attempt!

We walked up the steep little bit of hill on the far side to a house where a crowd of men were standing enjoying the strange sight of two women arriving at the post unescorted by white men. To our joy, one among them was Mr Evertsberg from Cwelei, and after we had introduced ourselves – Miss Bleek can get along fairly with the Portuguese when she has time – he came forward and asked if he could assist us, and acted as interpreter. Someone went to see if the Administrator, Senhor Madruga, was at the office. This apparently was a private house, and while we waited, a lady came out on to the stoep of the adjoining house and by signs and words invited us in. She proved to be Senhora Madruga and was most kind – regretted that she had not a room to offer us but invited us to have meals with her while we were at Menongue. Her husband was out on one of the *stradas* but she expected him back at any time.

HOSPITALITY OF THE ADMINISTRATOR'S WIFE

We waited some while, but as he did not return, Sra Madruga decided to wait no longer, so she took us into her spacious room with long mirror (whew! *what* a sight I saw in it – Rip van Winkle the younger!) and bathroom adjoining, and there we had a welcome wash and brush up, after which we joined the Senhora, her pretty little niece (both are very attractive in a dark olive-skinned way) and Mr Evertsberg at an excellent lunch to which we did full justice. In the midst of it a voice was heard, and in came a small

August 19. *Protea* (*P. busseana*) with 'delicate shades of pink and silver', Menongue

August 19. 'A lovely speckled red *Gladiolus*', Menongue

August 19. *Protea* c.f. *trichophylla*, Menongue

Fedde Report. XI 543 Allida — (not)

golden brown hairs except tip

① Smaller leaf 2·6 × 4·5 cm.

② larger leaf 3 × 5·5 cm.

seed

August 19. *Uvaria* sp., Menongue

August 22. *Gardenia imperialis*,
(previously *Randia physophylla*), Cwelei

Randia physophylla K. Sch
(fruit)

Randia physophylla
K. Sch
Riverbank, Cwelei; Angolan

stout man in spotless white, with a rhino sjambok in his hand. Mr Evertsberg whispered to me, 'The Deputy Governor'. He had come to see the Administrator – refused to sit down but strutted here and there chatting to our hostess – lips pursed out, a figure straight from comic opera!

THE IRRIGATION DITCH

After lunch, as the Senhor had not yet appeared, we departed to put up our camp. Our porters had been sitting outside in the street all this time. Mr Evertsberg showed us a suitable camping place a short way up the river. To get to it we had to cross a deep ditch, some six feet deep and about as wide, at which a train of men were working, carrying away a couple of handfuls at a time of the hard yellow clayey soil in flat bark holders. We were told that this was an irrigation ditch begun by the late administrator, over which thousands of escudos had been spent. Unfortunately they got their levels wrong and if ever water does come down the ditch it will run *below* the houses, and so back to the river.

While the tent was being pitched, rather hindered by two of Mr Evertsberg's large dogs, that chose a shady spot to lie in just where the boys wanted to fix the guys, I heard a voice say '*Doña!*'. I turned and found one of the boys gazing doubtfully at the dogs and appealingly at me. A word from the white person did the trick and the dogs moved off a bit. I photographed the protea and gladiolus. I hope the result will be good. On the screen both groups looked lovely but oh! for some colour plates. The beauty of colour all goes, and the beauty of line doesn't give any idea of it.

We were told to go back for tea at 4. At 3.30, after a scramble, we set out for the shop and on the way were joined by Mr Evertsberg who again kindly acted as our interpreter. With his aid we made our purchases and changed £10 to escudos at $100.00 to the £1, giving two £5 notes, which aroused much interest in the audience. The shop acts as a kind of club and there were three or four soldiers and one or two civilians in addition to the trader – a dapper little man in a shepherd's plaid suit – and his bearded, swarthy assistant, the latter evidently with a good dash of dark blood in his veins. The trader looked like the younger brother of the one at Longa – Sr Domingues – same small, rather pinched, features and longish nose, but probably is no relation.

It was past 4 when we got to the Administrator's, to find Sra Madruga instructing a boy in the art of ironing handkerchiefs, her own cook boy being ill. Sr Madruga had returned and by and by came in for tea.

When we started making camp Mr Evertsberg talked to the porters for us, and told them they must go to Cwelei with us but should return at once from there. They said they had no food, so by his advice we said we would give them food for the two days, whereupon they reluctantly agreed, and two of them went up to the shop with us where we bought two sacks of meal (34 kilos) to satisfy them. Having thus paved the way, Mr Evertsberg now asked Sr Madruga if our porters could go on to Cwelei, the *guia* being for Menongue only. 'Certainly!' he replied, 'if the porters are willing!' Needless to say, we did not say that they were willing merely because they thought they must, and that the government was behind us. This was a great relief, as we did not want another long delay and we were told, contrary to what the Cuito-Cuanavale *Chefe* had said, that Menongue district is sparsely populated, and porters hard to get – at Cwelei, next to impossible.

Having settled the question of these porters, we asked if we could have 16 the following Monday to take us from Cwelei to somewhere in the bush – Kaiundo possibly – where Bushmen were to be found. This, too, was granted and then he departed with our passports, *guias* etc., and we went into the garden with the Senhora and her little niece. It is a tiny plot. With all the veld to choose from, the dozen or so residential buildings have been built right on top of one another. There were a few flower beds, and another small plot devoted to vegetables – neat rows of beans, carrots, peas, *makovi* and other vegetables filled narrow beds – not a weed to be seen. A boy was busy sprinkling them with water and we were told that two boys were kept busy the whole day carrying water for the household. Looking over the fence between that and the next house, we saw some boys cutting up a roan antelope, and on the fence beyond four fine leopard skins. We asked if they were for sale and were told they belonged to the *Chefe* at Kaiundo, who was at present in Menongue. These skins are bought from the natives for $100.00, we are told.

DINNER AT THE ADMINISTRATOR'S

We departed to have baths and change and see that all was in order at the camp. Before we were ready it was quite dark and Mr Evertsberg appeared to say dinner was ready and escort us. Conversation at dinner was a bit difficult – neither host nor hostess talk English, though I think the latter knows much more than she admits and the few words she knows she speaks with a faultless accent. I couldn't remember a word of Portuguese – at least I could remember words but could not put them into sentences. Miss Bleek valiantly talked but each sentence took a lot of thinking before it finally came forth. In the end what happened was that Mr Evertsberg had to act as our interpreter most of the time. I was seated next to our host unfortunately, but he'd excused himself from much talking as he had caught cold and was very hoarse (I think this was rather by way of precaution). The dinner was excellent and towards the end the conversation grew easier and quite interesting – discussing *stradas*, rivers etc. The Cwelei *strada,* it appears, goes on westward through several posts to the coast – I think at Mossamedes.

Sr Madruga had been telling us of a small white boy across the river, who was ill with appendicitis, and just as we were finishing a boy came to say that the child was dead. The Senhora was much upset and wanted to go over at once, but her husband said no, later would do, and proceeded to have his cigar. We sat round the table chatting for a time and then bade them farewell with many thanks for their hospitality, offering ours should they ever come to Cape Town. Our host had given us our passports, again *visé'd*. There's no room on mine for more of these lengthy *visées* and *guias*! By the time we got back it was past nine and bed most welcome.

Departure from Menongue

Thursday August 20th—We were just about to have breakfast when Mr Evertsberg and the dogs appeared. He refused to join us but finally accepted a cup of coffee and a biscuit. We left camp at 7.40 and left Menongue (4800′) at 7.50. Mr Evertsberg walked out with us for about half an hour and then returned. I quite forgot to explain the reason for his presence at Menongue. He was on his way to Muié to represent the Cwelei station at the annual conference which is to be held there shortly, had difficulty in getting carriers and engaged a strange boy who happened to be at Cwelei to carry his box, which contained valuable papers, his money and all personal belongings. He had walked fast and had last seen this boy half way from Cwelei. When he arrived at Menongue on Tuesday evening, the boy had not appeared and search by one of the other boys and a government police boy failed to find either boy or box. It is practically unknown for a carrier to go off with a load, so everyone was much disturbed and worried, as it is not a thing one wants to start.

Passing the post the road goes downhill a short way and soon crosses another beautiful river, fringed with trees. This is the Luahuka, a tributary of the larger Kwevi. Here Mr Evertsberg said goodbye and turned back. He had quite enjoyed meeting two English-speaking folks in the midst of all the Portuguese, and we had most deeply appreciated his assistance. It made all the difference in the world.

The hard, wide *strada* wound up through open bush, where I found *Faurea*, with spikes of silvery fruits. Then we turned off from it along a narrower, more uneven road which crossed a small stream where a species of *Kniphofia* (?) was flowering. Then the road made its way up the side of a rocky valley through a village, then out into a more open valley, unfortunately recently burnt out, where groups of great granite boulders occurred at intervals. Welcome rocks! Such a change after all the sand.

A very broad-leaved *Protea*, leaf almost like *P. cynaroides*, was abundant here, but only last season's heads were to be found. However, at our first rest I collected several new specimens. After leaving this valley, our path, an old wagon road, led through woodland where I found a lovely white sweet-scented flower (Scrophulariaceae or closely allied family), a really handsome thing. We trekked in leisurely fashion, as we wanted to give the porters an easy day, stopping several times.

A meeting on the way

Finally after four hours' trekking, not counting stops, about 12.30 we stopped on the side of a valley down which we had been making our way for the last hour, to have lunch. I sent Maliti to get water and he evidently had some difficulty in getting it, as he was away some time. After nearly an hour we got our tea and were just about starting lunch when I saw a little crowd of people coming towards us across the slope far down the valley. From what we had heard of their plans, I knew it was Mr and Mrs Procter coming up to meet Mr Gale. We were very hungry, so hurried up and had our lunch – chicken bones and bread, besides the welcome tea. Just as we finished, Mr Evertsberg with three of his dogs walked past at express speed. Not seeing us in the trees, he went on to where our *tipoias* were, further down the path. There the boys showed him where we were and we all waited together, watching the approaching group, which by and by resolved itself into two donkeys on which were seated Mr and Mrs Procter, followed by eight or nine carriers.

When they arrived it was suggested that we should all camp there for the night. Mrs Procter wanted to turn back, but this we emphatically vetoed, explaining our proposed plans – i.e. a weekend at Cwelei, then a week or so in the bush, to be followed by a longer stay at Cwelei when Mr Gale's visit was over. Finally, she agreed and we all camped. We were reluctant to do so at first on account of the porters, as we had promised they should go back next day. However, Mr Procter had shot a small buck on the way and on putting the matter to the porters, such as were there – really only the *tipoia* boys – they unanimously agreed to stay and answered for those behind. The hope of meat undoubtedly decided them. As it turned out, the first of the porters, Samolova, Muyeye, Kaliye and the man with the pot, appeared three hours after we had got to the camping place, so that in any case they would probably not have got to Cwelei that night, or at any rate not till after dark.

Meanwhile Mr Procter's boys had been scouring the country in search of the missing porter and his load, which had last been seen nearby, and luckily they found the box, quite near the path in the bush. One end had been wrenched open and a few things extracted, such as a safety razor, but fortunately all the valuables were intact.

Mr Procter and Mr Evertsberg went out to try and get another buck but without success, and got back at dusk very tired, particularly the latter, after his hurried walk out. He'd left Menongue at 10 and got to us soon after 1, which speaks volumes as to our relative rates of travel! Then we all had dinner together in the Procters' comfortable tent. Just before dinner we heard a jackal or something of the sort bark not far off, so I made Muyeye and Samolova, both *very* reluctant, go and get some logs and make a fire in front of the tent and near my bed; so after dinner, as my fire had burnt up well, we all adjourned to sit round it and talk till past 9, when we retired to bed.

Friday August 21st—At dawn as usual we were stirring, got packed and had breakfast as expeditiously as possible, and arranged for Mr Evertsberg to take over Samolova to carry the recovered box. Samolova being a Muié boy, it suited all parties. Then we parted, each going their several ways – the Procters and Mr Evertsberg on to Menongue, where the former had business to do before going on to meet Mr Gale. They intended to get their business finished and then trek leisurely along the way to Muié till they met him. I wonder how far they will get. He was expected at Cwelei on the 25th (next Tuesday).

The river above which we camped was a small stream, which might have been the head of the Chimpompo. The way wound along the side of the beautiful little valley the bottom of which was unfortunately burnt out nearly all the way. Some way down we saw four large glossy raven-like birds feeding. From their call they are well known and the Procters had been telling us about them the evening before. Their native name is *mungombe*. They were described to us as 'black birds with huge beaks'. The female (*mpwevu*) says '*Wuchi, wuchi, wuchi u tovala*' (Honey, honey, honey, it's sweet). To which the male (*gale*) replies '*Pamba, pamba, pamba ngome*' (set up, set up, set up a hive), which is but in accordance with the prevailing native custom, which is to give the women the bulk of the work! Though making and setting up the bark hives in the trees is, I believe, not one of the women's multifarious duties. Our four however did not let us hear their call. I hope I shall hear it some time. A little later we saw three oribi playing in the valley – a game of follow my leader. One would dash off then circle round, the others following him. Then off at a tangent they'd go, to make another circle a little further on. It was very pretty to watch.

THE MISSION STATION AT CWELEI

By and by, the road left the valley and wound through treeland – the two deep ruts marked it as 'the old wagon road' we had been told to follow. We cut off several corners and my boys got rather confused, not knowing when to keep to the road and when to take a narrow path. After three hours' walking, broken by a half-hour's stop, after which, by the way, I gave my ankle a wrench, so took to the *tipoia* for the next hour, much to my boys' disgust, we came to a place where the road turned sharply to the right and a ditch of sorts had been dug. Here my boys went wrong and they and I followed the road till we were recalled by shouts from the boys behind: '*Mbonge!*', and looking round, we saw huts to the left below us – the mission station. It is beautifully situated. The houses are built among the trees just above the river which here comes down in a series of rapids, cut up by several islands into distinct streams.

We were welcomed by Miss Moors and Mr Pontier, the only members of the mission left at home. The loads were deposited at Mr Pontier's house where we were to have a room. Miss Moors had just moved out of 'Pontier's Hotel' and is at present occupying Mr Evertsberg's. Miss Moors gave us lunch and then we got our things straight and paid off the porters who were anxious to get back. We hoped to get a good personal boy at the Mission but were advised that this was next to impossible so asked the Kunjamba boys if they would stay on and they eventually decided to do so. As I wanted to sleep out of doors Miss Moors helped me choose a suitable place near her back stoep, or front perhaps, as it faces the river, and Ulombo was set to dig my sleeping place. I told him and Muyeye to get grass; result – a miserable handful. I had the tent put up nearby for convenience in undressing, sheltering my things in the daytime etc. Nearer to, the river was very attractive but terribly stony and they said full of snakes. After the warnings I received, I quite expected to find a snake in my bed when I went there at night. The rest of the afternoon was taken up with labelling and putting away specimens, of which I had found a good few new ones on the latter part of the journey.

'PONTIER'S HOTEL'

Our headquarters are in 'Pontier's Hotel', the famous house of which the roof was built first! It was built in the rainy season, so Mr Pontier first erected the high roof – grass on strong poles – then when that was all firm and watertight, he built up the stone walls underneath it – very slow work, he said, as only one of the boys knew how to lay stone, so he and this boy had to do practically all the building. The stones are rough-hewn and plastered together with the clay which they use for making the sun-dried bricks of which the other houses are built. Until recently, the Procters and Miss Moors have been sharing it with him – each with their own little kitchen attached. Now the former have their own house and the latter is in Mr Evertsberg's while he is at Muié. Her own house is in course of building and will soon be finished. We have our meals with her.

THE CWELEI

Saturday August 22nd—We spent the morning exploring the 'islands' under Mr Pontier's guidance. Just above the mission station the river divides into several streams between which is a whole series of islands. The channels between are narrow and rocky and the water rushes through them in a number of rapids and falls – very beautiful indeed, many palm fringed – the first palms we have come across since leaving Rhodesia. On some of the islands they have made vegetable gardens – rather marring the beauty of the scene, it must be confessed! Water is led down from the upper level of each island by small furrows and every garden is easily irrigated. Here they grow beans, peas, *makovi*, strawberries, carrots, radishes, parsley, sweetcorn – all kinds of 'garden truck'.

We went to the far side of the largest island where there is an impressive little fall and there, to my delight, I found one, or possibly two, species of the Podostemaceae flowering below the rushing torrent. Further up, we crossed again by a couple of small tree trunks, cut down and laid across the stream – on one of which were three plants of an epiphytic orchid – to a part where there was a thicket of dense bush. In this, leopard spoor was seen not long ago and here we found another orchid with a number of tubers half above ground. Coming back from here, we found a bird, rather like a small grouse, caught in a snare set by Miss Moors' boy. Mr Pontier cut it free and took it with us, but later on it slipped from his hands and flew away. Trying to give a realistic description of myself and Telosi scared by a snake, on the way to the Kusezi, I gave my ankle another wrench, much to my annoyance.

Exploring the islands took the greater part of the morning. We ended with a tiny one nearest the mission station on which were growing some bushes with a large magnolia-like glossy leaf and sweet-scented, large white flowers, which Mr Pontier thought were magnolias. We went to see if he or Mr Procter, who disagreed, was right. It proved to be a very handsome *Gardenia* with a large funnel-shaped flower. Unfortunately, the last two nights had been very cold and all the flowers were frost-bitten and brown in places, so no good for photographing. We collected some for specimens and I must wait till later for a photograph.

Soon after we got back, the porters we had asked for, for Monday, arrived – 15 of them, that is, not the 16 we wanted. Unpunctual Portuguese – when they are punctual, they are too early! We gave them a day's wage, $1.50, to buy food and they went off to the village, after trying to get us to sell them food. In the afternoon I had a good laze with a book, *Foursquare* – such a treat to read again!

The day here starts at 6.30 when the first bugle goes. A little later, about 7, a second bugle goes and there is a short service in the open air — round three or four very welcome fires this morning. I did not think I should be up in time as I felt really tired now the trek is over, but found that I have developed the early rising habit, and was up as usual with the sun. In the evening, just before sunset, I went across to the island with tubes to collect Podostemaceae and one or two other specimens, so did not get my bath till late. We were all dining with Mr Pontier, so I had to ask for a quarter of an hour's grace. After dinner we looked at some of the tiny bladderworts from the island and I showed my drawings.

Sunday August 23rd—Up at sunrise again, finished putting away my specimens and then took a photograph of the rapids. I took a second one a little later when the sun was higher. I wished I had more films to spare. If only my small camera, or rather the lens and shutter of it, would come from Kodak's. There is such a lot to take here. I'm reduced now to one roll of quarter-plate film, one packet of plates and two in the dark slides, so I have to go carefully.

As the porters have arrived we must go off tomorrow, so some of the day was spent sorting out and preparing. We are only taking what we shall want for the week and leaving the rest here. I am taking the small camera and films in case the shutter comes by the mail which is now due. We were going to Kaiundo, but on Miss Bleek's telling Ulombo that she would give $10.00 to anyone bringing Bushmen to her, he immediately said he had found out there were Bushmen at Kaiongo, about a day's journey down the river, between the Cwelei and the Kwevi, so we are going there first at any rate.

There was a short service in the morning and another towards evening, at which Ulombo read and preached.

I forgot – while we were sitting talking and looking at my sketches last night, a voice hailed us from outside. It proved to be a soldier, who gave Mr Pontier a letter to read. This was an authorisation of the soldier to collect certain men from the village to work. This was apparently rather a blow, as the people are dead afraid of the government and instead of going peaceably, run away into the bush and hide, then are hunted down, tied up, beaten and generally ill-treated, and probably made to work longer than they would otherwise be.

The morning proved events to have happened as expected. The people had all fled and consequently hardly anyone except the actual mission boys turned up to service. In the evening three of our porters came. Miss Moors is very kindly making some of our flour up into bread for us. She made the dough this evening and says the flour is very good, as good as any she has seen. They are very short themselves as it is difficult to get porters to go the four days' journey to fetch flour. It is a pity we did not get more.

ON THE WAY TO KAIONGO
Monday August 24th—A busy morning getting things packed, and loads (seven) allotted. We sent Ulombo and Kaliye on with the porters, with instructions to have camp

prepared for us. We are only doing a short stage today – about two-and-a-half hours to a village lower down the Cwelei, called Katendi. We got them off about 11am. Mr Pontier very kindly went out after blowing the first bugle (6.30) to try for a buck to set us up with fresh meat. He went to the north and by and by I heard a shot. Half an hour later he returned in triumph with an oribi doe of which we were given the hinder half – and very good it proved, tender even that evening when we roasted the 'tender loin'. The story of his hunting was a bit amusing – he saw two oribi feeding, aimed, thinking his magazine full, pulled the trigger – nothing happened! On examination, he found that he had left in the empty cartridge from his last shot, had forgotten to refill the magazine and had only one cartridge with him. 'No missing this time!' he thought, went down on one knee and aimed with great care, fortunately with success.

Miss Moors brought down the bread, three beautiful loaves, a cake, some biscuits and cookies she had made, and in addition the two of them gave us a good supply of potatoes, onions and garden stuff, so we left well supplied with eatables. After getting rid of the boys, we adjourned to Miss Moors' house for lunch. Mr Pontier undertook to store our goods for us, taking special care of my book box which contains most of our remaining supply of English (Cape) money.

The Cwelei people are finding it very difficult to get sufficient Portuguese currency for their daily wants. They can change English or Cape notes with ease, but not cheques, so I was glad to be able to help them a bit in this direction. We left Cwelei about 1.30, with eight *tipoia* boys and Muyeye carrying on his head the three basins containing all our bread and cake supply. The *tipoia* boys proved very inexperienced – were puffed after five minutes carrying, and besides, the path was hard and narrow and often stony – bad going for *tipoias*, so we walked most of the way.

It was an interesting little trek, more or less parallel with the river. For the first part the path ran just above but in sight of the stream, which here is fairly wide and unbroken by rapids, then through woodland where there were the most extraordinary conical mounds at intervals. Some mounds were quite large – ten or twelve feet high, others smaller, most with bushes growing at the summit and some with quite a variety of plants – several different kinds of shrubs, two species of *Sansevieria,* aloes, grass, were among others which I noticed in passing. The soil is very hard and yellow-brown, like the surrounding ground. I suppose they are old ant hills, but the effect is most curious.

Several times we crossed tiny tributaries of the Cwelei and almost invariably the strip of grassland along each of these had recently been burnt out. In these more open parts I found two species of *Euphorbia,* growing close to the ground, with underground rootstock, a yellow-green Thymelaceous plant (I found the same in bud along the Makongo and thought it was a *Thesium!*), a yellow-flowered shrub and, near the streams, a beautiful, light purple *Iris* or *Moraea.*

About 4 o'clock we came in sight of the village, Katendi, and saw beyond it a knoll covered with trees, where at sight of our advance, there was a hurried erecting of the tent. What they had been doing all the time I don't know – probably sitting in the village talking. However, Ulombo set the men (the porters this time) to work and they dug the beds. After a lot of instructing, their idea is to make a smooth flat piece and I find it very difficult to eradicate that and plant in its place my idea, which is that of a hollow, hammock-shaped hole. However, in the end I got somewhat like what I wanted. Grass and wood were brought in and we had quite a fair camp.

Tuesday August 25th—The day Mr Gale was expected. I wonder whether he and the Procters have arrived.

A CHANGE OF PLAN

We intended to get to Kaiongo tonight, but events altered our plans. The path turned inwards, eastwards, from the river and, as none of our boys knew it, a man from Katendi, the *Mueni* I think, came part of the way to put us right. He led us for some way, then gave Miss Bleek's *tipoia* boys minute instructions, which I could not follow in detail, and left us. For some way it was plain sailing, partly over riverland – a small, nearly dry river, the Dumbu – partly through wood. Then we struck the head of a small dry river where the path branched and my boys in the van took the upper fork to the left, which led steeply uphill. I caught sight of a fine piece of *Loranthus* (?) and stopped to get it, which brought Miss Bleek's boys up, very indignant and vociferous – the man had told them to take the right-hand path! Discussion as usual followed, while I hacked at my specimen. Then we cut down to the other path, which rounded the head of the valley, then cut up the hill to a large deserted village. The sound of hammering steel not far off indicated the presence of an inhabited village nearby.

Here again there was doubt as to the path – we followed one for some way, then someone in the rear called out, again much discussion and apparently angry recriminations, till finally Ulombo ordered one man to go in front and show the path. He did so, cutting back at an angle to our path and sure enough soon came into another path leading eastward from the village. The right path led on and on, chiefly through woodland, sometimes open – in one comparatively open patch I found the speckled *Gladiolus* again – and it grew very hot. Then suddenly we came to a group of store huts near a garden (*mayhu*) where a woman and small boy were. The former had a calabash of water at which all the men wanted to drink. Some drank too long and too deeply, so after some squabbling, the calabash was handed up to the woman, who was standing in the hut, leaning over the half door, and she dispensed water in a tin to those who asked nicely!

MUHERI'S VILLAGE

A lot of talk followed, in the midst of which a man strolled up and the talk began afresh. Finally, we had had enough and took our turn, asking the way to Kaiongo, distance etc., which brought forth the suggestion that *lelo* we should sleep at *his* village and *imena* go on to Kaiongo. We replied, 'Chahi. Lelo Kaiongo'. Then we asked if there were any Bushmen at his village. '*Wahi!*' So I got up and said that if there were Bushmen, we would stop and give salt. If not, then no salt, and we'd go on. To which Ulombo added that if they brought Bushmen, Miss Bleek would give *fueta* (pay). The man looked somewhat contemplative and then somewhat to our surprise, said he would bring a *Vankalla* (Bushman) if we would go and wait at his village. We asked, man or woman, and he said woman. Did she talk Bushman? (Illustrated by some queer clicks on my part) Yes, she did, so off we went led by the little boy, through a large clearing a few yards from the huts.

In other parts the huts have always been in the middle of the clearings. Here they seem to make them among the trees near the edge of the clearing. Then we went in and out of treelands and across gardens till finally, after about an hour, we came to the clean, apparently newly-built village, through which we passed, going on for another quarter of a mile, till we came within sight of the riverlands and camped among some trees nearby. The name of the village was Muheri (i.e. Muheri's village) and the river, such as it was, the Kandondo.

We made camp, had a badly needed lunch – it was nearly 2 o'clock – and then I went to look at the 'river' which proved to be represented at present, as far as water goes, by half a dozen or so water holes, of which only two had much water in them. Round them were myriads of butterflies, bees, flies and wasps sucking the damp earth and some even in the waterholes themselves. I returned for my killing bottle and net, and collected a few, but the bottle has grown rather weak and the specimens I most wanted were difficult to catch, so I contented myself with a few. After watching them for some time, I returned to the camp, where I found the porters just getting another advance of their wages (the third – they had one on Saturday and one on Sunday) to buy food, and our friend of the store huts with the *Vankalla* – not a woman but a little boy of 12 or 13 or thereabouts. Poor little chap! The tears were wet on his cheeks and he had evidently been hauled in much against his will. He was yellow-brown, about !Kõ's colour, but a little darker – partly from dirt – and the face rather broad. He had nearly forgotten his mother tongue, partly I think from fright, but Miss Bleek managed to get something from him. By the end of the interview he had brightened up and was even smiling. A present of beads and salt pleased him much, though I doubt if he was allowed to keep them, and the man who brought him got $5.00. Evidently there are more in the neighbourhood, but probably slaves.

A Bushman girl

Wednesday August 26th—We were up early as usual. I asked if Ulombo had found out the way to Kaiongo, and Miss Bleek replied that he had told her that someone was coming from the village to bring another Bushman and some goat's milk. As we are now reduced to two tins of condensed milk, the latter was welcome hearing. If they did not come soon, he would go up and make enquiries. Not long afterwards Muheri arrived, with followers carrying his chair, a fowl, some meal (the two latter a present), two mugs with a little milk in them and a tall Bushman girl, of the same general type as the boy and, like him, quite unadorned and clad only in a ragged bit of skin. Her head and face, however, were noticeably smaller – she looked far less intelligent than the others we have met. She

August 26. A young Bushman woman and her brother at Muheri's village. Captions on reverse of both photos indicate that they were 'slaves' in the village

remembered more than the little boy, but both left out clicks in many cases. She was able to name her parents, who she said were at Kaiongo, and said the boy was her *ché* (used by others for 'younger brother'). We stayed for some time while Miss Bleek talked to them and got what she could from them. Then about 9 o'clock we started, escorted by the *Mueni* of Kaiongo, who happened to be at the village. He is fairly tall and clad in a large skirt of striped cotton wound round his waist, an old overcoat and a felt hat, with a steel chain wound three times round his neck as adornment.

The trek was longer than we thought – for some half way the path led through the bush, then we struck the head of a small river – pretty, but as usual, burnt out. There, near a water hole, we stopped for half an hour's rest after two hours' trek. After that, we followed the valley till it joined a larger one, that of the Kaluli, which we crossed. The ground was very stony but looked interesting, and I looked forward to exploring it, as Miss Bleek said she understood Kaiongo was among the trees just beyond. Accordingly, on the far side I looked for a good place to camp but on enquiry, *Mueni* said Kaiongo was on the next river, the Mushombo, also a tributary of the Cwevi.

Camp at Kaiongo

On we went for an hour or more, then we came to a cluster of rough shelters which at first we thought were those of Bushmen, but turned out to be those of some of the Kaiongo folk who were clearing fresh ground for plantations. We passed several large ones – some manioc, but most round here are maize – and finally, hot and weary, (we had not used the *tipoias* much, and the late start caused us to trek through the hottest part of the day) after 1 o'clock we reached the village. I explained that we wanted to camp among nice trees near the water, and the *Mueni* led us through the extensive village – enclosures round groups of huts – to the far side above the valley down which was a lovely view looking towards the Cwevi – blue in the distance. We selected a spot for a camp and were promised Bushmen tomorrow. It is very open bush – tall trees with very little undergrowth. One of the trees has most beautiful young foliage – rich scarlet, and the red trees among the green on the far hillside are most beautiful.

Bushmen arrive

Thursday August 27th—The chiefs about here seem very reliable. About 9 this morning Muyeye announced '*Vasekele!*' We looked and there came Kaiongo, followed by a boy carrying his chair. After him came a crowd of his people and Miss Bleek had just exclaimed 'But where are the Bushmen?' when we saw the tail of the procession – a long row of men and women – actually 13 men and one small boy, and at the end three women, one quite old-looking, with a tiny baby.

We sat under the shadiest tree, and the chief sat opposite us with his people on his left and the Bushmen in a half circle on his right, the men in front and the three women behind them. It turned out that one tall old man with a broken nose, named Muba was the head or 'father' of all the rest. Some were his sons and daughters, or grandchildren. Others were married to his daughters, or sons of his brothers and sisters – in fact one large family party.

Miss Bleek had a busy time getting names and interrelationships as far as possible, while I measured them against a nearby tree. Most of them were tall, indications of mixed blood, though one or two of the women were quite tiny. A large number have dreadful colds – one or two, including Muba himself, could hardly speak. The youngest woman had practically lost her voice. They are of !Kõ's type of colouring, i.e. yellow-brown but considerably darker. Several of the men are very similar in type to Golli-ba, though not so dark. Anyway, there is a good rich field to work in.

The people began to get rather noisy, but as I had finished measuring, I took the three women aside to show them what pictures I have here. Immediately the centre of attraction shifted and the whole crowd of the Bantu people made a big circle behind my three little 'Red People' and watched the show with interest.

About noon everyone was getting tired, so Miss Bleek proceeded to dismiss them. She gave Muba a liberal present of salt, saying it was for them all, and explained that while we are here she wants four to come every day, two men and two women. Evidently, however, they weren't satisfied, and Kaiongo explained, first that he had promised that if

August 27. Miss Bleek interviewing Bushmen, Kaiongo. The scanty huts are winter (dry) season huts. Note the calabash hanging from fetish sticks

they came we would give them *each* a present, perhaps only a pinch of salt, but into each one's hand – and secondly that they were afraid to let the women come. To the latter we replied well, then let the men come alone. The former was met by giving each, man and woman, a present of beads and a good length of cotton. They all departed happily, after which we had lunch and a much needed rest.

Friday August 28th—We decided to send Muyeye to Cwelei with letters today, so yesterday afternoon and evening were taken up with letter writing – to the Cwelei people, and home. I also sent Uncle John an account of our trek from Kunzumbia to Cwelei. It was past ten before I had finished. We are very short of small change, though we have plenty

Forest people

The Bushmen we saw east and west had some slight difference in appearance and habits, but all spoke one language, all called themselves by one name, !ku or !kun. This word means 'person, people', but when used without an adjective signifies 'Bushman' to them. Their various black neighbours are designated as... dzu !ku 'black people'. They sometimes call themselves !gei !ku 'red people' or !o !ku 'forest people'[1].

of $20.00 notes, so sent in eight of these to Mr Procter to get change for us if possible, also to Mr Pontier for sugar. We gave Muyeye instructions to come back quickly if the mail were in, if not, to wait till it arrived. Yesterday afternoon we told the porters they were to build us an *nsinge* and that tomorrow they could go. They set to work without a murmur, and with a certain amount of instruction, they erected us a very comfortable and most welcome shelter of the same type as the Kunzumbia one, though much smaller. When it was nearly finished, I set Ulombo and Muyeye to build a table, and everything was finished in time for us to have dinner in comfort.

This morning we paid them off – hadn't enough small change so had to give them three $20.00 notes from which each of the fifteen was to get $4.00. They understood quite well and again took everything without a murmur. They really have been a most docile, willing lot. The young boy who was, I thought, going to be cheeky, dug me a most comfortable bed, giving a willing *'Si, Senhora'* to all I said to him, whether he really understood or not. Some of the Bushmen came back from the village with them last night while I was writing. The village folk were singing and dancing, and our porters struck up a chant of their own and went in a body to the village, returning about 9.45. As they passed the *nsinge* I heard the peculiar pitch of Bushman voices among them.

Later on today I walked past where they had been camping, and further on I found a series of fires still smoking, where the Bushmen had evidently been sleeping. They turned up in a body again this morning (all the men at least) and told us they had slept here at Kaiongo. Again we sat under the trees and, while Miss Bleek talked to them, I tried to sketch one or two, not very successfully. I am trying pencil sketches at present. Later I want to do one or two paintings if possible.

A GOVERNMENT MESSENGER

They had not been here very long – Kaiongo did not come this morning, but one of the headmen did and sat beside us listening – when one elderly small man, one of the best types, suddenly pointed to the village and said something about *'Municipale!'*, and sure enough, a government messenger had come to the village to collect porters. The village folk cleared and we told the Bushmen they'd better go too, as we did not want them caught while with us. However, they seemed quite happy where they were, and even after having been given presents of salt and tobacco, were in no hurry to depart. Miss Bleek explained again that she did not want them all at once, but would like three or four each day, and this time they seemed to understand.

Before letting our porters go, we asked the *Mueni* if we could get enough porters here to take us back to Cwelei, and he said yes. In the afternoon (yesterday) five fine strapping young men gave in their names as porters, not only to Cwelei, but wished to be written down at once for Bihe. Today, at a fitting opportunity, we mentioned to both Ulombo and Kaiongo that if we could get sufficient porters here (27) to take us to Bihe, we would be glad to enlist them, instead of writing to the government. I hope we have sown seed which will bear fruit. Signs at present are hopeful if the *Municipale* doesn't in the mean time take too many of the porters.

This village is large and seems prosperous – heaps of food – we have already bought meal (mealie meal), eggs, manioc, sweet potatoes, beans, a carrying basket and a nice young cheetah (?) skin. There is a great demand for beads but as the supply is small the sale is strictly limited to one small teaspoonful of beads for one *good* egg, i.e. one which does not bob up on being put into water! The money value here is two eggs to a *piastre* ($0.50).

Saturday August 29th—Soon after dawn I heard the fowls (four now, as Kaiongo sent us a present of one) talking hard in the *nsinge*. Turning over, in the tree beyond I saw silhouetted against the eastern sky the shape of a large bird – a large hawk, I think, and wondered whether the talk was about him or the mealie meal in the basket. I've a shrewd suspicion it was the latter but was too lazy to get up and see. Besides they are welcome to their share. By and by the hawk left the tree and flew with great powerful slow strokes of his wings past me so that I had quite a good view of him, and the fowls left the *nsinge*.

Various things were offered for sale yesterday and today, and we had to explain the position of affairs. We had only $7.00 small change left. I showed the *Mueni* the $20.00 note yesterday, and rather to my dismay he went off with it, accompanied by Ulombo. However, in half an hour or so he came back with change, so we were able to get a couple of men to do the remaining bit of work – building a small shelter a little way off. They worked well and steadily, and in spite of one or two interruptions finished in the afternoon.

This morning two large baskets of meal were brought. Kaiongo had promised us a large bag, and we did not know whether this was what the chief had promised or another lot. This to the best of our ability we explained, whereupon the young men with the meal departed, leaving behind the elderly man (with his hair plaited into a topknot) and little boys who formed their escort, to return shortly with the meal in a nice bark bag and word that the *Mueni* was coming. So we all waited and when he arrived and had sat down, business proceeded. This was the meal he had promised and the price was $15.00. We again explained the small change difficulty, and asked the vendor (he of the topknot) if he would give us $5.00, and we would give him the $20.00. He did not see it at all! However, when Ulombo and the *Mueni* had both explained the system of 'change' he grasped the idea and departed to find change. He could only find $4.00, and promised to bring sweet potatoes (*musambi*) for the odd $1.00. Someone else brought in $20.00 in small notes and asked to change it. *Pour encourager les autres* I gave each one an envelope to *kusweha* the precious $20.00 note.

THREE BUSHMEN

Today three *Vankalla* (Bushmen) arrived – the chirpy, fat Puck-like one, Kambinda, with two large front teeth and a frightful cold on his chest and the two Gandus – Uncle Gandu Kameya, and nephew. The former always has his mouth open and looks most unintelligent – until his interest is aroused, when his whole face brightens. After about an hour and a half, they began to say they were hungry. First Kambinda asked to go and have a smoke to relieve his cough, then Kameya said he was so hungry, his inside was crying out for food! As we felt something the same, we decided to give them some meal to cook and eat, while we got our lunch.

Then came the difficulty of a pot – the boys were using theirs to cook beans, so finally I lent them our big pot on condition they cleaned it. Then there was too much water and Gandu came to announce it to us, i.e. to ask for some more meal! That given, they were happy and proceeded to cook and eat their very lumpy porridge, which they stirred with a stick. Finally, a procession returned carrying plate, cup and pot, and Kameya came to me as dispenser of pots, to ask couldn't they buy a pot – they'd bring meat for it. It was hard to resist such appealing requests, but we really can't spare one, not even the one with the hole, which they would probably not appreciate.

Ulombo has shown signs of restlessness. Before we left Cwelei he said his child was sick and he wanted to go back, but would come with us for a short time. It turns out that she had smallpox some time back and has not been well since, so apparently it is nothing fresh, and today he said he wanted to return.

August 29. Gandu Kameya and Gandu Ndende (the younger, i.e. nephew), Kaiongo

In the afternoon I walked to the south-east past the porters' and Bushmen's camps and climbed the kopje to the east. It was most lovely: glorious glimpses of the distant blue valleys, woods of pale green mixed with the flaming red of the young foliage – a leguminous tree I think, whether *munyumba* or another rather like it, I don't know – some are coming into bloom, but all the buds were out of reach. As I turned to come back I heard 'tap tap tap!' and, after a little search, found two woodpeckers – the red crest showed up well – it's the same as the Kutsi one. There are a lot of birds about. The second day I saw two turquoise-blue birds chasing a hawk. I suppose they had a nest. They got above him, chattering away and then dashed down at him, worrying him till he departed, though it looked as though he were more than a match for the two of them. I suppose he did not like wings flapping about his head.

The wind has a way of coming in sudden whirling gusts. Once at Cwelei, during our last lunch, we heard a curious clattering roar and there was a quite respectable little whirlwind screwing its way along through the trees, across the ditch and down towards the river.

LETTERS AND NEWS

About 5.30, in the middle of my bath, I heard Muyeye's voice. He had returned with the post, including, to my joy, my camera. What has happened to my letters, I don't know. There was nothing except two letters from Munnie to Grace, one written on her arrival, the other a week later, forwarded by G.V. without a line. It is rather worrying and very disappointing after the long wait for letters. My camera must wait till tomorrow to be put together. Mr Procter sent the change ($160.00) and Mr Pontier the sugar. There were notes from Mr Procter and Miss Moors, and oranges and cookies from the latter. Mr Gale had arrived as planned on the 25th. I made bread after Miss Moors' plan, all the flour in at once. The yeast was a bit tired, but the bread is very nice and tasty, though somewhat tough.

Sunday August 30th—No Bushmen arrived this morning. Ulombo wished to depart, and said he was going via Cwelei, not straight to Menongue (a day off) as we had expected, so we wrote hurried notes to Mr Procter, and to Mr Bailey, Mr McGill and Mr Dempster, to the three latter to ask them to redirect letters to us to the Cape, should any come, and I also put in a very short one to Munnie. These we gave Ulombo to take to Mr Procter, so that they could go by the return post, and I had already written a letter to Mr Muir for Ulombo to take with him. He received my old pink silk pyjamas – the jumper as a *chikovelo* for his daughter, the trousers to be cut up into a sash to go therewith – with which he was mightily pleased. Muyeye now succeeds to the coveted post of cook boy.

A VISIT TO ' BUSHMAN TOWN'

The morning passed in odd jobs, then we had an early lunch, in case the Bushmen came. Finally, at about 1.30, we decided to go in search of them. A small boy from the village was visiting in the kitchen and we got him as guide, and at a quarter to two we set off,

Kashanga in advance, carrying Miss Bleek's camera and chair, with a small bag of meal on his head, we following behind. Three quarters of an hour through woods brought us to a Bushman camp – a large one, consisting of a large circle of huts with one or two in the centre. To one side were two groups of huts, each with a fence round them, and in the centre a rough shelter under which three men were doing iron work – one working a bellows, a second heating and hammering at an axe head, while a third made a handle. Sitting by was a black man. Miss Bleek has never before seen Bushmen doing iron work. All the huts were made of leafy branches, the stocks planted in the earth, the tips bent over and interlaced, more or less oval in shape, some quite open on one side. In some the beds were made of grass, while others had mats. There were various baskets, pots etc. about, several fowls and dogs, including two families of pups. One of the tiniest, by the way, got into trouble with an infuriated mother hen who nearly pecked the life out of him. I rescued him and he sat and wept, too bewildered to find his mamma.

Old Muba, his wife and adopted son (a brother's child, very short – a dwarf – legs small, arms long, big head and broad shoulders) were in the first shelter to the right as we entered, then came one in which a very old dame, Muba's sister, was sitting smoking a calabash pipe. There were several women and children about, but most of the younger men, and some of the older ones too, and some of the women were away getting food. We spent two very interesting hours, photographing, measuring and looking round. Miss Bleek collected a good deal of information, while I tried vainly to sketch some of them. Two or three of the babies are rather sweet. The tinies are simply clad in a string of beads, sometimes with a brass anklet or bracelet as additional ornament. Finally, signs

August 30. Kaiongo woman, children and dog at their hut
made from interlaced leafy branches

August 30. Kahorta the blacksmith working at the forge, Kaiongo

of weariness began to appear, so we presented a gift of meal and salt. I gave Muba a box of matches I happened to have, and his nice old wife a safety pin which at once went through the hole in her ear, where I fastened it for her.

Altogether it was a most interesting afternoon. By the way, I set my camera up this morning and found, to my great annoyance, that Kodak's have sent it back still defective – sometimes the shutter works all right, sometimes not. I think I can manage to make it go, by dint of poking up the spring each time, but it is disappointing in the extreme. I spent two rolls on the Bushmen and shall take several more, as I had a dozen left. The walk back through the wood was very pleasant.

The *Mueni* brought his small boys to the *nsinge* in the evening. The youngest was dressed in a bootlace-like strip of leather tied round his middle. The others, rather older, had tiny leather aprons. He brought a small basket of meal which he wanted to sell for beads. Miss Bleek was having her bath but, when she had finished, explained that beads were only for eggs. I offered him some safety pins, but he sent the eldest little boy off with the basket to go and ask someone. By and by his old mother arrived, very disappointed as she had no eggs and badly wanted beads. She was so delightful that Miss Bleek finally gave her some beads and told her they were not for sale but a present, whereupon she put two handfuls of meal on a plate and said that was a present for us.

On the way back from Bushman Town, we again saw two woodpeckers working together, while nearby were a dozen or so of the blue birds, their plumage shining in the rays of the declining sun.

Arranging for porters

Monday August 31st—The *Mueni* came early, bringing a fowl for sale, and then followed a long conversation re the porters. A couple of days back we had a long talk on the subject, and he wanted us to send Ulombo and Kaliye with one of his headmen to Menongue with a letter to the government. We explained that we must have the names of the porters first, and after much discussion the subject dropped. Today he wanted to know what the pay to Bihe would be, and then tried to explain something to us. At last we made out that the *Municipale* was returning today and would take five porters – that when he had done so, then the porters would come to have their names written down. After much effort on both sides, to which finally Muyeye was added (we told him not to talk so fast and his efforts, very wide mouthed, to talk slowly much amused them), and finally we understood one another. Kaiongo had his second in command with him. We then had a little desultory conversation before parting, telling them some English names, e.g. bull, cow, pig etc., and asking some of their words. Finally a present of a small piece of soap each, left over from the washing, pleased them much, we having previously explained that soap was *muito caro* and we had very little, *ndinde ndinde*.

Photographing Bushmen

About 11, just as I was about to start mending stockings, *machila* etc., three Bushmen appeared – Gandu Kameya, Tepa and Mulingelu. I tried to get Gandu full face without success, then tried the other two, the old three-quarter view again, with a little better result, though neither is very good. I must try some painting soon. Soon after 12 they

August 30. Ndala, eldest son of Tepa, Kaiongo   August 31. *Mueni* (Headman), Kaiongo

August 31. Clockwise from top left: Gandu Kameya, Kuyella, unknown, Tepa, Mulingelu and Gandu Ndende at Kaiongo

got hungry and were given sweet potatoes and manioc to roast and eat, while we had lunch. Miss Bleek gave them such a lot that they were too sleepy to do much after lunch. Finally, she sent Gandu and Tepa away to smoke, and kept Mulingelu alone, whereupon he brightened up and had much more to say for himself.

THE TATTOO PLANT

After the Bushmen had departed, we had some much-needed tea, and then walked down across the valley, over the river and up the kopje on the far side. It was getting late so we could not go far. The red trees were particularly beautiful with the low sun rays shining through them. Yesterday, on the way back from Bushman Town, I found a large white bell-shaped flower, purple inside, very handsome, growing on a broad-leaved shrub – the one with the large fruit, the juice of which is used in the tattooing of faces. I think it is a *Gardenia*, or very close to it. None of the blooms was very good. I want to go back tomorrow and look for a better one to photograph. Down by the river are numbers of irises, which I must also photograph, and then try to collect some corms. It is really a very beautiful one.

Green and white *Eulophia*

Tuesday September 1st—Soon after breakfast I took my camera and went in search of the *Gardenia,* hoping to find some flowers open – vain hope. I could not even find the bush – presume I did not go far enough. There are plenty of plants, but apparently it is not the flowering season. I found a rather nice, tall slender green-and-white *Eulophia* which I photographed. There I was joined by two little boys from the village, who proceeded to accompany me on my rambles – evidently thought I was quite mad, turning round to gaze about, then climbing to the top of the kopje and mounting a rock to look about again. The small boys have quite nice manners as a rule, and these were rather dears. I got one of them to climb the red-leaved tree to get me some blossom, unfortunately still in bud.

The Bushmen and their women

Then I returned via the village and gave the folks some entertainment – put up my camera and let them look. To my surprise, all the Bushmen and their women appeared to be in the village. Some came out of huts, others were obviously just passing through on their way here. My camera show rather delayed matters. We had asked the women to come and about half a dozen appeared, with a strong escort of men. Some of the women we had not seen before. I tried to sketch two – one a middle-aged woman called Kakongu, wife of Kuyella, I drew on drawing paper and later started to paint. She was very difficult to get, constantly moving and altering her pose, and then she was hungry and wanted to go. I stopped then, and later, after she had eaten, started painting it. I did not get very far, however, when I had to let her go. After they had departed, I went on

September 1. Kaiongo women and children

with the painting. I've got her squatting down with her bark blanket wrapped round her – as much of her as I could get on to my paper.

Two Bushmen girls

While Kakongu was eating, Miss Bleek got a pretty little girl, Kaku, very slight and dainty, with curiously shaped breasts, and more decked-out than most – sticks in her ears, a blue bead on the front of her hair, two osprey-like things in the back, an elaborate bead necklace, some beads quite large, and bits of china, with the china stopper of a beer bottle at the back, and a small brass bell in front, and a large skin kaross. It turned out that she and her sister, both young unmarried girls, are Muba's grandchildren and with their parents live at a village some way to

September 1. Muba's grand daughters, Kaku (left) and her sister

the north called Chikomba (Chikomba's village). They are here on a visit. The sister, too, was very much dressed. I did a quick sketch of her and took a snap of her standing with another girl. We have arranged that the Bushmen are to dance for us tomorrow afternoon, and we are to give them a goat.

This morning I cut up the peel of the oranges Miss Moors gave us. They have been soaking all day and now practically fill the larger of our two enamel basins. Tomorrow I must boil it and make bread, besides which I want to do some painting. I shall have a busy day.

While we were having dinner, the *Mueni* came to know if we wanted to buy *ngulu* (a pig). He was decidedly the worse for something – this time I think from smoking, probably dagga. Once before he was somewhat 'above himself' – the afternoon we arrived – but that time I could distinctly smell beer or whatever alcoholic drink it is they take. He then becomes very Portuguese, is 'Bernardo' and talks very loudly. The first time he proceeded to inform us that he had been baptized at the Catholic Mission and all his children too – giving the names they received. Fortunately, he does not get very bad, and remains quite respectful. This evening he departed almost at once. A man from Kunjamba arrived tonight and wishes to enlist as a *tipoia* carrier to Bihe. After our experience of Kunjamba boys as *tipoia* carriers, I have my doubts.

A letter is written

Wednesday September 2nd—A busy day, as I expected. I was just preparing to make bread when the *Mueni,* with five or six other men, arrived. After a pause to collect ourselves, we started in. He explained fully what he wanted and we tried hard to understand. Finally, after much effort, Miss Bleek gathered that he wanted her to write a letter for him. What with his strange pronunciation and her scanty knowledge of Portuguese, it is doubtless a gem. He and the four porters and the two other men then departed.

Miss Bleek gets lost

I proceeded with marmalade, (first boiling only – it has been on all day and isn't really tender yet, whereas the lemon was ready in two or three hours). While I made bread, and some baking powder rolls and biscuits to go on with, as we finished the bread at breakfast, Miss Bleek went for a stroll and got lost! About 11am Muyeye asked where she was. I hadn't the least idea, except that I had a vague notion that she had gone towards the river. By noon, as she still had not returned, I began to get anxious and to contemplate sending out search parties. However, by half past she turned up with a very gory finger which she had cut on a piece of grass, having thoroughly lost herself. She must have been walking for close on three hours.

Termites at work

In the intervals of cooking, I worked at my drawing of the Bushman woman. It's better in some ways, but I am not at all pleased with it. About midday the dreaded *Municipale* arrived. Soon after lunch Muyeye brought three *piastres* with a request to buy food, as all the women had fled to the bush. As soon as possible I baked the bread. Early in the morning I examined the things in the tent. Last night I thought I heard the knocking of termites at work, and this morning found I had heard only too well. Of the tent bag, the top and the edges are left, the rest – *wahi!* The only other thing that has suffered much is my knitted sweater. It has huge holes in the front and back, alas! I am much annoyed with myself for not being more watchful. I heard puppies squealing this morning and there in the kitchen was a *camba* with five pups. I got tired of the squealing and feared that Samolova was going to be emulated, so made enquiries, only to find that they, too, were taking refuge from the *Municipale*. Evidently the owner feared his pups would arouse the cupidity of His Frightfulness, and so brought his five puppies to the supposed safety of our camp. There were one or two others running about, I think.

Return to 'Bushman Town'

At 3.45, as soon as the bread was finished, Miss Bleek, Kaliye with a bag of meal, and I set out for Bushman Town. We took a wrong turning and found ourselves in a *mayha* (plantation) which we had certainly not passed last time. We met two men and asked the way, receiving a vague direction. We went on for some time till I found a vague cross-path like the one we took last time, struck it and followed it till we were near open ground. Kaliye stopped and said it was no good going on that way. Just then we saw a figure in the distance and he ran off to ask the way. I looked round, saw dead branches to the left, and there was our goal. Miss Bleek went on, while I waited for Kaliye and his guide, one of the Bushmen, to return. We found they had pulled down several of the huts, making a larger space, and had built four or five new ones farther out. Miss Bleek explained matters, and arranged that the Hottentot-looking man is to come tomorrow morning to fetch the goat. I hope we'll be able to get it as things are! Then we are to go over at about 3 and they'll dance for us.

PLAYING THE BOW

Miss Bleek was busy with Muba and Hottentot (Kayata) and I was trying to sketch one of the older women, when music started and Miss Bleek called to me, 'Oh look, he's playing the bow!' I turned, and there, sitting on a bark blanket and holding a long shooting bow, was Kuyella, playing away and singing a fascinating little song to his accompaniment. I went across, stool and all, and tried to do a quick sketch of him, much to the amusement of half a dozen men and women who were watching me. The bow was an ordinary shooting one with the string tied down in the centre. The stick was held by the left hand, pressed against a calabash held mouth down against his chest. The first finger rested against the string like a violinist's left hand. The lower end of the bow was held between the toes of his right foot, and the right hand, elbow on knee, played a slender rounded stick against the string. The notes were rich in tone and of fairly wide range. Later Muba played on his of rather a different type, evidently for playing only, small bell-shaped calabash at one end, much smaller and more bent bow. The note was much poorer.

My little musician evidently thought the song was necessary for my drawing, and went on good-naturedly playing and singing for some time while I sketched him. Then his small son, about two I should say, strolled up, dressed in his bead necklace and leather waist tie. His father put down the bow, one of the other men took it up and Papa gave his tiny son his first dancing lesson. The dancing is done largely with the shoulders, and he started to teach him the shoulder wriggle. I did so wish the light had been suitable for a snap. The baby is fascinating, quite unafraid, he walks up and starts slapping his father,

Playing the bow

This is not a special musical instrument, but the singer's own hunting bow with the string pulled in and tied back at about a third of its length. The player sits on the ground, the bow held by the left hand at the tied bit against a calabash resting on his bare chest, the lower end of the bow held between his right toes. In the right hand he holds a short stick with which he taps the bowstring, producing a twanging sound. By slightly altering the position of the left hand and calabash, he can vary the note a little. The calabash is an ordinary household utensil not specially kept for playing. The performer sings as he plays; the time kept is good, the melody very slight, merely an accompaniment to the voice[2].

September 2. Kuyella playing the bow, Kaiongo

and the other day picked up a broken piece of pot and offered it to Miss Bleek.

Vishaka, Muba's old wife, was much exercised by the absence of the veil from the back of my hat, and asked what had become of it. She also pointed out that she hadn't a safety pin in her right ear. It was certainly an omission on my part not to take another! My suggestion that she should come to our *nsinge* to get another was received with scorn. It was pointed out to us that the sun was getting low and if we didn't go soon it would be dark. We took the hint and reluctantly left just as the sun was setting, having given presents of meal and tobacco.

The walk through the woods both going and coming was delightful. Coming back there was a fine sunset – fiery gold, and the colours all round were lovely. Muyeye tells us the *Municipale* has tied up the *Mueni* Kaiongo, in the large hut. Apparently he demands ten porters and Kaiongo only produced five, so is tied up till the other five appear. I'm afraid we are done for, so far as porters are concerned. Hope we'll get enough to take us back to Cwelei. The other evening, Sunday I think, Kaiongo and two or three men appeared at dusk with a small boy carrying a small oribi round his neck – such a beautiful little thing, with a thick soft coat. They wanted us to buy it – we refused. Again I wished for daylight that I might have photographed it. Madame has been ill for the past two weeks. She has a bad cough, poor little thing, and is in a bad way, I'm afraid.

VISITORS

Thursday September 3rd—A day of visitors! I put on that dreadful orange peel to boil (only did half yesterday) and then started painting. I put in my lady's other eye, and left that. Just as I was finishing it, one of the *Municipale's* soldiers came with a small boy to sell eggs. Then I started to sketch Kuyella sitting singing and playing on his bow, to try using pastels. Then three red-fez soldiers appeared with wounds to be dressed. One was the boy who brought our porters to Cwelei. He had a nasty open wound on his shin, dressed with a piece of bark. The next had a huge swollen ankle, and wore a piece of string tied tightly round his leg above it. The third wanted medicine – when in doubt, give something nasty! So we cut up an iron jelloid and gave it to him to eat. Whereupon all the others wanted to try it too. Several brought eggs – our stock of eggs is now excellent. Most were fresh – a few bobbed up and were rejected.

BUSHMEN DANCE DELAYED

Hardly were these gone, and they were in no hurry, when Kayata (the Hottentot) arrived with six other men including Kuyella (my musician), the two Gandus, Kambinda and a couple more whose names I don't know, to fetch the goat. Unfortunately, owing to other pressing business, Kaiongo has not yet been able to let us have the *mpembe* so the dance must again be postponed. Fortunately, the 'People of the Woods' (one of their names for themselves) quite understand the position. They sat around interrupting, till I took the six off to the kitchen to show them the camera, leaving Kayata to Miss Bleek. They thoroughly enjoyed the camera and then, Kameya having made a pipe out of a couple of leaves and a hollow reed, they proceeded to have a smoke, while I retired to draw Kayata, a difficult task. Soon they were round us again – having come at nine they were hungry long before 11.

THE *MUNICIPALE* AND HIS SOLDIERS

By and by, Kayata peeked round the corner of the *nsinge* and greeted someone deferentially. Miss Bleek looked out, and there was the *Municipale* and his soldiers. However, our friends not only did not mind them, but are evidently on the best of terms with them. Kayata greeted 'our little soldier' almost affectionately by the name of Dondi, and started talking to him. Dondi or one of the others wanted to know why he had not been to Menongue lately!

The Bushmen sat in a half circle and chaffed the soldiers who stood behind, making one laugh so much that he choked. Then the *Municipale* (who has rather a fine strong face, though not very pleasant) bought some string from Kuyella, giving a metal *piastre* for it. Even he smiled at their jesting! He was most polite. I suppose his job is far from pleasant. By and by, they took themselves off and we continued our business, till by 11.30 the people were so clamorous for food that we sent them off to bake and eat sweet potatoes while we got our lunch, a very excellent onion omelette.

KUYELLA'S SONG

Before we had finished, the six went off to the village, leaving Kayata for Miss Bleek. Later they returned and again came and sat round. They evidently consider me a huge joke and quite inexplicable. Miss Bleek I think they wonder at, but give some explanation for in their own minds. They are most interested in the drawing and all the outfit – particularly Kameya. The drawing pins attracted them, and as I have heaps, I gave them each one. Kuyella tried to fix his in his hair, so I suggested he should put it in his wooden comb, which he did. Kambinda stuck his on the end of the stick in his ear. I took Kameya's, made a paper windmill and stuck it on the end of a stick, and blew it to show how it would go. Whereupon he tried, without success. All this, I'm afraid, rather interrupted Miss Bleek's work. Then Kameya suddenly remembered the *dzo* (honey), and begged for some, whereupon I gave them a drop each, except Kayata who, poor old thing, gets toothache if he eats it. He's lost all his front top teeth and looks as though he had pyorrhoea. Then Miss Bleek asked Kuyella what his song was and got this:

/nwa se //kabama //gu a ii
/nwa se tatana //gu a ii
/nwa se talana //gu a ii

Which is a kind of invocation for rain – the first line is 'Come out dance for rain for us'. The other lines are variations – different names for the dancing or its accompaniments – clapping and the wriggle of the shoulders.

The two older men started singing it and then the others took it up. Suddenly Kameya, putting on his beseeching look and gazing hard at me, (he evidently thinks I'm the softie) substituted *songe* (beads) for *//gu* (rain). The others took it up and Kuyella started dancing. It was most amusing. I made them shift out into the sun and while

A song and dance

In the dance I saw, the women and girls stood in a group on one side singing and clapping. The song accompanying this dance of the !kū was a repetition of the following three lines[3]:

New moon, come out, give water to us,
New moon, thunder down water for us,
New moon, shake down water for us.

Mary denoted Bushman clicks with standard symbols, / ≠ // & !, already in use in the 18th century[5] and standardized by Bleek in her *Bushman Dictionary*.

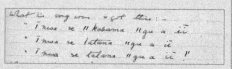

September 3. Words of song

Miss Bleek notes that the Bushman language and traditional practices were already heavily influenced by their Bantu speaking neighbours and predicted complete acculturation before long: *I fancy another half century will see the end of their existence as a separate tribe*[6].

Kuyella danced to their chorus, took two snaps. Then of course, having danced for them, they had to have a few beads each, and finally departed in excellent spirits. Miss Bleek told Kameya that if he went to Bihe with us, he should have the big pot. Thereat there was great excitement and several seemed keen to go. We went on to say that if two went they should have the pot and, Miss Bleek was going to say blanket, but hesitated for the word and they chorused *'chikovelo!'* (dress). I don't suppose, however, when it comes to the point, that any of them will go. I want to try to buy the musical outfit if they'll sell it.

My pastel of the musician is rather funny but does give some idea of the scene. The one of Kayata is less successful. The highlights have come out green instead of yellow! I have three really good pastels. The rest, *Reeves' Greyhound* pastels, are awful – very hard. The paper does at a pinch, but is really too soft. Oh, for a big supply of drawing paper! I ought to paint heaps of the heads, doing quick simple sketches, but the paper is so scanty, only four sheets left, that I have to go carefully. The pencil sketches are good practice, of course. I had the marmalade boiling all day, till about 2.30 when I put in the sugar and boiled it hard for two hours. Finally, I got tired of it and took it off. It is not nearly so good as the lemon marmalade, but better than nothing. Then we walked down to the river and found a huge patch of iris. Brought back a blue composite, which I painted, and a river fig. Today we have given sanctuary to a cock, three hens and one small chicken.

To 'Bushman Town' again

Friday September 4th—Another full day – directly after breakfast, about 8am, we started out for Bushman Town. I thought I really had marked the path, but again we found ourselves at the gardens. It rather suited me, as I wanted to photograph the sweet-scented white flower which I had noticed in bloom last time, so we cut across the plantation. Then, leaving Miss Bleek to go on the path, I went to find my flowers, telling her to take the first path to the left and then look out for a grass path. I took my photograph, and started to cut through the wood, when I heard voices and the sound of pounding of meal in that direction.

September 3. Kuyella (centre) and Mulingelu (right) started dancing an invocation for rain

Then I remembered I wanted a piece of castor-oil stem to make a bubble-pipe, went back to get it and decided I'd better go the way I had sent Miss Bleek. I did so, and the first path to the left led me straight to Bushman Town, which was alive and humming – little boys running around, women pounding meal, the whole community present and busy beginning the day's work – no Miss Bleek! Enquiries confirmed the evidence of my eyes – they had not seen her. I explained and they asked anxiously if Kaliye was with her. I was about to start out in search of her, with two of the men and some little boys, when one of them spotted her down near the stream, and in a few minutes she arrived. She'd gone the same way as last time, instead of the short way.

It was very interesting getting there so early and seeing the whole community – a large one. While Miss Bleek started work, I showed the younger folk the camera, much to their amusement, and then the sketches which I had taken for them to see. Then they mostly returned to their respective jobs and I looked round for somewhere to sit, whereupon a little boy was sent after me with the stool. I chose a shady spot and tried to do some sketches of various individuals including Friendly Baby, his Mama and Papa (Tepa?) and the group at the forge.

VISHAKA'S TANTRUM

Suddenly the scene was enlivened by Vishaka, Muba's wife, who lost her temper and started scolding all and sundry – quite what started her I don't know. I thought at the time that she was annoyed because several men and women were hanging round us instead of fetching water, pounding meal etc. Anyway, she gave them all a good scolding, besides stamping and ramping round in the centre of the clearing, giving vent to her rage in a rapid flow of angry talk.

Miss Bleek says she was started by a black man who was proceeding to remove a lady dog and her family – presumably they'd been boarding with the Bushmen for a few days. We'd seen them there the first day we went, and it was one of those puppies that was so badly henpecked. Next time we went I saw mother dog standing up, poor thin thing, with all six pups hanging on to her breasts eagerly feeding. The black man preparing to remove the family drew down the vials of her wrath upon him, whereupon he presented her with one pup. This, however, failed to appease her and she proceeded to the general scolding. Even the three men working at the forge had a share in it. The girls near me looked rather amused.

Finally, after raging around, lifting up and shaking a couple of empty calabashes, scolding the men at the forge and the girls nearby, she dumped herself down near me, still erupting at intervals. However, just then Miss Bleek came to speak to me, and I suggested that another safety pin might have a soothing effect, so Miss Bleek presented it and it worked like magic. Her mutterings ceased and she clapped her thanks and appealed to me to put it in her ear, which with some trepidation I did – I was awfully afraid I'd prick her.

A GERMAN LUNCH-GUEST

Soon after, we took our leave and departed, after Miss Bleek had given round some tobacco and salt, and safety pins to two or three other old women, and had told one of the four brothers, the eldest, Mushambi, who was working the bellows, to come this afternoon to fetch their meal. The path back is quite easy to find. We left at 11am, got back at 11.40, and I was just frying sweet potatoes and eggs for lunch when someone came to fetch Kaiongo's chair, which had also taken sanctuary here in our kitchen. Muyeye told us the *Municipale* (*Mu sepoi!*) had gone and *Cindele* was in the village. On enquiry, we learnt it was someone from the Catholic Mission, guessed it would be 'the German', and sent a message by Muyeye to ask if he'd come to lunch with us.

We waited some time, boiled some eggs and mixed the preparation in the pan, and finally with a swish of skirts the Father arrived. He was in a white gown, with a white sun helmet and riding gaiters. He uses an ox for transport. He is a tall man with aquiline features, rather long black hair and a full wavy black beard, rather like my mask of St. Francis. As we expected, he was the German, or rather Alsatian, of whom Mr Bailey had told us, so Miss Bleek was quite at home with him. He stayed till half past three, talking about the people, the work at the mission, Bushmen, our journey out, etc.

A 'WAR'

Incidentally, we learnt somewhat of the true history of the last few days and heard to our surprise that we had just been through a war! It appears the *Sepoi* went to the next village, Kakunda, to get porters. They had just had the boys' initiation month and the final feasting was about to take place. Not only did they refuse to go, but they beat the *Sepoi*! Whereupon he came here to get assistance, which the chief gave him, but on the

September 3. 'A blue composite (*Vernonia* sp.),
which I painted', Kaiongo

iongo, Angola

September 4. Hand-coloured photo of iris (*Moraea schimperi*)

September 6. Hand-coloured photo of *Gardenia imperialis*, Cwelei

way back the helpers took to their heels and fled. Then the *Sepoi* fetched soldiers and returned to find Kakunda deserted! So he came here to make up his quota of porters. The Father heard all this the other side of Menongue from one of their workers, who declared there was war at Kaiongo!

Then too, he told us this chief is one of the Kuchi men. (The mission is now on the Kuchi. It was at Cwelei, but they moved, as several 'hunger years' had completely depleted the Cwelei district). When he was invited to become chief here, he (the Father) advised him not to accept, pointing out the difficulties of the position. Despite this, however, he accepted. Apparently he has been here only a short time, a year or two. They started a school here, but hadn't anyone very good to put in charge as catechist. They put a man in temporarily, telling him he must build a house, make a garden and get a wife, none of which apparently he has done, hence the present visit, to put someone else in charge of the school. He confirmed our impression that this is a prosperous part. There is never lack of food here at Kaiongo.

Altogether we had a very interesting afternoon. He's going to see if he can help us at all over the porter question. He mentioned that a German farmer from northern South West Africa is at present at Cwelei with his wagon and family, on his way to Bihe. Miss Bleek thought it might be a way out for us, but further enquiry rather knocked that little scheme on the head. He is trekking north to try and find a good spot to farm.

Soon after he left, two of our Bushmen, Mushambi and Kayata, arrived. They had come earlier and heard we had a visitor, so waited in the village till he had left. I had a try at drawing the latter, while Miss Bleek talked to them and got a song to the new moon. Meanwhile I'd put the marmalade on to boil, and in the intervals between the sending of the invitation and the arrival of our visitor, and his departure and the arrival of the Bushmen, started drawing and painting my white flower. I wish I could get the seed of it – it's most beautiful and sweet. When the Bushmen had gone, we walked down to the river and I photographed the *Iris* patch, and two or three plants close up. I changed my plates last night – a hot job under an eiderdown.

A short time ago we sent Muyeye up to the village – no need to ask if the people are back, the village is humming! We saw some of the women returning, complete with blankets and meal baskets, from down the river. Muyeye went to ask Kaiongo if the goat could appear tomorrow. It is to be here about 8am.

The moon has just risen (8pm), a golden globe near where the sun, apparently the same size but fiery red, appeared this morning at 6.10am It was full the day before yesterday, and most beautiful.

Most of our visitors have departed, except the chicken which I hear cheeping – the cock made the night hideous during the early hours before dawn. Finally, in desperation, I got up, seized him, to the consternation of the hens among whom he was roosting, and put him in the *camba*. The shock to his nerves was such that never a crow crowed he till well after sun up! One of the hens paid for her board – before she departed she went into the tent and laid an egg. Or possibly it may have been Madame, who seems to be recovering and has been saying she wanted to lay since noon yesterday.

WHY THE HUTS WERE MOVED

Saturday September 5th—Miss Bleek heard yesterday the true reason for the moving of the huts. In winter (the dry season) they build open shelters, more for shade than anything else. In summer the huts are more rounded, close and dome shaped, and they are beginning to rebuild in preparation for the rainy season. The new huts are built with fresh branches, of course, and probably the less dry leaves help to keep out the rain. Later they will thatch them with grass to keep out the wet. The work of demolition and rebuilding had proceeded quite a lot yesterday. The old people's shelters were untouched, but nearly all the younger folk had moved to their new ones and removed the old. Tepa and his wife were still in their old one just behind the forge, and she spent the morning reclining on the bark blanket in front of it.

THE GOAT IS OBTAINED

This morning we sent a message to the *Mueni* asking for the goat, and he brought the goat's owner, who wanted stuff for it. Told there was no stuff, he retired – a message later brought back the reply that he wanted $40.00! In the end, after we thought he had agreed to take $25.00 (according to Muyeye), the goat was brought, proved large, and as the owner still wanted $40.00 we gave it. In the circumstances it seemed the only thing to do. Then, having at last got the goat, the Bushmen failed to turn up, so at 1.45 we went ourselves.

A BUSHMAN GAME

This time we found the right path and got there about 2.30. As we approached, I heard sounds of shouting and laughter, and, on arriving, we found the young men and boys playing a game with slender sticks tipped with clay, which they were using as darts. Resting the dart in the crook of the left elbow, they put the middle finger of the left hand on the end, and with the thumb and middle finger of the right hand shoot it out at one another. They aimed and shot very well – two or three specially so, the object being to hit someone who does his best to dodge the dart as it comes.

'REAL OLD BUSHMEN'

Then Vishaka came up and drew our attention to the fact that they had visitors, 'real old Bushmen'. They were three men, two brothers whom Muba called younger brothers – really cousins of his, and one slightly younger, Gandu Kameya's father, very like Gandu. He had his wife, a tiny little thing, very Bushman in feature but exceedingly dark, and younger son with him. Their names were Gongu, Tshala (brothers, Muba's cousins), and Gandu with wife //Kaba and son Tshivemba (about 13 or 14). He was probably Muba's nephew, and cousin of the other two. We took several photographs. Miss Bleek got some information, and I tried some sketches. There was much noise and excitement. Some of the men started dancing, but eventually we arranged that they should fetch the goat and the meal tonight, and tomorrow morning we'd come and see the dancing, so we start early. Then Kayata presented Miss Bleek with a chicken, and finally we left, escorted by four Bushmen and the three boys from the village.

Shuttlecock game

I did not see any games save one play or practice of aiming with pretence arrows. These darts are made of very slight reeds about 40cm long with blobs of clay at one end. These were laid across the open left hand, the blob being just free of the hand. The other end is hit with the right hand, sending the dart flying. The players were wonderfully quick both at aiming and avoiding the missiles[7].

September 5. Shuttlecock game, Kaiongo

Our favourite baby was again very sweet. He marched up to Gongu, who was sitting next to Muba, and presented him with a long piece of bark. Old Gongu took it, put his arm round him and thanked him, clapping his hands, whereupon Baby too clapped his tiny hands in imitation.

While waiting for goat, porters, Bushmen etc. this morning, I went down to the river, armed with axe and digger, and with some difficulty got three or four *Moraea* corms. They grow deep in the damp clay soil among tussocks of grass, so are hard to get out without injury.

Now (6pm) we are waiting for Father José to arrive. As yet no porters have appeared. I hope we get enough to take us to Cwelei on Monday!

PÈRE JOSÉ

Later. About 7pm Père José arrived. After dinner he stayed talking until 9.30, German, which I find rather hard to follow. Miss Bleek, of course, is quite at home therein, more so, it seemed to me than in English. The Father is very interesting and knows a lot about the people, their customs, beliefs, etc. He has, with difficulty, managed to enlist eight porters here, and hopes to get the remaining at Kakonda tomorrow. It is very good of him to take so much trouble. Poor man, he is evidently hungry for his mother tongue. He is Alsatian and has the soft 'ch'.

AN INTERESTING AND VARIED DAY

Sunday September 6th—An interesting and varied day. A walk, a dance, buying and giving of gifts, a tragedy and a dinner party were all among the events of the day, to which, on my part, must be added photographing and drawing the red-lined *Gardenia*, cooking a dinner, and a sketch, while Miss Bleek mended my *tipoia* cover for me, packed up and generally prepared for tomorrow.

Well, to begin at the beginning: by 7.45 we had had breakfast, got our things ready and started for the Bushman camp. Just after we had started Muyeye ran after us, evidently anxious to witness the dance, but Miss Bleek sent him back. Judge then of our annoyance when, after ten minutes with the Bushmen, the Kunjamba man, Sasali appeared. However, we could do nothing.

It was a cloudy day, but the light on the trees was lovely and I thoroughly enjoyed the walk. As we approached, we heard strains of music and song, and thought we'd be late for the dance. However, it was only a preliminary, and most of the people were still in their huts, eating. The goat was tied to a stump in the centre of the clearing, where he protested whenever anyone approached.

BLOWING BUBBLES

Miss Bleek went to talk to the Gandu family, while I produced the bubble apparatus and started blowing bubbles, much to the amusement of Kameya, Tepa and Kayata and several others who joined in. Tepa sucked up once and got a mouthful of soap, which he did not like. They exclaimed '//gu!' (water) as I started, and were most careful to avoid the bubbles as they sank.

CHASING THE GOAT

In the midst of all this, the goat broke loose and started scampering about, and first one then another threw down their pipes to go and *quatta mpembe* (catch the goat). A most exciting chase followed, women and children scattering with shrieks of merriment as the goat dashed in and out between the huts. Finally, we heard a dismal bleating from the direction of the river towards which the chase had passed, and in came Kameya in triumph, the goat held over his shoulder by its hind legs, while Tepa followed behind keeping the head taut by the twice severed rope – composed partly of our rope, partly of bark.

THE DANCE

After this, they began to prepare to dance. The women stood in a half circle on the left, the men on the right, all clapping their hands and singing their little chant, at the same time bending at the knees. Then first one, then another of the men would come out doing steps, the 'shiver shake', sometimes alone, or by twos and threes or even fours. The two most to the fore were Kameya and Mushambi. Tepa and Kuyella also did a good deal, while Kambinda specially fancied himself in the 'shiver shake'.

Sometimes the dancers danced up to first one then another of the ring of men clapping, shaking and shivering. It looked as though they were begging for food. Only occasionally did they dance towards the women. Mushambi, the eldest of the four brothers – very small (comparatively), and one of the most Bushman-like of all, who works at the smithy and has particularly muscular little arms – was wearing the most absurd little straw hat. We took several photos. Unfortunately, I didn't take my roll of quarter-plate films to Kaiongo, so only had the small camera.

TRADING

After some time, it began to get hot and, as there seemed to be nothing more to see, we started to try to do our buying. After several attempts, I got Kuyella and made him understand I wanted to buy his musical outfit. '*Ambe!*' was his reply – the local negative. In Bushman he said he couldn't sell it, he had only the one. Then I pulled out my 'sunshine' cretonne at which he looked, and then with a disgusted shrug of his shoulders he turned away. He was tempted and fell – a case of 'the woman tempted me' again! He called to a small boy who brought him the bow which was strung for playing (which was really the one I wanted) and his own slightly smaller bow. He transferred the string

September 6. Preparing for the dance; a calabash resonator is attached to the musical bow
Note fetish sticks against a tree

September 6. Mushambi 'wearing the most absurd little straw hat' while dancing

September 6. Bushman dance, Kaiongo

from the former to his own, tightened it, tried to tune it, then walked out into the wood away from the noise, and with his ear close to it tuned it to his satisfaction, came back and sat down and played, singing his little song as he did so. I had unfortunately used up my last film, but Miss Bleek had one left, which she let me use to photograph him. I do hope it comes out well.

Finally I got the bow, ready strung, the calabash with etched edge and a couple of sticks which were brought in from the wood and peeled by one of the old visitors. He, Kuyella, was using a dry grass stalk. Then Miss Bleek bought the younger Gandu's knife for a large blue and white linen square, about a yard square, I think. I got another etched calabash for the old blue silk kerchiefs and a box of matches from the pretty flat-faced little woman, mother of the fascinating baby and wife of Tepa, I think, while Miss Bleek bought another from Tepa for an old towel. She bought a pipe from someone else for a red kerchief, and the two other red kerchiefs bought two pots, one for each of us. Mine from Kahongo has four little raised parts on the upper curve of the bulge.

The other musical instrument, with the small calabash at one end of the bow was not forthcoming. Muba said the 'black man' had taken it, probably in exchange for something. I wanted some of the darts they were playing with, but 'wahi!' said Tepa, with an expressive shrug, indicating that he was sorry but they were scattered in the wood. Several little boys rushed to fetch grass stalks and clay, and made some which we bought for matches and beads.

Gifts

Then came the giving of gifts. I forgot – before the dance, we gave out the safety pin earrings, beads and buttons which I had threaded. The white cotton buttons on coloured cotton proved surprisingly popular and were quickly dispersed and tied on, either as earrings, round necks or in hair. My old broken aluminium fork made a very nice hair ornament.

The real gifts were a yard and a half of blue stuff (the very last of the *tanga*) to Muba, Miss Bleek's old plaid shawl rug to Vishaka, towels or some bits of old stuff to the old women, beads and G.V.'s old red and white hat to the visiting Gandu family (the little old mother was delighted when I clapped it on her head), Miss Bleek's old rucksack to Kayata, and red bead bag to Kameya.

Kahorta was evidently ill. He sat on a stool, looking very miserable and very pale, not taking any part in the gaieties, and the others were most anxious that he should have something, so besides tobacco etc. which all the men got, he also got a towel, but was beyond taking much interest even in that. There were still a good many who had not had anything in the stuff line, so we told them if they liked to come back with us and fetch the meal and salt, we'd try to find something. Kuyella begged for a piece of soap to wash himself to honour the dress. This, cotton and needle for Muba, and cotton for several others we promised. To our consternation a train of three men and two little boys followed us. However, we thought we could manage.

September 6. Old Muba and his wife Vishaka (left); and Kahorta the blacksmith (right)

GARDENIAS

I had a nice little instance of their quickness on the way back. I was looking for *muvanguvangu* fruits and flowers as we went, and once left the path to look at a largish bush a few yards from it. Some way further on Kambinda, who was behind me, called '*Doña!*' and pointed to the left. I did not see at first what it was. He led me to a large *muvanguvangu* (red-lined gardenia) which had a large number of fruits on it. He'd spotted what I was looking for, without apparently taking much notice. Growing just next it was another *Gardenia*, either the same as, or very close to, our one at home. It had several flowers, but all were just going over and very yellow. A little later, to my joy, I found that one of the buds I've been watching was open, so picked the branch with it and another bud, and one with smaller buds, and took them back to photograph, draw and press.

Back at the hut, Miss Bleek cut in two the old tablecloth, giving half to Kambinda and half to Mushambi. Kameya looked very crestfallen at not getting anything more, and to our surprise offered his beautiful long knife for sale. This I bought for my bronze silk *chikovelo,* which pleased him much. To make it complete, I put a handkerchief in one pocket. Then I cut up two pairs of stockings and gave a pair to each of the small boys. Then they departed with the remainder of the meal, salt, a reel of black cotton and the needle. The soap I forgot! However, half through the afternoon came two more with an escort. They had not had anything. By this time our stock was nearly exhausted. However, we each unearthed an old petticoat to make them happy and I sent the promised piece of soap to Kuyella.

Before the first lot were disposed of, the chief, another man and a crowd of women and children arrived. We were really too weary by this time to feel inclined for another crowd and on the *Mueni* departing suddenly to speak to someone, I got lunch and then suggested they left in peace, and we spun lunch out as long as possible.

After that I took my photo of the lovely *Gardenia*. Hardly was that finished when the chief returned and I gave a little time to entertaining him, and took a snap of him. Then I started drawing my specimen and after watching for a time, chief, men and boys finally departed. We were left in peace for a time, until the second lot of Bushmen arrived. I'd told Muyeye to catch and kill two fowls before we left. He caught them and put them in the *camba!* It was only by chance that I discovered about 1 o'clock that he had not killed them. I had them killed and put in the pot as soon as they were ready.

MADAME IS KILLED
This left us with Madame only. Miss Bleek wanted to give her to the chief's old 'mother' (not really his mother at all – possibly an aunt, possibly not even that) and as the old lady came to see us, I told Muyeye to catch Madame, which he proceeded to do by running her down. Sasali, the Kunjamba man, must needs join in, and the clumsy beast stamped on poor little Madame's head as she tried to pass him, killing her on the spot. However, we explained to the old woman that we had wanted to give her the hen as a gift, and showed her what had happened. She was quite pleased to have her dead. Poor little Madame – I felt quite sad to lose her, and I'm not sure yet that Sasali didn't do it on purpose. Anyway, he didn't get her.

PACKING UP
About 3, Miss Bleek sent a note up to the Padre to say we hoped to see him to dinner. Then she spent some time packing curios etc. in the basket and mended my *tipoia* cover for me while I drew and pressed the *Gardenia,* baked bread (half flour, half *musambi,* which I had set before we went out this morning) and some rolls with the remaining flour. These turned out beautifully and were delicious. The bread, too, rose well. I got the dinner (roast chickens, chicken and onion soup, sweet potatoes, rice and fresh tomatoes) well forward. While Miss Bleek bathed I went to the knoll just below the village and did a hurried sketch which I must finish later, then I bathed and started packing while waiting for the Padre.

A DINNER GUEST
About 7 he arrived. He had been able to get some more porters for us and thought we should have enough to leave in the morning. He was very interesting and told us about the people, Kangali. Among other things, leopards are taboo for them. They may not shoot them or eat them, and if anyone were to shoot one in the village, the whole population would at once migrate down river so that the spirit of the leopard might depart from the village without troubling them. But they may dig spiked pits to catch and kill them on the spikes at the bottom without being themselves harmed thereby.

MARRIAGE CUSTOMS

Then again, as regards marriage, the custom is one man, one wife, though it is allowed to have more. In the whole village he could only think of one with three wives and one or two with two. The women are definitely not the slaves of the men. If the woman does not wish to go into the fields, she does not go. She stays at home and the man must go to the fields. When he only buys one piece of cloth, it is always for his wife! When a man marries he has to pay certain things for his wife – goats, pigs, fowls etc. If she dies (I suppose within a certain time) he has to pay the same again. And if she goes to visit her mother, when he fetches her back he must take a hen or something of the sort as a present.

FETISH WORSHIP

Their religion is a form of fetish worship. There is usually a witch doctor or doctoress in the village, and he or she is consulted in case of illness, receiving payment in kind of course, hence the Christians are far from popular with these folk, since through them they would lose their easily obtained livelihood. If anyone makes a kill, buck or so, part has to be laid on the little fetish stick altar. Then certain bits of this may be eaten, after which the hunter is free to proceed with the rest of the buck.

Should anyone die, his spirit would haunt his people and the witch doctor is sent for and consulted as to who caused his death (no-one dies 'from natural causes'!). If the witch doctor bears anyone in the village a grudge, he is named, or the witch doctor is primed beforehand by the man's friends as to who his enemy in the village was and the latter has to pay, a large part of the payment of course going to the witch doctor. Incidentally, we had the fetish worshippers largely to thank for our difficulty in getting porters. Some were keen to go, but the fetish authorities were vetoing their going.

The Padre looked tired and as if he had a touch of fever. He said he had managed to get us eight *tipoia* boys and five others, so we hope we shall be able to get off tomorrow. He left early (8.30) and we betook ourselves to bed as soon as possible. I had to do a bit of sorting and packing first, but did not spend long over it.

I was just getting to bed when I heard the Bushmen singing and clapping, and then the chorus of '*wah!*' as something specially pleased them. At first I thought they were coming down to our camp – it sounded as if they were coming over the hill behind the village, but I suppose the sound carried well in the still night air. Perhaps it had a specially carrying quality this evening. Now the sound seemed quite near, now far away, but soon it stopped. I suppose they proceeded to the feast of goat. By the way, late in the afternoon Sasali came up with a piece of meat in his hand and announced that the Bushmen had sent it for him. We think probably it was sent to us not to the kitchen. We thought that was the reason for the anxiety to be present at the dance.

I quite forgot to mention yesterday, that after receiving the news of the goat's arrival, and just as we were about to leave the camp, there was a fowl chase and Kayata came up to Miss Bleek with a dingy black hen in his hands, and presented it to her. Poor old hen went into the pot today with the one bought from Kaiongo, for the Padre's dinner and food for our journey. Kaiongo's hen I called the 'Charlady' – she had a wild-looking black topknot from which a line of black feathers led down on each side under her – chin, I was going to say – beak, for all the world like a dilapidated bonnet with black bonnet strings. She was a fine large bird with a lot of fat, and was either laying or about to lay. It may have been the 'Charlady' who laid the egg in the tent and not one of the visitors.

FAREWELL TO KAIONGO

Monday September 7th—We were up early, had breakfast and had everything packed and ready by 7am and were waiting only for the porters. Last night we told the Padre about our promise to the Bushmen – Miss Bleek said that if two of them came to Cwelei with us, she'd give them a blanket. 'Oh, if you said that, they'll come!' he said, when we expressed doubt as to whether they'd come or not. Our stores being strictly limited and nearly exhausted by now, the blankets were to be one of Miss Bleek's cut in half. Imagine our consternation when, in the middle of breakfast, I looked up and beheld what looked like the whole Bushman community (I'd counted at least 80 adults at the dance yesterday) led by Muba and the three visitors, coming down the path behind the camp. There were about a dozen men and small boys, and at the back six or eight young women. When they saw we were eating, they sat down round the kitchen fire. The women stood in a bunch on one side till someone brought over some fire, when they sat down round it, well away from the men.

When we had finished breakfast, Muba and the other men gathered round the *nsinge*. It turned out that four of the young men wanted to go with us: Kameya, Tepa, Gandu the younger and Mulingelu. We explained that we had blankets for two only, but the other two could have dresses. The rest had come down to say goodbye – the women had come in hopes of further presents! We were very glad to have the Bushmen, as when the porters arrived soon after, we were one short, so Kameya carried the tent, Tepa my black kit bag, Gandu the basket with the curios and Mulingelu the big pot and axe. The other axe I intended to have in my *tipoia*, but was told that Sasali had it and was going to carry it. The Padre came down with the porters and then returned to the village. All ready, as we thought, the porters said they wanted to go back, get their blankets and eat. We vetoed it, but however, *tipoias* and loads as well as ourselves and the men went up to the village. There all the carriers vanished to eat, while we were speaking to the Padre, looking at his school etc., where a class of some half dozen youngsters were kneeling before the native catechist and reciting after him something in chorus.

Muba and Kuyella followed us, carrying respectively the calabash and Miss Bleek's brown kit bag which had been left behind. Incidentally, the carrier to whom they had been allotted came up 20 minutes or so later and said he'd left his blanket too, and it had vanished. We had our suspicions, quite unfounded possibly, of Sasali. Others blamed the Bushmen but I'm pretty certain they at any rate had nothing to do with its disappearance. I put in time mending my shoe (one of the stitches Mr Pontier put in for me had broken) by means of my knife spike, half a hairpin and a pair of tweezers I borrowed from the Padre, and taking snaps of the stages in the process of bark blanket making. An old man was beating out one, while another pegged out one already beaten and dyed, a third being already pegged out in the sun.

September 7. 'Our *nsinge* at Kaiongo'

Finally, after many attempts to collect the men, we started – time 8.40am, not bad for a first day's start – Muyeye leading, Kaliye and the four Bushmen following close behind. The rest of the porters did not come up till 9.30. The way led along a broad, well-trodden, hard clay path – old wagon road it looked like – first up the valley and later through the bush. It was a lovely part of the country – valley with slopes of soft bright green patches of colour (a leguminous geophilous herb with sensitive leaves), woodland the most glorious shades of red, bronze, green and gold, all but the latter being spring colours, mingled with a few brown-leaved trees, most of which were shedding their leaves.

After the first bit of woodland, we came out on a tiny river valley, some small tributary of the Cweve, which we followed for a bit and again entered wood. About 11.30 we reached a deserted village on a rather large valley, the Kaluli, a good way above where we crossed it on our way to Kaiongo. Here Miss Bleek stopped, thinking it was the luncheon spot, but, after some difficulty, we made out that it was better further up, so after the men had had a smoke, we went on for another half hour or so along a narrow sandy track along the side of the valley. Then we passed a pool of water, a kind of bay, where we dipped down to the river and crossed a tiny bend of it, beyond which we stopped under some trees for lunch.

Afterwards we followed up the Kaluli for some time, then crossed it and continued up through bush until we were 5200 ft up, then down and up through the woods for some way till about 4pm when we reached a village, Isambulu, situated some way from its river, the Dumbu Yalaha. We passed through the village, Bushmen always in the van, nearly always in front even of the *tipoias*, and went about five or ten minutes nearer the water, then camped among the trees. On the way we had passed a great many gardens with storage huts, very like the spot where we met the woman, boy and man on the way over. Our *tipoia* boys were excellent, the best we have had since Mr Dempster's, and I thoroughly enjoyed the trek. The walking was delightful, so that I was reluctant to get into my *tipoia* and ride, but when I did, the going was so pleasant, and the colours around so pleasing to the eye, that it was hard to leave the comfortable hammock when my allotted time therein was over.

HERBARIUM
DES UNIVERSITY

Lepidagathis macrochila

MP.

lukombokombo

Acanthaceae 7971

Englerophytum magalismontanum

? Heeria

Stephelis narrow deciduous

Musakala

In camp, Gandu Kameya, Kaliye and Sasali put up the tent. Miss Bleek started to show them but the latter was well up in tent erection, while the other Gandu departed with the calabash for water. The Bushmen were given some rice. Result, shortly after, as I was putting my specimens in the wire press, one of the older of the porters came and squatted down in front of me and asked for '*belela*' (relish) as he had *jalla* (hunger), caressing his front with his hand. Miss Bleek asked him if he wanted part of his wages to buy food, and that did the trick. After waiting a few minutes longer, to try the power of the human eye, he departed amid little spurts of laughter from the kitchen.

Tuesday September 8th—We left camp at 7.20, a quarter of an hour later than we should have, as I had to stop and darn my stockings, both of which had developed holes. A short pleasant trek, chiefly through woodland, brought us in a little over two hours to the Chimpompo which we crossed by a bridge. The porters, however, mostly just stepped across. On the slope beyond the river we stopped for a drink and rest, i.e. a smoke for the men. Our four friends again led the way and entertained us with information concerning the trees etc. we passed by the way. I longed to take photographs several times, but the beauty of the distance consisted mainly of colour, so did not, beyond one snap of the Kaluli yesterday and one today of a large *Protea* with Gandu Kameya, carrying the tent, standing at its foot.

RETURN TO THE MISSION AT CWELEI
The last of the porters having come up, we left the Chimpompo and in about half an hour reached the mission village, and a few minutes more brought us first within sight and sound of the rapids, and then to Mrs Procter's house, where we were welcomed by her and Mr Procter, and the German farmer, Herr Weiss, of whom the Padre had told us. After chatting for a few minutes, we followed our porters to our old quarters, and there paid them off, and gave them the rest of the salt. Then, after vainly trying to persuade the four Bushmen to wait with us till tomorrow, the coveted blankets and frocks (my green linen and the mauve and yellow striped silk I bought at Lewis, Oxford Street in 1921) were presented. The first of the two frocks was chosen by Tepa, the other went to Mulingulu, who evidently thought Tepa's choice preferable.

Then I bought Gandu's sheath knife (Kahorta's workmanship – he is evidently the chief smith among them) for an old night gown, and the belt on which it hung for the last of my 6d knives, quite a good one with a large blade. Kameya and Tepa had bags made out of the skin of a wild hare or some such animal, and these too we bought – mine for an old flour bag and two little coloured handkerchiefs, Miss Bleek's also for a bag and an ancient pair of woven pants. The latter bag belonged to Tepa who was not keen to sell – he pointed out that there was a hole in it. Then when we said that didn't matter, he objected to the two small handkerchiefs offered and said he wanted something for Kahorta. Finally the pants were found and accepted. Then, thoroughly happy with their blankets and *chikovelos*, matches, salt, etc., and finally satisfied that they had nothing more to sell, nor we more to give, they departed, saying they intended to return at once, and I think that they would get home tonight – quite a possibility for them. Before the giving of gifts and buying started, I photographed them all separately, using my Vest Pocket Kodak and portrait attachment, which I had unfortunately not taken with me to Kaiongo, and then in a group together.

Everyone disposed of, we went over to the Procters' for lunch. Mr Pontier was away for the day, and Miss Moors, who had come down to say how do you do to us, seedy with her first experience of malaria after five years in this country. In the afternoon came a welcome read, stretched out on Miss Bleek's bed. After that, tea, and then I made my bed in the old hole, and we had our baths just in time for the evening service, held under the trees at the top of the station. Just behind are two tall proteas, and on the top of one is a silvery bud which will soon open.

In the evening we chatted for a time and then Mr Procter started to write a letter to the Administrator for us, asking for porters to Bihe. We said we had planned to start early next week. They pressed us to stay longer if we could, and finally it was decided to wait over for Miss Moors' birthday on the 17th (Thursday), and to ask for the porters for the following day, Friday. The suggestion that we should stay over the weekend, we negatived. It is delightful here, but it will, at the best, take us a month at least to get back to the Cape, quite possibly longer, and we both feel that we ought to get back, though I am sorry to miss the spring flowers.

In the midst of the writing of the letter, the typewriter went wrong and Mr Procter couldn't get it right, so finally left it till morning. We decided to wait a day, instead of sending Muyeye in tomorrow. He received the news that he was to go with evident reluctance, though he didn't actually refuse. He made a great parade of sore feet however, which was quite wasted on us! Now he and Kaliye can do our washing tomorrow morning and the following day (Thursday) he can go to the post, Serpa Pinto, with the letter re: porters and one to the trader asking for a load of flour. He won't be very keen to carry the latter, I'm sure!

When I took stock, the small axe was missing. On enquiry, Sasali had left it at Kaiongo. We doubt it – think he probably put it away safely somewhere on the way. It appears he is an old soldier, a regular mischief maker and generally not to be trusted. He did come in with Mr Gale and was supposed to take a load back to Kunjamba, but instead came on to us at Kaiongo. We hope to get rid of him but unfortunately Miss Bleek wrote him down for Bihe.

End of Volume IV.

THE DIARIES
Volume Five

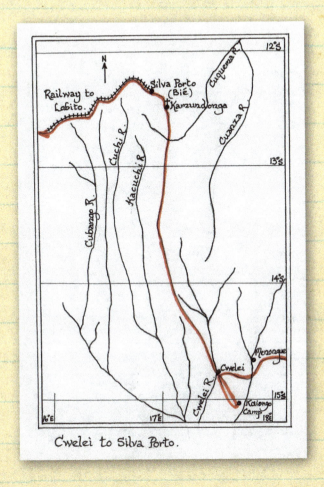

Cwelei to Silva Porto.

Cwelei Mission Station to Cape Town
(September 9 to October 22, 1925)

'I found that there was already something hanging on that tree — a yard long, emerald green snake, stretched out on the branches overhanging the river, just where I wanted to bathe. Needless to say, I did not hang anything else on that tree, nor did I bathe there.'

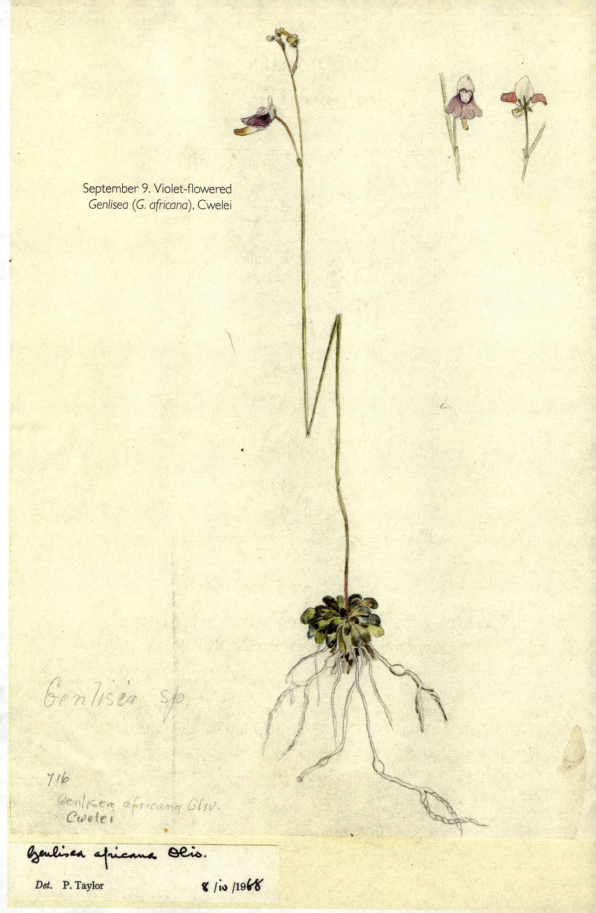

September 9. Violet-flowered
Genlisea (G. africana), Cwelei

Genlisea sp.

716
Genlisea africana Oliv.
Cwelei

Genlisea africana Oliv.

Det.   P. Taylor   8/iv/1968

Cwelei. 'Pontier's Hotel'

Wednesday September 9th—After breakfast I intended to label and put away the Kaiongo specimens – I did not take enough labels with me to finish labelling there. Instead I spent the time putting away things and generally clearing up, then went down to the river and collected some small sundew in flower and a new species of *Genlisea* with a pretty violet-coloured flower, after which I had to come back and change my shoes and stockings, by which time lunch was ready!

The program here is:

6am:	First bugle.
6.30:	Second bugle, followed a few minutes later by service, then breakfast and the morning's work.
11 or 11.30:	Lunch.
12.15:	Bugle for workers to come in.
Afternoon:	Work, rest, read as necessity or fancy dictates.
3.30:	Tea, then odd jobs till 4.30 when baths etc. precede the 5pm service – held on alternate days.

The German settlers

This afternoon we all, except Miss Moors, walked over to have tea with the Germans (Mr and Mrs Weiss, Greta and Mrs Weiss's younger brother and sister, whose name is Reidel) who are camped about a mile from the station. He is a German farmer from S.W.A., who prefers to keep his German nationality. Two years ago they left S.W.A. and trekked to southern Angola near Diriko. There they started farming but failed to obtain from the Portuguese the title to the land they had taken up.

They decided to trek northwards to Bihe where it is said that settlers are encouraged, and have now been nine weeks on the way. They have two carts, one large, the other small, some 26 head of cattle – trek oxen, cows etc., dogs and two monkeys! On reaching Menongue, they found the traders absolutely out of everything except sardines and vinegar! Fortunately they were allowed to get flour from the government stores, then came on here to rest the cattle for a day or two while waiting for their passports, *guias* etc. Since then they have been waiting, expecting the necessary papers each day. Instead comes an official letter asking one day how many cattle they have, next day, what kind of cattle? Herr Weiss had goats as well, but sold them in Menongue. Next day it is some other absurd question.

Mrs Proctor's two boys went ahead of us with boxes of cake, scones, coffee etc. It was a pretty walk – they are camped a short way up the river in a grassy clearing. Just as we were sitting down to tea, the young man and woman brought out the bread, two huge tins, and started putting it to bake. I wanted to watch and all the rest followed. The square hole just big enough to take the tins had had glowing cinders put into it. These were now removed and the tins put in, and covered with two pieces of sheet iron on which the cinders were spread. There the bread had to bake for two hours. While we were at tea the delicious scent of baking bread was wafted across to us. It is half mealie meal – Frau Weiss mixes the mealie meal with yeast etc. overnight and next morning stirs in the flour.

A BUSHMAN BOY

Some time ago a small Bushman boy turned up at Herr Weiss's farm – a *Sepoi* had stolen him (and some others I think). He had escaped and had walked four days through lion-infested country, for over 100 miles I think, to Herr Weiss's farm, where he arrived nearly starving. At night he had dug a hole in the ground and crept into it. He is now about eight or nine I should say, very dark and of rather a different type from any we have seen. Unfortunately he has quite forgotten his language. Mrs Weiss has dressed him in a little butcher blue overall and as he is well fed, his Bushman figure more obvious when clothed makes him a comical little object.

After tea we were amused by the two monkeys with which Greta and Mrs Reidel played. The larger hated the little Bushman boy and the latter in return feared it and ran away. Then they went for a walk, the monkey jumping on to and riding on the dachs-hund's back, till he decided Greta's was more comfortable and jumped up, riding on her back like a baby on a native woman's. Mrs Weiss was born at Claremont – she was the eldest of ten, of whom the brother and sister are the youngest. They went to S.W.A. 15 years ago and though Mrs Weiss talks English well, the brother and sister have quite (or nearly?) forgotten it. We left soon after 4, Miss Bleek staying behind to talk to Mrs Weiss a bit. On the way back we noticed several large proteas in bloom or with buds nearly out.

Before we set out for tea Miss Bleek was sitting on the stoep reading and I was inside, when Sasali came up. He returned the $1 Miss Bleek had advanced him to buy food, and which she had forgotten to deduct when paying him, and $0.50 for the salt, and said he was going back tomorrow to look for the axe! We couldn't understand it, till Mr Proctor told us he had given him a good talking to! We are rather sorry, as it will make it more difficult to fire him. In the evening we played Word Making and Word Taking.

Thursday September 10th—Directly after morning service, Muyeye was given the two letters and departed for Menongue, none too willingly! I quite forgot to note that on arrival we found a mail waiting for us – I had letters from England to my relief, as well as some forwarded by G.V., a *Punch* and a *Courant* with interesting accounts of the meeting at Oudtshoorn of first, the Botanical Society and secondly, the Southern African Association for the Advancement of Science (S^2A^3). The English letters were dated June. In them or in those to G.V., M. mentions the boat by which they will probably return 'in October'. I suppose I'll get back before they do, but quite possibly not!

About 10am Kaliye who was pounding monkey nuts to make nut butter came to me with a smile to announce the arrival of the *Cindele* from Kaiongo. Later when we went across to lunch we found not only the Padre but Herr and Frau Weiss, Greta and Miss Reidel. Mr Proctor had asked them all to come to lunch. They were hoping to start in the afternoon, but the expected messenger from Menongue had not yet arrived. The

Padre had waited at Kaiongo as he said he would, but the *Chefe* not arriving, he could wait no longer. On the way he met several of our porters leisurely returning home, also Sasali in search of the axe! After lunch I showed my drawings and sketches – hope it did not bore people too much!

Miss Moors spent the day in bed, but said we were to dine with her in the evening. After protest we went but she really ought not to have got up and before long she had to go back to bed and leave us to finish dinner alone. The attack of fever is a light one which makes it in some ways more difficult to lie up as she should do. I found another specimen of my violet-flowered *Genlisea* – hope to find more. I must go collecting up and down the river. Mrs Proctor suggests going a couple of hours down on Saturday for a picnic.

Friday September 11th—Early in the morning Mr Weiss came over with the expected letter from the government, this time a command to take all his cattle back to Menongue to be passed by the deputy inspector of cattle (or something of the sort) – this after he had passed through Menongue, and stayed there three or four days, three weeks since! Really the Portuguese are as annoying and obstructive to deal with as they possibly can be. One feels very sorry for the man and his family.

DEVELOPING FILMS

I spent the morning developing films, chiefly Bushmen. The negatives are excellent but the differences in temperature of the various washes nearly ruined them. Fortunately Mrs Proctor had some alum (of which I had none) and I hope I've saved them – must wait till morning to see, when the films are dry. To my delight, two snaps of 'the baby' have come out particularly well.

I should like to get as many as possible done here, if they are all right. So far I've done six and have nine more powders. The washing of course is a difficulty but we are so near the river that there is plenty of water. I'd like to put them to wash in the river! Kaliye has made us excellent nut butter. It looks very good. Today he was kept busy carrying water for me. It is delightful on our stoep in the mornings – the river rushing below is a constant source of joy. I wish I could do justice to it in a sketch – I know I can't, but nevertheless hope to try. After supper we listened to the gramophone and talked till past 9 o'clock.

Saturday September 12th—Mrs Proctor had planned a picnic down the river but as Miss Moors was still seedy and staying in bed another day, it was decided to postpone it. I went on with the developing. Using the alum stops the blistering and the films are quite good. Some of the negatives look excellent. So far out of 64 exposures there are only three utter failures – one the shutter did not work, one I knew I had spoilt by taking two exposures on one negative, the third was in the shade and is underexposed. Others are far from perfect but ought to be worth printing.

After lunch I went over to the island to get flowers for Mrs Proctor. There is not much to get besides the pretty, wild grasses. The few flowers there are quickly wither in water. Muyeye got back in the morning having left Menongue yesterday afternoon. He brought a letter from Sr Madruga promising us our porters, but probably later than the 18th – they would have to bring a load back for the government. But – alas! – no flour – the trader was out of it, the government could spare none. I wish I had insisted on getting a good supply on our way through Menongue instead of letting myself be overruled by Miss Bleek! The mission people themselves are very short. The Catholic mission at Kuchi where they usually obtain theirs has finished last season's wheat, and have not yet begun to cut this season's. Our friends have sent carriers further afield but they have not yet returned. Miss Moors has only enough left for one more baking!

Kaliye heated the water early so that I had my bath when I came back from the island, and had not quite finished when we were summoned to tea. Mr Pontier left for Menongue yesterday directly after tea – he was hoping to get a new back wheel for his bicycle as well as to change some money – some for himself and some for us. Saturday is a holiday for the workers so it is his free day. They are hurrying to get Miss Moors' house finished but are handicapped by the shortage of workers. Mr Pontier is anxious to get it finished as he wants to get south on a 'prospecting' trip just inside the border, and in South West Africa. There were several stations run by the Finnish Mission. These have been deserted since the war, and both Mr Gale and Mr Bailey are anxious to take over the Angola stations. So Mr Pontier has been chosen to go and spy out the land. He would like to have gone at once – it is ideal for trekking at present and in another month the rains will probably have started, but he may have to stay to finish the house, of which part of the wall is finished up to the roof but not all.

The days slip by very quickly. I haven't entered up or packed any of my specimens yet, nor done any sketching or specimen painting, and there is plenty to be done here. After supper we all went across to Miss Moors' house, chatted for a while and then had the usual Saturday evening prayers. Miss Moors is better and got up in the evening. I've been feeding her with arrowroot the last two or three mornings, and we all put our feet down and said she must stay in bed and not attempt to get up. It has been a light attack but has hung on for a long time, a couple of weeks I think, and she has only just started treating it drastically, i.e. taking doses of 30 grains of quinine on two successive days and then decreasing the dose on successive days till normal.

Unexpected post

Sunday September 13th—No early bugles or drums (the latter is in use during Mr Pontier's absence) this morning, so we treated ourselves to a late morning, and did not get up till 7 o'clock! Just after service, which is held under the trees on the hillside just above the station, two men with loads arrived – another post from Muié! It proved to be the quickest they have yet had. Mr Evertsberg had got in just before the mail left, having taken 11 days on the journey. This evidently spurred the post boys on to do their best and they got in in 13 days. My share was a long letter from M. telling of a delightful motor weekend they had spent – Bedford, Stratford-on-Avon, Warwick, Cheltenham etc. At the latter they had visited my 'second home', Rosenhoe, seen the gardens etc. and been of course warmly welcomed by Mrs Duckworth and Gladys. In addition I had a letter and *The Courant* from Uncle John. The paper contained my account of our first experience of *machila* travelling. The post, being unexpected, was doubly welcome. We expected Mr Pontier back today, but he did not arrive. He will probably get in early tomorrow. Miss Moors was up and about, looking a bit washed out, though she says she is 'fine'.

Monday September 14th—Mr Proctor was showing me his tools etc. and was just explaining the working of a 'magic' steel square used in timbering etc., with most elaborate calculating figures, when we heard voices in the house. Mr Pontier had returned. He had not got his bicycle wheel but had changed the money etc. – for us he had got $110.00 to the £1. Miss Bleek and I walked out to see Mrs Weiss. While the latter and Miss Bleek talked, her sister, Miss Reidel, Greta and I went collecting down towards the river. There was not very much to get, but I got a new umbellifer – yellow, finely-branched umbels at the head of a long, thick grey stalk, a flowering leguminous tree – the flowers very like a fine wisteria in colour and arrangement, and a fine yellow-and-red orchid, the latter at the margin of the river bank. The river here is wide, deep and slow – a great contrast to what it becomes a hundred yards or so further down where it divides and forms the islands separated by two or three lines of rapids where the water makes its way over and between rocky ledges.

We took a message from Mrs Proctor saying that we would take tea out to their camp in the afternoon. It was noon when we got back and very hot. We went out again about 3pm taking tea as arranged. Mr Pontier met Mr Weiss near the Luahuka yesterday. He had made good speed, as he had only left in the morning, by ox cart. The cattle started the previous noon.

Tuesday September 15th—Mr Weiss returned last night and came down early this morning – all was well, the cattle had been passed and he had his passports etc. Mr Proctor's theory that they wanted to persuade him to settle nearby was confirmed. As, however, they won't give him title to the land he takes up, he is trekking into Bihe (which is just across the Cwelei) and may possibly settle about a day or two on the other side. Everyone here would be very pleased if he did stop close by – it would mean a regular supply of milk, butter etc., besides other farm produce. They were leaving about midday and Miss Bleek, Mrs Proctor and Mr Pontier went down to see them off. The latter returned with two riding oxen which he had purchased for £4 each. I did not go again as I wanted to get on with various odd jobs. I've started a sketch of Miss Moors' house (Mr Evertsberg's really) from where we sit for morning service. The light changes so quickly that I can only work at it for about an hour in the morning.

Wednesday September 16th—Mr Proctor, talking of the natives, their names, languages etc., mentioned the three names they have for white people: here they always speak of

Nâla Proctor, *Nâla* Pontier etc. – *Nâla* means a crab. *Cindele* means a great devil and was first applied to Portuguese traders! While the third term *Lifoolah* (*Mafoolah* plural) means foam, scum, and is applied to white people in general. A choice selection of names!

I have started packing away my specimens – have taken the pot box, nailed the lid on and taken off one side. I hope they will all go in, but I doubt it. Some are not yet dry.

We had supper at Mr Pontier's, and then all six did word making – we managed three rounds. It is usually enough for one evening.

I now have an evening sketch of the view looking up-river on hand, as well as the morning one of the house. The red trees are rapidly turning bronze and green, and in the burnt patch across the river, black when we got back, tufts of bright green grass are showing everywhere. There are such a lot of hawks about – their flight is wonderful; it is fascinating to watch them soaring in great spirals, the broad fan-like tail used as a rudder.

I went over to the island and took some snaps, including one (which won't come out!) of a diver sitting on an old branch just above the water with a background of *Osmunda* and water ferns. I have also taken photos of the lovely sweet-scented gardenias which I have been painting – back, side and front views and the fruit, all most interesting to do. The only other flower I've done here is the yellow-and-red orchid.

BIRTHDAY CELEBRATION

Thursday September 17th—Miss Moors' 30th birthday! We all had lunch with her at midday and in the evening Mrs Proctor gave a dinner party for her with a 3-tier birthday cake as table centre, with candles round it. I did the flowers for her – palm leaves with some grey *Helichrysum* and grey and gold umbellifer in one vase which stood on the machine stand under the window, and grasses and blue lobelias in the small one. There isn't much to be found in the way of flowers but it is interesting searching the islands to see what one can find. Incidentally, I found a fine collection of Podostemaceae growing on an old tree trunk in the rapids. The erect part of the plant was quite long, much better developed than on the rock. The day was memorable for my first jigger in the sole of my left foot. I did not realise what it was – it looked like a tiny thorn, so I was lucky to get it out egg-sac and all complete. Our evening again ended with word making – it's a fascinating ploy.

Friday September 18th—Muyeye has not turned up to work for the last two days. On enquiry we were told that he was 'sick'. However, he did not come for medicine. This morning we told Kaliye to tell him to go to Miss Moors and get medicine. He did not do so, but Miss Bleek had occasion to take Kaliye to Miss Moors for some interpreting, and at the same time got her to make enquiries about Muyeye. Again 'Muyeye is sick' was all she got. 'Oh, I suppose he wants to go back to Kunjamba', she said, whereupon Kaliye said he was going back on Monday anyway! This being the first indication we had of any such desire on his part, we felt annoyed. However, Miss Moors sent him some strong medicine to take – croton oil, two drops. As he has not come near us, we haven't heard anything more.

September 16. Yellow and red orchid
(*Eulophia angolensis*, previously *Lissochilus angolensis*), Cwelei

September 19. 'We saw a number of orchids (*Ansellia africana*) growing on hollow trees, some with fine sprays of flowers', Cwelei

September 19. 'An emerald green snake, stretched out on the branches overhanging the river', Cwelei

September 21. 'Two large black birds looking very like turkeys in size and gait', Cwelei

I managed to put in an hour or so morning and evening at my two sketches, neither of which is very good. In the river one the trees on the island came rather nicely but the rushing water is frightfully difficult, it is so light, almost colourless yet full of movement and life — far beyond my feeble attempts. In the morning one I cannot get the light I really wanted, with the sun low down and deep shadows on the roof of the house. Also the paper seems to have got greasy or something at the top and unfortunately the line of demarcation comes right across the sky!

The long awaited fresh supply of flour came in — two loads (45 kilos) from Kubango Catholic Mission, five days off by carrier — not quite as much as was ordered, as the supply of last year's wheat is running out. No sign of the carriers yet, for which I am in a way rather thankful, as I haven't got half my specimens packed yet.

A PICNIC AT THE RIVER

Saturday September 19th—Soon after breakfast we started for our picnic. Mrs Proctor and I mounted on two of the donkeys, the third — small, brown and woolly with white nose — going along unloaded for company. Mr Proctor tried several times to ride it, whereupon ensued a bucking exhibition. The first time he finally slid off over its tail, the second he tumbled sideways, the third was thrown full length and the donkey almost put its foot on his knee, whereupon Mrs Proctor put <u>her</u> foot down and forbade any more bare-back riding! Kaliye and Litome, Mrs Proctor's smaller 'help', carried the provisions. Mrs Proctor felt worried — sure she had forgotten something. Before we reached the village she realised she had left the camera, so Mr Proctor went back for it. I offered my black donkey (really his mount) but was told it would not go back.

We went on through the village and along the path we took to Kaiongo — some way out Mrs Proctor remembered what it was she had forgotten — matches! As no-one else had any, Litome was called. I took his load, a basket, on the front of my saddle and he ran back for the *foo-foo*. It must have been a funny sight — me nursing the basket, the reins tied to the horn of the saddle and dangling down, and the donkey picking his way slowly and deliberately along the path avoiding the stones and following every turn and twist even when the path was hardly visible. As he doesn't understand English and I couldn't get at the reins my 'whoas!' were ineffectual when I did want to stop.

It was a lovely day with a fresh cool breeze, and the path we took kept close down near the river bank till we reached the Chimpompo. At one spot we passed an old trading station on the far bank, made conspicuous by a large clump of prickly pear. After crossing the Chimpompo we left the river bank and crossed the plain with its anthills, little and big, and scattered trees — at first burnt grassland with innumerable tiny pointed cones, between which dark green grass was springing, then came a belt of very dark green acacias followed by open woodland with huge old anthills covered with vegetation. We saw a number of orchids growing on hollow trees, some with fine sprays of flowers. On the way back we collected a few, leaving the larger plants for another time.

Then we left the path and cut down through burry grass, between large kopjes to

September 19. Mary in front with picnic basket followed by
Mrs Procter on their way to the picnic

the river. It is a lovely spot – the river divides into two main streams with some smaller
branches, spreading over rocks down which it rushes in a series of rapids. We took off
shoes and stockings and spent the time wading about from rock to rock crossing the
nearer stream, till Mrs Proctor called us to lunch.

When Miss Bleek took off her stockings Miss Moors exclaimed that she had jiggers
in her toes and there and then sat down and started digging them out with a safety pin.
I returned in the middle of the operation and offered my knife, then found I fortunately
had a needle in my bag, so with these two instruments the process was satisfactorily
completed. Some had been in a long time, weeks probably – one had hatched out and
only the egg sac remained, the other two were smaller. They had probably started at
Cuito or soon after and had gone on ever since!

After lunch we went back to the river. Mr Proctor went downstream, Miss Moors
and I started up, crossed our wide stream and then made our way up a small side stream
between banks of *Schizaea* (*Gleichenia*) and bushes of sweet-scented *Myrica*, gardenias and
other flowering shrubs. At the top it led us to the other, western, main stream which
proved too deep and rapid either to cross or to go down.

We were making our way back to the other stream when slip! splash! down I went. I
was hardly down before I was up again in fear for my camera. Fortunately I had left my
watch in my bag in anticipation of some such event. The water got into the case, but I
do not think it got into the camera itself. I had been longing to bathe and now having
got half wet thought I might as well go right in. So Miss Moors found a shady spot while
I prepared to go in, hanging my wet clothes over various stones and bushes to dry as I
went. The last garment I decided to hang on a dead tree just at the edge of the rushing

stream where I intended to bathe but just as I was about to reach up I found that there was already something hanging on that tree – a yard long, emerald green snake, stretched out on the branches overhanging the river, just where I wanted to bathe. Needless to say, I did not hang anything else on that tree, nor did I bathe there. I retreated and made my way to the main stream a bit lower down, but all the time I was bathing (not a long time!) that snake stayed and watched me with a steely gaze – he watched me and I watched him. When I retired to dress he also retired, where I do not know. This is the second green snake I have seen here – the other day, going up-stream collecting just behind the house, a slender dark green grass snake crossed the path and disappeared into the grass. Just beyond it I found some specimens of a purple-flowered *Genlisea*, rather a beautiful little thing.

To return to the picnic: as soon as I and my clothes were dry (on the hot rocks even my wet tunic dried quickly) we started back, cutting across the rocks to the main (eastern) stream and making our way down it, enjoying every step, now on hot dry rock, then on flat rocks in the stream and then knee deep in the rushing water. It was past 2 when we got back and we were sorry to find Mr Proctor preparing to return. He, meantime, had tried to cross the far stream on the lip of a fall, missed his footing and gone down with the water over the slide, getting a thorough wetting (with a roll of films and a packet of paper money in his pockets!) and several bruises. However, a secluded spot in the warm sun soon dried him and his clothes.

Mrs Proctor says the most interesting event of the day was the rescue of a bad egg. She peeled the hard-boiled eggs in preparation for the lunch and, finding one which was not so young as it might have been, she threw it away. It lodged on a rock at the side of the stream, making a blot on the landscape and I went over to remove it, found it was

September 19. Picnic on the Cwelei River

only slightly passé and said it was a waste to throw it away, as the boys said they had *jalla* (hunger) and so brought it back, whereupon I was greeted with derision!

Even the afternoon trek back was pleasant – the breeze was so fresh that the afternoon sun was not unpleasant. As we left the river, a whirlwind started some way back and the column of black ash and dust came whirling along towards us. Fortunately, its path was parallel with ours just there and it swirled past us. As it passed, we felt the anticyclone rushing in towards it, taking a few golden leaves in with it. The whirlwind crossed our path just in front of the leading donkey and then disappeared in the distance. We spent some time collecting our orchids, climbing a tree for some and got back about 4 o'clock, all very ready for the delicious tea Mrs Proctor gave us. There were such a number of hawks along the river, some so close that one could see the broad tail twist as it changed its direction. The flowers too were interesting – the yellow *Hypoxis* was fine, and several small red, yellow and blue flowers. We had dinner at Miss Moors', and after that we did not stay long – soon after prayers we all went to bed, after a very pleasant, enjoyable day.

Sunday September 20th—Got most of my specimens packed. Have also repacked suitcase and book box, but there are a good many odds and ends still to do. It isn't much good doing them until the porters appear.

Muyeye managed to come for his wages yesterday! He looked sulky, probably still felt a bit seedy, but I do not think he has really been ill. However on enquiry he said Miss Moors's medicine was *celi celi*. Today he started asking for a present, 'Like the one Miss Bleek gave Telosi'. Miss Bleek could not make out what he wanted, so took him up to Miss Moors to get her to interpret. It was service time, so I called them away in the middle. Afterwards we got Miss Moors to explain: first that there was only one such suit; secondly, that we had nothing to give as presents now; thirdly that at Bihe we should probably have something to give away. To which she added a reproof to him for leaving us in the lurch like this. Result: he decided to stay on! On the whole it is a relief – having had this fit of the sulks or illness, whichever it was, he will probably be cheerful and willing for a week or so.

Monday September 21st—Early this morning I saw something dark moving in the woods across the river; what it was I could not see, so ran indoors and got out the glasses. After some search I found my quarry – two large black birds looking very like turkeys in size and gait, walking solemnly through the bush in Indian file, turning aside now and again to pick up some food – looked like a frog or lizard. I watched them for some time. The one in front seemed slightly smaller than the second, with a white blaze at the base of the heavy black beak and a white line in the wing, both of which seemed missing on No. 2; down the side of the head and front of the throat was a red patch looking in the distance not unlike the wattles of a turkey. The legs were short and stout.

The mail for Muié was despatched directly after morning service. The Kunjamba boys were to have left too, but Mr Procter decided he must go in to Menongue to see about Mr Muir's parcels, so held them back a day. It was rather nice not to have to write a big mail! Sent M's last letter and a short note to G.V., also a line to Jeannie, which will probably not arrive before I do.

The day was taken up so far as I was concerned with packing specimens, walking up the river to get flowers for vases (by the way I found a new specimen – an orange-coloured, long bell flower on a shrub about 3-4 feet high near the river) and working at my two sketches.

Tuesday September 22nd—Mr Procter left early for Menongue, though not as early as he intended, as he was kept waiting nearly an hour by his old friend Sasali. When at last he and the other Kunjamba boy did appear, they didn't want to take the load they had promised to take! However, in the end they took it and Mr Procter at last got started. He expects to return on Thursday. Mrs Procter came over here and sat chatting with us while I mended my ant-eaten sweater, a ticklish job which I am glad to have done. Kaliye helped her boy Makai as Litomi has gone with Mr Procter, while Muyeye was set to make a carrying frame for the pots. He has taken all day over it and is not yet finished. However, apparently there is no hurry, as there is not a sign of the porters yet.

I collected five *Gladiolus* bulbs on the kopje between the river and Miss Moors' summer house. I hope it is the speckled Menongue one, but all are fruiting so I am not sure. All my specimens are now packed, except a few in the press, which are not yet dry. I washed out my mended jumper and some frocks etc. I'm afraid I shall have to descend to an iron once more!

Wednesday September 23rd—After breakfast Mrs Procter let me use her ironing board and iron. I pressed out the sweater and three frocks. How long they'll stay presentable is a question! We all had lunch with Miss Moors and in the evening, being Wednesday, dined with Mr Pontier. About noon clouds came up and the afternoon started dull – rather a relief after the burning sun of yesterday. There was a slight shower, nothing much but enough to pock the ground, in the afternoon. The marks of the rain stay on the sand for days.

Thursday September 24th—Mr Procter got back about 9am having slept at the head of the Chimpompo, or rather its tributary, above where we camped. His business had been satisfactorily accomplished and the trek was very pleasant – the Chimpompo, he says, looks lovely. As regards porters, however, the position appears fairly hopeless. The Administrator is 'hundreds of porters behind' and he does not know when he will be able to let us have the 27 we asked for. Mr Procter suggests sending to Kuchi and trying to get them there, so tomorrow we shall send Muyeye with letters to the *Chefe*. We all lunched at Miss Moors' on the remains of Mr Pontier's dinner, which was more than ample!

Friday September 25th—Mr Procter spoke to Muyeye yesterday. At first he was reluctant to go, but when it was suggested that he should not go alone but with a friend, he became quite keen. He and another Kunjamba boy left early this morning. Mr Procter gave him the letter directly after service. He, his travelling companion and Kaliye then came over here, got salt and their money envelopes. Then followed a great counting of money. It appears that they plan to buy *tanga* with it. Kaliye handed over all his to Muyeye to buy food and *tanga*. Hope it doesn't mean that they are planning not to go on to Bihe with us! The boys have to try to get food on Mr Procter's behalf on the way. They go round by the mission, I think, and took a letter. There is great difficulty in getting sufficient food for the school boys and workers. After breakfast we went to see Miss Moors's house. It is very nice, has lovely views of woods and river from the various windows. The walls are all but completed and the framework of the roof, three great A-pieces held in place by long cross bars, in place. Scarcity of helpers is delaying the work very considerably.

THE MISSION CATS

On leaving I walked up past my bed and there reposing in the centre of my grassy hole was the greater part of a very dead, very malodorous fowl. Skeez, the yellow tom cat, in the offing, grinned at me and at first I allotted him the blame, but the consensus of opinion is that it is Kaboseta's doing. I gingerly removed it and carried it a few hundred yards into the bush where I threw it away. In the late afternoon however, when I went to make my bed, I found it – or rather all that remained of it, considerably less than in the morning – again on the bed, covered with ants. Henceforth Kaboseta is not only not encouraged, but strenuously discouraged from coming near my bed! She and one of the other cats nearly always come to greet me in the morning. They too are nearly always present at morning and evening service. Bobs, another black and white cat, and Skeez don't often appear. The other evening however, they were both there, and Skeez discovered some fish (*belela*) which one of the boys had put in a hole in a tree. He managed to fish out three in succession, which he much enjoyed. Just as he was trying for a fourth the preacher, Maliti, finished and moved to his seat. Guilty conscience read in the move the coming of retribution, and at the first step Skeez took a flying leap from his tree-trunk and fled for safety! Naturally the congregation had difficulty in hiding their amusement!

Today Mr Pontier called us to see what he had – his tortoiseshell cat had a family of four kittens, two tortoiseshell, one yellow like Skeez (the father) and one white with a black splodge on its hindquarters. Funny little mites, with such pink, wrinkled paws! The happy event took place in his 'dirty clothes basket', i.e. paraffin box!

In the morning I went across to the island and wandered about collecting, right to the far end of the main island. Then on the way back I met Mr and Mrs Procter who asked if I'd seen a flower which had come out in Mr Pontier's furthest patch, a head of deep red flowers some six inches across, *Brunsvigia* or some such genus. I must photograph it. It's the first I've seen.

Saturday September 26th—Just as morning prayers were finishing yesterday we heard voices and steps, and a young Portuguese on a riding ox, with a train of followers, appeared.

September 26. *Siphonochilus puncticulatus*, 'spent a couple of hours painting it', Cwelei

'A lovely mauve flower (*Siphonochilus puncticulatus*)... after lunch I photographed it'

HERBARIUM
RHODES UNIVERSITY

No.

September 26. An orchid (*Bulbophyllum sandersonii*) with the
flower on a long, flat, tongue-like process, Cwelei

Bulbophyllum Sandersoni
Reich. fil.

? Angola.   Orchidaceae   17

He turned out to be a trader going to open a store on the Cweve between Kaiongo and Menongue. He had come through from Bihe in seven days – slept the previous night at Makengi's. He had a long train of carriers with loads – these passed the house a little later as we were finishing breakfast. He asked Mr Procter for a guide to show the way to Kaiongo. Tapelo, the head man, showed him the road to the next village where he would be able to get a guide.

Exploring the islands in the river

This morning Miss Moors and I, in tunics etc., started out soon after 9 and spent an hour and a half wading about the islands, trying to cross into Bihe. We crossed several branches of the river, but were stopped by the last large one. We had not time to try for a crossing as we were lunching at Mrs Procter's and did not want to be late. As it turned out we could have stayed half an hour longer. We both thoroughly enjoyed the outing and are planning to go again.

At midday when we went over to lunch the Procters drew our attention to a lovely mauve flower, three inches across, growing right out of the ground just in front of their door. After lunch I photographed it, then began to dig it out.

We called to the boys Makai and Litomi to come and look at the camera. Then Tapelo and Maliti and two or three schoolboys came and we had some fun with them, showing them 'Nâla Procter' standing on his head on the screen, etc. Then Mr Procter brought out half an old pair of binoculars. They looked through the broad end at Mr Pontier on the roof working at his chimney and said he was just like a *kasila* (bird), at Mr Procter and said he was *ndendi-dendi-dendi-endi* – nothing could be smaller than the way they said it! Looking the right way at a chicken, the boy put out his hand and tried to grab it!

Finally I got up my specimen with its curious tuberous roots, and then spent a couple of hours painting it – two views, the first poor, the second fairly successful, though giving absolutely no idea of the lovely crystalline quality of it. That finished, I started on the local form of the red river clusters, with its dainty four-winged fruits and silvery leaves. In the middle, Mrs Procter came to summon us to tea.

Then I went to the island to get flowers for Miss Moors' little bowl. It was lovely there in the evening light. Halfway across, I heard Mr Pontier call, saw his hat wave and went across to the falls just above the shelter. He was on the other side and threw two specimens across to me – one new, an orchid with the flower on a long, flat, tongue-like process, the other the climbing bell-flower. Leaving him to cross the river and get home, I went on to get my little 'fire lilies' (*Cyrtanthus?*), grey *Helichrysum* and white grass, which made a very pretty bowl for the centre of the dinner table. After evening prayers we went for an hour's moonlight walk, up the *tapalo* and down through the village. The cool white evening was delightful.

Sunday September 27th—Only three more days before October is here! A good thing I refused Prof. Compton's work.

I finished my red-and-silver specimen, pretty thing, and went on with the orange-belled shrub, at which I was working when the bugle for morning service went. My mauve cup-plant is putting out another bud – there was a purple point showing at the top of the white wrappings. Perhaps it will be open tomorrow. Kaboseta is so tiresome. She slept in the tent last night. She is very affectionate but very nervous and far too pressing in her attentions. I hate discouraging her but still more do I hate ancient chickens etc. on my bed! After tea, went across to the island to get some fresh flowers. It was lovely there. Mr and Mrs Procter and Miss Moors came over too – the latter I met as I was coming back. In the evening we had a little service of our own in the dining room – sang hymns, then Mr Procter read and gave a little address.

Monday September 28th—Went across to the island and photographed the red *Brunsvigia*-like bulb (dug it up and took it with a background of waterfall with a yellow Malvaceous flower). I intended to do the bell creeper too, but the flowers weren't out sufficiently, so left it till later. I took the yellow flower and red bulb back to paint and press. Both are very difficult. I enjoyed doing the bulb and leaves of my red friend but couldn't get anywhere near the rich deep yet bright red of the flowers. Mrs Procter came along on her way back from the garden where she had been planting seeds, and stopped to watch the process – I was just drawing in the flowers. The yellow one she did not know well. After lunch I went on with the painting and was just thinking of having a rest when I heard voices, went out and found Mrs Procter had come across with her crochet, for company.

A BUSH FIRE THREATENS THE MISSION

There was a big bush fire which had started over near the gardens behind the village and was sweeping downwind towards the station. As one thing and another has hindered the burning of a fire path round the station, Mr Procter and Mr Pontier with all the workers had turned out to fight the flames. I went and joined them – got a leafy branch and joined the beaters. Quite interesting, but very hard and exceedingly hot work! I soon found myself breathless and panting – badly out of training after three weeks of the soft life, little exercise and much food! The fire came quite close to the schoolboys' huts but was prevented from coming closer. It swept across, more or less parallel with the river. After it was past I left them burning a strip down to the wagon road, went back and as everyone was out of the way, slipped into my bathing dress and had a dip (head included!) in the river, most refreshing! I had it all to myself till just as I was coming out when I met Chindongo (Mr Pontier's funny little cook boy) coming to fill the water bag for the thirsty beaters. Mr Pontier had been out shooting in the morning and had returned with an oribi – and a headache. Beating the fire cured the latter.

I went to look at my mauve cup this morning and behold! It was gone. The basin in which I had put it still stood in the path, water and all, but I could not see a sign of it. I suspected Kaboseta and looked all round in vain. While I was dressing after my bathe, Mrs Procter noticed Kaboseta strolling round the corner of the house with a bit of stick

September 26. Fire lilies
(*Cyrtanthus welwitschii*), Cwelei

Cyrtanthus

velet, Angela

September 28. The red-flowered bulb and the yellow
Malvaceous flower photographed together, Cwelei

A head of deep red flowers
(*Scadoxus multiflorus*, previously
Haemanthus zambesiacus)

A yellow malvaceous
flower (*Hibiscus* sp.)

aemanthus ?zambesiacus
Bak.

Ysani
Cwelei
28.9.

in her mouth, suddenly thought that she would probably take it where she had taken my specimen, followed her and found it, much battered and minus all but one bud. No sign of Muyeye as yet. Mr Procter expected him back before this.

Tuesday September 29th—Miss Moors and I set out about 5.30 and went for a walk through the bush – walked for 40 minutes along paths, then turned back and cut through the bush, passing a spot where a log was mounted for sawing planks. I got several specimens – one new, a purple-flowered leguminous shrub, my lovely 'bell' *Gardenia*, and several mauve cups. By the way, none of the boys could give Mr Procter the name, except a little Chokwe schoolboy who called it *Kavalanganja* (something 'cup'). We got back just in time for morning prayers. Among other flowers, the fragrant white 'scrophs' were abundant and I took back a bunch for Mrs Procter's room. During our walk, we heard 'punk! punk!' – like a drum being beaten. For a moment it startled me. 'Woodpecker on a hollow tree,' said Miss Moors, and so it proved. We saw him and his mate a few minutes later.

After breakfast Mr Procter mended the hole in our pot with some metal cement. I was watching him finish it when six fine tall men arrived with loads of corrugated iron. Mr Evertsberg had bought it at Bihe some months ago but for some reason, lack of porters I think, it had been left at a mission some way from Bihe – a mission run by American Negroes! The porters were Ovimbundu. They were very welcome, as the iron was needed for the building of the chimney of Miss Moors' and Mr Pontier's houses. The latter has been building his in his spare time – Saturdays. After the men had put down their loads, we noticed one had an apron of a particularly pretty buck skin – cinnamon brown with spots and bars of white, unknown to the Cwelei folks.

THE PORTERS ARRIVE AT LAST

Then I went across to the house and was just about to start drawing my new specimen, when Mr Procter came across to announce the arrival of our 27 porters from Menongue. As we had quite resigned ourselves to several days more, and I'd made my plans accordingly, this was quite a disappointment. Owing to the scarcity of food for them, we must get off right away. Accordingly, a busy morning packing followed, while our kind friends made bread and cooked a joint for us to take on our way and generally busied themselves on our behalf.

Packing was nearly done by lunchtime but I still had my specimens to pack. This was hastily accomplished after lunch and then (2.30) all six of us, dressed for wading (more or less!) started out for the islands.

A FINAL VISIT TO THE RIVER

There, under Mr Pontier's guidance, we spent a delightful afternoon wading through the various branches of the river, right across into Bihe. We explored several islands and had some lovely views of the station, bits of river etc., and wound up with a most exciting crossing in the middle of one of the small falls. It was fine to feel the force of the water against one. On the way I sat down once very thoroughly in deep water. It didn't matter, as I had my bathing dress on. Miss Bleek, Miss Moors and Mrs Procter all had one slip apiece but did not go right down. We found a number of both kinds of epiphytic orchids, and two kinds of fig – one is a very fine tree, the other is the little river fig. We got back about 4.30 – met two green snakes – one Miss Moors almost trod on after we left the river, the other I saw below the house. I put my specimens in the press before changing so did not get up to evening service. Our day finished with dinner for everyone at Mrs Procter's, after which we made words from 'franchise'. No sign of Muyeye.

Mr Pontier, not content with his little oribi yesterday, went off again this morning hoping to get roan, which had been reported as seen on the Chimpompo. On his way he saw a leopard on the far side of the valley. He shot but just missed it and it went into the bush. Shortly after, he saw a roan and as he was stalking it, a herd of wild pig crossed his path. He shot the leader, wounding it, and they trotted off. He followed, shot again and got a small but fat porker, with which he and two boys returned. The latter were going back after the wounded leader which they did not think would get far. It is unusual to see leopard by day. Lion have been reported across the river.

Departure from Cwelei

Wednesday September 30th—Up early and by service time every last thing was packed and ready for the porters. They had been given beans and salt, *tipoia* boys chosen etc., last night after evening service, at which they turned up in force. Mr Pontier helped us but they gave us no trouble. They had been working in the afternoon carrying bricks and had got their sticks and tie bark ready for the morning. Directly the first bugle blew,

September 30. Leaving Cwelei: Miss Moors, Mr Pontier, Mr and Mrs Procter and Miss Bleek in *tipoia*

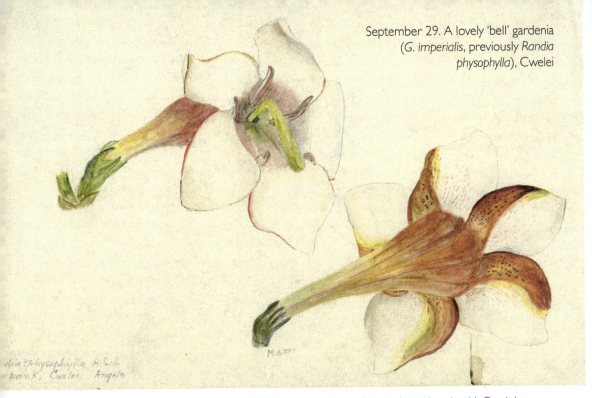

September 29. A lovely 'bell' gardenia
(*G. imperialis*, previously *Randia
physophylla*), Cwelei

September 29. *Genlisea glandulosissima*, 'in swampy grass fringe along river bank', Cwelei.
The painting is accurate enough to be treated as a botanical specimen and is annotated as
such by an authority on the group

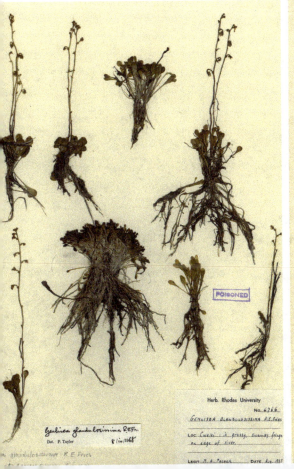

September 30. On the road leaving Cwelei with Miss Bleek (left), Mrs Procter, Miss Moors and Mr Pontier (right)

September 30. Reaching the road to Bihe accompanied by Mr Pontier, Mrs Procter and Miss Moors. Miss Bleek on right

Kaliye was sent to call them, and Miss Bleek gave out the loads. After breakfast and after taking some snaps – Mr Pontier's house, me in my *tipoia* etc. – we started at 8.20, accompanied by all our Cwelei friends who walked with us as far as the bridge which we reached after an hour's walk. There we said goodbye, after having our last drink of the Cwelei's waters, and our friends turned back.

A fine forest

For an hour our way lay along the *tapalo*, then, after half an hour's rest, we turned off to the right along a woodland path which led through some of the finest forest we have seen – really big, fine trees, well spaced – then through thicker bush to a small river. After this there was a succession of small streams – very pretty country with many flowers. About 12.30 heavy clouds gathered, there was a little thunder and for some while scattered heavy drops fell – not enough to wet us, though over to the west on the high land it seemed to be raining heavily. Soon after, we reached the Muiambei, and followed up the stream for some little way till we reached the home of the trader, Ferreira, on the right bank of the river. On the far side was a large village.

Halt at a trading station

This trader is the younger of the two brothers Ferreira who had tea at Cwelei with us a couple of weeks ago. We were rushed up to his store through the grass gate into his courtyard and our things dumped on his little back stoep. He wanted us to camp right there, but we said we preferred to go among the trees and fortunately he did not insist. However, he made us have tea, very welcome, and insisted on our having dinner with him. His establishment includes the inevitable native wife, Lisa, who waited on us etc., besides several other attendants. Fowls, pigeons, dogs and cats were all about, and a couple of hawks made repeated swoops after chickens and pigeons – without success, but completely undeterred by the gun which he fired once without hitting. He's quite a young fellow, and was very sorry for himself, as he had burnt his finger and it was very painful. We doctored it as well as we could, but hadn't much to give him.

We made camp on the hill above his store and I was just writing up my diary while Miss Bleek bathed when Ferreira brought the old chief to ask for medicine. He had a pain across his chest just over the heart, probably indigestion. We gave him some tablets of bisurated magnesia as the most suitable, as well as the most harmless, of the few drugs we have. Miss Bleek proposed cascara, but in the circumstances I thought that a bit risky! While waiting, the *Mueni* expressed his admiration of my hair! Then Ferreira stayed talking till dusk when we went down to dinner – such a dinner! Vegetable soup, almost solid, of which Lisa gave me two great cupfuls. I struggled hard but had to give in before the end. Then roast chicken and potatoes, followed by fish done with onions and vinegar, then tea and cake, beautifully light and delicious. Whew! It was a struggle! With difficulty we prevented our host from giving us three times the chicken we needed. Soon after 9.15 we left and went to bed – very welcome after the previous too short night (sleep from 12 to 4.30 not enough for me!).

AMONG THE HEAD-WATERS OF THE RIVERS

Thursday October 1st—We left Muiambei at 6.10. The morning was cloudy, but later it cleared and was hot and sultry. We crossed the river, passed through the village and then, having had the way pointed out, we trekked through very ugly country, half burnt, all its natural beauty destroyed, for an hour and a half. Then we got to unburnt country where we had lovely glimpses of blue valleys to the right, the Cwelei in the far distance. About 8.10 we reached the Luasevi (*mwana* Cwelei), a beautiful open green valley where we saw a large buck feeding. As we approached, he looked immensely long-necked, like a giraffe. My boys said he was *ntengu*, others *mpulu* (wildebeest), but the latter I am sure was wrong. He watched us for a long time, then moved off to the woods.

After a rest, (I found a lovely blue-flowered succulent plant which I photographed) we went on, crossed the river, and then followed a succession of small rivers. At 11.25 we reached the Mavombo (*mwana* Chisolo, *mwana* Kweti, *mwana* Cwelei), followed up it for 20 minutes, then crossed it and stopped under trees on the far side for tea and something to eat. The water was poor, very full of sediment.

At 1.05 we started again through woods. We trekked till 5.15 when we crossed the Luasevi (*mwana* Kakuchi), and camped at the edge of the woods on the far side. Between 1.05pm and reaching the Luasevi we crossed the Lisolo and the Kangola (*mwana* Kuseki).

AN EARLY START

Friday October 2nd—We left camp at 6.10, 15 minutes after I first saw the sun's rim rising above the far side of the plain, here very wide and flat. I forgot to note that yesterday I saw several hoopoes in the woods, the kind so abundant round Oudtshoorn. Today I heard a *piet-my-vrou* calling. It is interesting – I suppose they take a couple of months to get down to the Cape.

It was a lovely fresh morning – night quite cool – the eastern sky lightly flecked with cloud. To the north-east cocks were crowing in a village not far off. At 7.50 we reached the elder Ferreira's store on the Chitotomena. We crossed the river and went up to the large, newly built store. Nearby, ten head of cattle were grazing. A shaggy dog came out to greet us, and at the back were numbers of fine big hens, chickens, goats, pigs etc., altogether most prosperous looking. The trader pressed us to have coffee, wanted to give us two bottles of wine etc., and was much disappointed when we refused. We tried to get some information about our route, without much success – he appealed to our men for confirmation! About 8.15 we got under weigh again following the Chitotomena up to its source, a pool at the head of an amphitheatre which we reached after half an hour. Then there was a climb up a very steep but short hill above the river, followed by a long gradual slope through low brush.

At 10.10 we reached a small river with a large village near it, below which we stopped till 11.20 for lunch. After that we trekked for some way till we came to another Vachivokwe village, Comboio, on a small river. It was only 1.10 when we reached it, but the next stage was said to be long, so we camped on a slope opposite the village.

Then followed a wait of an hour or more before the porters came up. As soon as they did, I started putting away the last three days' specimens – a long job. A huge crowd of villagers, men, boys and women collected to watch the operation, and I got a number of Chivokwe names – slightly different from Mbundu or Luchazi in most cases. Meanwhile Miss Bleek tried to buy a pig for the carriers. After much explaining of the price, it was finally demonstrated in cash and the owner departed – but the pig did not materialise! We bought eggs and were offered fowls which we foolishly refused. I managed to get a much-wanted meal in train and made scone-loaf while Miss Bleek got the tent etc. up and had her bath. The bath was almost as welcome as the meal.

After dinner Miss Bleek said she had a suggestion to make – I waited expectantly – that we should make a long stop in the middle of the day, cook a meal and then go on! It quite took me aback, as I thought the long stop was just what we were trying to avoid, and had been advised against. Bed was good – most welcome, but on a skew slope, heels rather higher than head and left side lower than right. In future I'll see that it is more nearly level. As it is so warm at nights we have returned to our beds and given up digging graves. Miss Bleek will persist in sleeping without a mosquito net. I think it is foolish, though as a rule there are not many mosquitoes; however, I cannot do anything more.

To the source of the Comboio
Saturday October 3rd—Left Comboio 6.15, crossed the river, then turned up it without passing through the village and followed it to its source – a large pool in an amphitheatre of hills. A good many of these small streams seem to rise in this or a similar manner. Then, as on the Chitotomena, we climbed a short steep hill followed by a long gradual slope, to the high land where there were numerous gardens and some new clearings for the coming season. At 8 o'clock we reached and crossed a small river, the Sando (?), and an hour later the Achizinga (*mwana* Kakuchi). We rounded the head of this, putting up several birds, rather like plovers from their cry – standing they looked greyish brown, in flight grey with white bars on body and wings, feet stretched out behind. Here we stopped for half an hour when our *tipoia* boys were hailed by the porters, and came and asked if we would start. They had taken the wrong path, so we had to re-cross the river, follow it down on the far side for a mile or two and then cross again, the porters meanwhile having mostly got on ahead.

By the way about 7.30am we passed some very fine specimens of the lovely white-and-red cup-shaped flowered twiner. I stopped and photographed one spray, much to the interest of my *tipoia* boys who had not yet seen the camera at work.

At 10.35 we crossed a tiny clear stream, the Lizungu, probably flowing into the Matendi, and after that our way lay along an overgrown *tapalo* through woods. About 11.10 we stopped at what was evidently a regular halting place for carriers. It was about five minutes above the Matendi, and we were told we'd better have our lunch as there was no more water for some time. I tried some of the *mukola* fruit in my tea instead of lemon, and found it excellent. I wonder whether it is tamarind – it might be from taste and appearance.

A DESERTED VILLAGE

Here we were delayed by one of the carriers, Kalosi, with my book box, containing, by the way, the bulk of the remaining cash. One of the others went back to find him and brought in his load. He had complained yesterday that he had painful legs. We gave his load to the *tipoia* boys and tried in vain for another porter in the next two villages. The second was Amatendi on the River Ndalashamba. Here again we were on an old *tapalo*, much overgrown and running along the hills above the Kutundu. About 4.10 we turned off to the right to Chilundu (Vachivokwe) now deserted. We passed through the village and camped below it above the narrow deep stream. Kalosi had medicine, and another boy came to have his eyes doctored. I bathed them with boracic and permanganate (the latter unintentional – there was some in the pot). The men seemed very nervous of something, '*vangonde*' I think they said. At first we thought they meant the herd of five or six cattle (*vangombe*) which were feeding near the village, but I think they meant the spirits from the deserted village.

SEARCHING FOR ANOTHER PORTER

Sunday October 4th—A day of stopping in villages, very tiresome, the excuse being the search for a porter. We left Chilundu at 6.20, crossed the stream and then followed the Kutundu (*mwana* Kuseke, *mwana* Cwelei) till 8.10, when we went up to a Vachevokwe village Somakambwande, where we stayed till 8.55 trying to get a porter. Finally we got a guide to the road, but no porter. He led us through the wood for some way to an old wagon road along which our way led for some time. Here I first saw a yellow *Loranthus* growing on a leguminous tree.

At 10.05 we reached Kawololo where we again spent a quarter of an hour on the vain search for a porter. At noon, yet another village, Kapaia. This time, however, we refused to stop, asked the way and went straight through down the slope to the river Kapendu which we crossed, and climbed the slope up the far side to some trees where we stopped for lunch. This hillside was very interesting – open wood, chiefly of *Faurea*, *Protea* and a fine-leaved leguminous tree with a *Philippia*-like shrub, orchids etc. – particularly the attractive three-horned *Eulophia*.

Lunch interval lasted from 12.20 to 1.50, when we went on up the side of the river for some way, then took to the woods once more, passing through gardens galore, where the women fled at the sight of the *tipoias*. Then again came to open river slopes and finally our path led straight down the side of a wide valley and up the far side. The men were doubtful of the way and Kaliye was sent racing on ahead to ask a woman who was carrying water up the far slope. She too was so frightened that she could hardly tell him the way. Kaliye imitated her most realistically with gusts of laughter, '*munakasi kukina*', shaking himself all over.

We went up the slope on top of which was the village of Sobe, and camped a short way from the store where we had hoped to replenish our stock of food. However, the trader was away and though his people came and greeted us, everything was locked up and we could get nothing. A woman brought us a present of five or six eggs, and we gave her some tea. Then we bought a fowl – a small boy brought a miserable sick chicken which we refused, eventually having to pay $6.00 for a hen – quite a good-sized one, however. It was 4.15 when we arrived, so that the time was short to get my specimens put away as well as getting camp settled etc.

Monday October 5th—We left Sobe at 6.25 by a wagon road which led through the trader's post. Two of his boys requested to be allowed to carry Miss Bleek in her *tipoia* as far as the *tapalo*, and set off in grand style at a rapid rate, which soon decreased. In half an hour we reached the *tapalo*, crossed it and went on along a woodland path which led down to a river valley, the Mavuka (*mwana* Kusehe) which we crossed at 7.30. The path then led along the side of the valley, crossing several clear small tributaries, sometimes cutting through a tongue of woodland but always following the line of the river, till at 8.15 we reached a village, Katentambundu through which we passed, leaving the river. When I was taking a rest in my *tipoia* a very noticeable ladybird joined me – polished red with several fine longitudinal stripes of black edged with yellow – a beautiful little thing.

About half past nine we reached another river, the Cindemba, with a large village, Mutandu, on one bank and another larger one, Machosi, with an old store formerly (?) kept by the younger Ferreira, (now at Muiambei) on the other. The valley between was deep and we rested among the trees and rocks near the river for 40 minutes, before we went on through Machosi. Ferreira's store was whitewashed and, besides his name in large letters, there were quaint designs and figures riding bicycles etc., painted on the front wall in red, the figures very reminiscent of Bushman drawings! About 11.30 we reached the head of a small clear stream in a very pretty shady little valley, flowing down

October 5. Angolan woodland

to the Kusehe not far off. A woman was fetching water and here I found two orchids –
one the dense head of orange-red flowers of which I had already found one specimen at
Cwelei, the other a new one – a handsome tall red fellow, with a big head of numerous
rose-pink flowers.

The little valley led down into the wide, open grassy valley of the Kusehe which we
crossed (11.45) getting on to a wagon road again. Miss Bleek was on ahead, left the path
and struck off to the right attracted by some bush which, however, was far off and much
burnt. Eventually she stopped at a tree in the middle of the burnt patch. Fortunately
there was an only partly burnt log to sit on. Here we stopped (12-1pm) for lunch, then
went on along a path through the wood, leaving the wagon road behind on the left.

The path went on through bush, crossing one or two small streams and then suddenly
came to an irrigation ditch near to Kakula. Here we found a charophyte and a very
beautiful red composite. We had some difficulty in finding a good camping spot.
Eventually we went past the post, through an old village site, surrounded by the great
circle of trees common in this district, and camped beyond, just above a small stream.

Dinner with the trader

Camp made, Miss Bleek went to the store to buy salt, flour (if possible), 'belela' for the
boys (dried fish was all that was procurable) etc., and came back committed to dinner
with the trader, Silva. It was unfortunate, as we were a long way off and had a lot to do.
However, we had to go. Miss Bleek went on while I finished what I had to do, and by
the time I left it was almost dark (7pm), and I had some difficulty in finding the way.
However, we had some time to wait for dinner.

Meanwhile the trader occupied himself in making up two beds behind us – much
to our consternation – rain was threatening and, rightly or wrongly, we concluded he
meant them for us. We told him not to worry, as we certainly could not stay. He pressed
on, saying rain was coming. But I'm not sure that the beds were not merely for himself
and his native wife, a rather pretty little thing who brought in the dinner eventually,
assisted by a small boy. Others who assisted at the meal were four dogs and a cat. The
latter and one of the former both had new families of four each – these we did not see.

The meal, quite good, was very long – and to poor sleepy me seemed interminable.
Finally it finished with 'extra choicest' green tea (Lipton's). The pot, coffee pot really, of
boiling water was brought in from the kitchen across the yard and our host put in two
or three large teaspoons of the tea, covered the whole with a serviette and let it stand for
a quarter of an hour, after which we were allowed to drink it – very good too. Finally
at 9pm I felt we really must get away to bed or we'd never get started in the morning.
I hurried poor Miss Bleek through her second cup of tea, and showed my watch to Sr
Silva who agreed that it was 'muito tarde' and let us depart after another offer of beds, (not
much pressed!) Poor man, it's a bad life for a lonely man and I don't think we could have
enlivened it much!

It was very dark getting home – the lantern helped a bit, but Miss Bleek, who was
leading at first (as she had been over the path three times by day and I only once by

night, I thought she'd know it better), wandered up to the old village. I had the bearings however, and we soon got on to the right path. It was cloudy but did not rain. We got to camp about 9.20.

Tuesday October 6th—We were astir fairly early but it was hard to get up after the late evening. I had a narrow escape – my watch slipped off my wrist and I thought I had lost it. The ground was covered 6-8 inches deep with one of the geophilous legumes – terrible ground for losing anything. I searched for some time, then to my great relief Miss Bleek found it by my suitcase just outside the tent.

We left camp at 6.20. The morning was cloudy but larks were trilling on every side. Our path ran down along the valley of our little stream for a mile or so and then along the larger Kutupu (*mwana* Kwanja) of which it is a tributary. Crossing one of the streams by a rough little bridge of logs, I took the snap I had wanted of the porter who wore his hind skin neatly folded for trekking – bag hanging over it, bark bag of food with two carrying sticks on his head (7.15am).

After an hour and a half we reached another of the characteristic Vankangale villages built inside a circle of large shady trees, reinforced by a stockade. These trees they say are called *linongo* (plural *manongo*) and are planted from slip, a species of fig I think. They have large buttress roots growing from the lower branches. Possibly the hedged villages are a relic of more troublous times – trouble caused either by raiding tribes or marauding animals, I don't know which. Here we stopped for a quarter of an hour (7.50-8.05) while some of the *tipoia* boys bought food.

WILD FIGS AND ANTHILLS
Quite early, soon after leaving our little valley, I went into the wood after specimens and was just picking an orchid when I put up a rabbit, cinnamon brown with a white scut. There are such a number of ground orchids – sign of spring coming, I suppose. They and the various species of *Gladiolus* are harbingers of the rain season flora – far richer, I am sure, than anything I have come across. Soon after leaving the village, we passed some magnificent fig trees – wild fig, like those at Cwelei with the old branches simply covered with the curious short, fruit-bearing outgrowths. The figs are bright red when ripe, the tree a beautiful shape, throwing a fine shade. It is quite different from the fig of the village stockade. Here, too, are numbers of huge anthills, usually covered with vegetation – two or three large trees perhaps. On several were young erythrinas ablaze with blossom, and a curious plant with large geranium-like leaves and snaky spikes of bloom. They were all too high up to get specimens, unfortunately. It is frightfully difficult to get my specimens put away in the evening – I expect they will mostly get spoilt.

Later we got on to a ridge and had lovely views of blue distant valleys. At 9.40 one was pointed out as the River Vissumba (*mwana* Kakuchi) and 20 minutes later (10am) we reached a village, Kasina, on the slope above the young *donga* Vissumba. We passed through the village and crossed the river. I had stopped to collect and photograph a very fine rose-pink orchid in a swampy bit of ground near a small river; incidentally I found a new *Utricularia*, a lovely little thing, palest mauve in colour and only four to six inches high. Consequently, I was far behind Miss Bleek, and took a path leading towards the palisaded village, again tree-circled. Just then a man came along on the upper path a few yards to the right and, his tone full of surprise, forgetting even to greet me, he called out that <u>that</u> was the path. I asked if the *Doña* had gone that way and his answer at any rate <u>implied</u> 'of course'! Followed by my *tipoia*, I transferred myself to the right path which led above the village, passed one or two buildings outside the stockade and then down a wide grassy *tapalo*-like stretch down the valley to the river, across which I could see Miss Bleek and her *tipoia* boys. It was evidently an old *tapalo* – the Vissumba was shallow but fairly wide and crossed by a plank at one side, the road crossing it in a kind of 10 ft wide, sandy drift. I filled the big water bag and went up the hill to join Miss Bleek, who had got the fire going and the billy on. Above was another large village, Chineni, with a trader's store.

Villagers were continually passing back and forth. One arrived with a pig on a string (fastened to its back leg) and the *tipoia* boys called him and started bargaining for the pig. By and by, one came up and said the man would sell it. Miss Bleek said she would buy it for the porters. Accordingly, the owner, not the man who was driving it but an older man in a blanket, came along supported by one or two others, pig and driver nearby, and a couple of *tipoia* boys watching the deal, and Miss Bleek counted out the amount we understood was asked and handed it to the old chap. He spent some time counting it over, then handed it back and departed, pig and all. It was most annoying as we wanted Mr Pig for the porters who have really been very good, and would willingly have paid twice as much, had we known it was asked. Everyone was disappointed, but everyone took it quite philosophically. Pig and party departed uphill, presumably to the traders, but as they all returned forthwith, evidently there too no sale took place. This is where we feel the lack of a good boy who understands both us and the people. Kaliye does very well but he isn't much good for this kind of thing.

Miss Bleek went up to the traders to replenish our supply of sugar. I don't like the long midday stop. The extra time would be so much more valuable in camp, and the boys are nearly always quite ready to start after the hour. However, Miss Bleek prefers it. Today we stopped from 10 to 12, at least an hour longer than necessary. At noon we got under weigh, went on uphill past the trader, and along a wagon road through the woods.

It was hot and about 1.15 the sky grew overcast and thunder rolled round about us. The porters lagged behind for some time – they don't keep up as well as at first, I think because they have found that our pace is slow and for some time only the four boys actually carrying the *tipoias* were with us.

I was taking a ride and our path led through wood when suddenly I saw a glorious ground orchid, hopped out and told the boys to put down the *tipoia*. It stood about two feet high, the slender stem coming straight from the soil, below which was a large horizontal, jointed tuber. At the top were three beautiful rose-purple blooms – inside the spur there were dots and lines of deep red-purple. I photographed it, but the light

October 6. A very fine rose-pink orchid
(*Eulophia cucullata*)

October 7. Elaborately
painted mask

Diospyros batocana, a large
bush or small tree in forest.
The fruit is used to make
meal in the lean season

October 10. Carpet shrub (*Parinari
capensis*), Kamundongo

was rather poor and I only had a film, so could not focus on the ground glass. I hope it will be successful but much doubt it (exposure 1/5 sec., aperture 1/16).

In the distance we got views of the blue valley of the Mitekiteki (*mwana* Kuthi (?) *mwana* Kukema, *mwana* Kwanja). About 2 we reached the village of Menonga on the Njamba. The porters wanted to go to the village but we insisted on going on beyond, though they complained it was very far from water. We camped among a number of *Protea* trees. *Protea* wood gives excellent coals for my sweet bread baking, the best I've had. A larger tree gave us some shelter, of which we were glad later. As evening approached, so did the thunder, and it soon became evident that we'd get some rain before the night was past.

By and by a deputation from the porters came and squatted down by us. At first we couldn't make out what they wanted, and called Kaliye to interpret. Finally we understood: 'Would we please not wish rain. If we did, it would come, and the porters didn't want to get wet!' We told them that rain didn't depend on us, and advised them to get plenty of *vikuni*!

Just as I was ready to go to bed, the rain came – a heavy thunder shower – thunder and lightning close round us. Fortunately for us, we only had one or two heavy showers and then it passed off. Unfortunately for me, however, I hadn't put my bed ready and got a bit wet in my efforts to do so. Once in, my groundsheet and coat, with Miss Bleek's macintosh on top, and my Burberry and 'Victoria Falls' sunshade over my head, and I was warm and comfortable, well protected from the weather and far happier than in the tent.

The specimens and other loads were put under the *tipoias* and covered with Miss Bleek's old piece of oilcloth. At the first drop, Kaliye rushed over to put his *wunga* under a *tipoia*. Incidentally he put himself there too and frightened the other men when they came to put their meal too in shelter. He stayed till the rain was past and had a good sleep. Finally, after nearly an hour, he reluctantly crawled out and betook himself to one of the fires. The water was far off, and the porters very reluctant to go and fetch it. Apparently it was just water holes or a very small stream near the village.

WITCHDOCTOR'S HUTS
Wednesday October 7th—We left camp at 6.10 after inspecting two very interesting little huts a couple of minutes' walk from our camp. They evidently belonged to some witchdoctor or worker of magic. In one was an elaborate mask, the front roughly triangular, painted white, with eye, nose and mouth holes and net for the back of the head. Below was a wide fringe of stiff grass or reeds, thick and standing out stiffly all round. Hanging nearby was an apron of similar make. In the neighbouring hut was a huge white structure painted in black with jazz patterns, made I think of very thin skin or hide stretched over a light framework, probably a lantern or some such thing. It was altogether an interesting camp. Nearby, roots had been dug up for weaving or basket work, or more probably for the making of these contraptions.

CHISEYA

Our route again lay along a wagon road, and the erythrinas were fine, blazing with blossom. The fig trees, too, were grand. About 9.40 we reached Chiseya. This was our first long waterless stretch. We didn't pass a single stream in the three-and-a-half hours, and even at Chiseya we were far from the water. Here we got on to a wide *tapalo* and just outside the village four heavily-laden wagons were outspanned, while a bit further on we saw the cattle belonging to them.

Chiseya is another village inside a hedge of trees, though it has outgrown its hedge and there are quite a number of huts outside it. We did not stop here, as they told us there was water nearer the road a bit further on, so on we went for nearly half an hour (10.10am), when we reached a store. There we stopped for Miss Bleek to buy some mealie meal. I'd used the last of our flour the night before, and as flour seems unobtainable we thought we'd better get what we could in its place. Then we went on till we found a decently shady tree and there camped. The porters had another $1.00 each to buy meal at the store. Again we stayed nearly two hours till 11.55. After two hours' trek our overgrown *tapalo* cut across a wider, better-kept one where again there was a store. We went straight on, however, and our road kept more or less on a ridge of ground, river valleys to right and left in the distance.

About 2 my oldest *tipoia* boy, who knows this route fairly well, pointed out a *missão* on the ridge parallel to ours to the east. We had just passed a big tree-encircled village, either deserted or almost so, when he pointed out the red mission buildings among the trees. He said it was connected with Kamundongo and the name was Mungola. Not much further on, a small valley led down to the right, and here the porters said we must camp. Not knowing whether there was water an hour or so further on, we finally did so, though the water here was far off and of poor quality. Women were clearing the stony slopes of the valley and there were two villages not far off – a tree-circled one across the valley and another up behind us.

I'd decided to try sleeping in my *tipoia*, and managed to find two trees conveniently placed, so that it could rest on both. Trying it as a hammock it was delightfully comfortable but proved disappointing as a bed – too short and too curved. I spent the afternoon transferring my specimens. The press is too full and they are not drying at all well. They don't get enough sun. It is a labour – by the time night fell I was not finished and had to go on by candle light.

Thursday October 8th—We left camp at 6.05, a clear cloudless morning with a cool breeze, and expected to reach Chiunge about midday, leaving a short half day to do tomorrow. Instead, we never went near Chiunge!

After an hour our wagon road branched – my *tipoia* boys took the left while Miss Bleek's, who were way ahead (I'd stopped to put away specimens) took the right. My path led down a steep hill to a little valley with a stream in it, not very clear or swift, but still with plenty of water, and this would have made a much better camp than last night. The slope on the far side also was steep, the bank rocky with trees among the rocks. Just beyond were large gardens so evidently there is a village somewhere near. It's a pretty spot. A short way on we re-joined Miss Bleek who was having a rest. After ten minutes' rest (7.35-7.45) we went on again and at 8 reached a wide, well-kept *tapalo* with marks of a motor lorry's wheels. Along this we went, hating its shadeless width. I was just about to explain to Kaliye what the marks in the road were, when I heard strange noises behind me – the porters explaining the ways of the motor car to Kaliye the uninitiated!

AUGUSTUS THE CHAMELEON

Just about here a movement in a small shrub in the *tapalo* caught my eye, and held my attention. It was quick but not the lightning-like dart of the lizard and not too quick for me – a fine big chameleon who showed fight when I tried to catch him! However, not being afraid of his aspect and fearsome attack, I caught him, much to the horror of Kaliye and adjacent *tipoia* boys, and after trying him on my arm gave him a branch to sit on, and there, much against his will, Augustus sat. He was just the colour of my *frou–frou* jumper, golden brown with a pattern of silvery grey. Kaliye told me to take care – Augustus would spit in my eyes and make them sick! Later, if I passed one of the men, they would draw to one side lest Augustus should touch them.

At 9 we came to a spot where the *tapalo* forked, and on the hill above the three roads was a large mission station, the name of which we could not discover. The church, built right on the road, had a very ornate front, but the walls and roof were collapsing in places and were under repair. It looked like a Catholic Mission and, though our porters evidently expected us to stop, we did not do so. We enquired of some bystanders which was the road to Kamundongo and the one to the left (west) was indicated – accordingly, down it we started. Then the porters started arguing – some wanted to start down a path leading off to the right, which they said led to water. A youth who was walking along with us, showing us the way, argued indignantly that that was not the way to Kamundongo, and as he pointed southwest when asked where Chiunge was, he was probably correct.

However, down the path we went. It led down into a valley, past some scattered huts, and there we found a water hole, quite unusable by us – the water was thick – so on we went some way further and presently came to a spot where the path dipped down into a deep ravine, very wild and rugged, and there (10.15) we camped above the dip among trees. The scenery reminded me of a rugged bit of the Yorkshire Riggs.

My first job was to photograph Augustus, who made himself very dark and refused to pose nicely on the branch of mauve wisteria-scented pea flowers which I got Miss Bleek to hold for me. Augustus refused to pose and Miss Bleek held the branch too near, so I do not think it will be much good. Then Augustus, to his great relief, was allowed to go and by and by we saw him doing gymnastics from one tree to another, his skin two most lovely shades of brown, one dark, one light. He's the biggest chameleon I've ever seen, except the green one I saw in the middle of the Zambesi. The porters were very much more interested in the camera and spent a happy half hour looking at one another in the view finder. Only, when one called out that he could see the poser, everyone crowded round in front of the camera to see too! Naturally it was quite a long time before anyone did see anyone else.

Going on again at 11.50, we descended the steep slope into the ravine down which was a winding stream, hidden on the right by trees and undergrowth, but from the sound evidently flowing swiftly over rocks. Just above, we crossed it by a log bridge, and here were numbers of a very handsome pink pea-flowered plant, the heads of flowers rather like lupins, but a true pink, not purple- or rose-pink, very attractive. Out of this glen, the path dipped again and crossed another smaller ravine. Soon, however, (12.30) we once more joined a *tapalo*, old and overgrown, but nevertheless far too open and sunny. We passed a village surrounded by trees at 12.45 and another large one, Kavatali, which had outgrown its hedge of trees. There were a lot of huts outside the circle and women standing about watching us. At my request the *tipoia* boys (Kangambi, who has been my chief informant as to villages and rivers, doesn't know this country) asked the name of the village, but apparently the women did not know it! Eventually, however, they obtained a name which, as near as I could get, was Kavatali.

We stopped for a short rest at 2pm (ten minutes) and then on again till 3.30, when we reached a store at the road side and stopped to buy a fowl, quite a good one ($5.00). The trader was sitting in front of his store with his baby on a chair in front of him. The latter set up a howl at sight of us, and kept it up. The Senhora was out, but we were offered coffee, which we refused on the plea that we were anxious to go on and camp. Yes, there was good camping ground at the Kukema, the river the boys had already mentioned and which apparently was only a few minutes further on. He seemed a very decent little fellow and while the fowl was being caught we chatted to him. Mr Gale had stopped at his store on his way through – he was much interested to hear of our journey. The fowl bought, the porters also wished to buy, so we went on slowly, leaving them to follow.

AT THE KUKEMA

Soon we came to a large river valley which we thought must be the Kukema – no water, only a mud hole or two. Enquiry of women returning from their gardens brought the information that this was not the Kukema – the Kukema was further on. By and by, we met a coloured servant-girl with a bath of newly-washed clothes on her head, and a few minutes later saw the Senhora who had evidently been superintending the family wash and at the same time washing her hair etc. Then we passed the washing pool – still not the Kukema – and it was not till we had been going for a good hour that we came within sound of rushing water. Then the road dipped down a steep hill and there at last was the Kukema.

On the left was a large, deep, rush-encircled pool. From that led a deep passage 4-5 feet wide, cut through the sheer rock and down it all the water of the river rushed roaring and foaming to empty itself into a second wide pool, this time cut out of the rock itself, whence, after a short rest, it again rushed through a rocky passage. A grand spot. On each side great carved rock masses lay and the upper passage was spanned by a light bridge of poles laid side by side and lashed to cross pieces with bark. In the rock were many pot-holes, evidently scoured out by stones in flood time. My *tipoia* boy, Sachingi, (the one who knew the villages and rivers so well the first part of the journey) drew my attention to the largest of these and informed me that it was made by '*Kalunga, Njamba*'.

It was 4.30 before we reached the Kukema, all weary after the long trek. Just across the river we made camp, but had to wait nearly an hour before the porters arrived with their loads. In the meantime I went exploring, first to the upper pool, then down the river. The vegetation of the rapids is very interesting – three or four Podostemaceae I think – one a curious branched coral-like structure, another with long thin leaf-appendages and one or two very close-growing flat ones. I collected what I could but I've used up practically all the formalin and all the large tubes. It was dark before we had finished our baths and had supper, and bed was very welcome.

LAST STRETCH TO KAMUNDONGO

Friday October 9th—I hastily put things together and then went off to collect Podostemaceae, leaving Miss Bleek to start on with the head of the train. Our friend the trader told us Kamundongo was six hours' trek, so we had a full day before us and not a half day's trek as we had planned. My collecting did not take long. I passed a tree simply covered with Mr Pontier's purple-tongued orchid. I hastily scraped off what I could of the curious plants and got back to camp just in time to see Miss Bleek agitatedly trying to make my *tipoia* boys understand that they must wait for me. She started on with the rest while I stayed behind to put on my shoes and pack away my specimens, finally getting away at 6.25, (by my watch that is. I think it has lost a good bit during the week, but as the sun has never been visible at rising, from our camp at any rate, I have not been able to correct it).

Kaliye was rather amusing last evening. I was helping to peel potatoes while he dressed the fowl and he started asking questions. He said the porters said Miss Bleek must be my *mukuluntu*, which I think means either aunt or uncle, at any rate one having authority – wanted to know if it was so, where we lived, where the various missionaries came from etc. It was rather difficult to explain.

The morning was glorious. Leaving the Kukema, the *tapalo* struggled up the steep side of the ravine, then straightened itself out somewhat, though still going rather too much to the east of north for my liking. The *tapalo* dipped into another ravine, which we crossed by a rude triple bridge, then climbed steeply up the far side. It was as wide as a main road but at a steep angle. A short way on, gangs of men, women and children were working at it, clearing it and cutting side drains. About 8.45 we came to a spot where another and evidently a main *tapalo* came in from the south-east at a sharp angle. A few hundred yards back from the junction of ours with it was a store and here the boys asked leave to go and buy food. They received $1.00 each, one went up and returned. The rest sat around till we began to get impatient and told them if they wanted to buy they must go quickly. Then they explained that the *Cindele* was in bed, so Miss Bleek went up in her *tipoia* and knocked his lordship up. However, he had no meal, only beans and such like. Kaliye spent the $2.50 which he had of his wages on a small mirror.

It was 9.30 before we got started again, along the interminable, hot, uninteresting *tapalo*, of course far from water. About 11.30 the air suddenly became tainted and I was wondering what it was when the boy walking in front of my *tipoia* suddenly swerved to the right edge of the *tapalo*, putting his hand over his nose, and my *tipoia* followed him. There half in the ditch at the side, half on the *tapalo* itself lay a disintegrating human body. It must have been there for weeks – the ribs were partly bare – and there I suppose it will remain for weeks more, it being nobody's business to remove it, and this on one of the main highways, only a day's march from Silva Porto.

Soon after 11.45 we came to a group of buildings on the road, and just short of this left the *tapalo*, cut down a path to the left through plantations, and rejoined the road further on. We crossed a small stream which we were all glad to reach after the long waterless stretch, and stopped on the rise beyond it in the first decent shade we came to (12-1.25). I'd managed to get a fine big *mukola* fruit early in the day and this we enjoyed in our tea. An hour later we passed another store (2.25), where two *tapalos* met. The one coming in had a sign board which told what places it led to. Here again our boys stopped to buy – the storekeeper said Kamundongo was an hour further on.

THE MISSION AT KAMUNDONGO

We were all getting very weary when after an hour's trudge we came out on the side of a fairly extensive valley on the far side of which we saw the mission buildings. Dipping down to the valley, the *tapalo* crossed the stream by a fairly strong bridge and then we struck into a path to the right which led up to the mission, which we finally reached about 4pm. We walked up to the first house, which had a grove of orange trees to the right, and there we saw an elderly grey-haired lady gathering eggs in her back yard. We introduced ourselves to her. As we thought, she proved to be Mrs Sanders, and she took us inside, called the doctor and gave us oranges to refresh ourselves with. By and by, she fetched the other members of the mission – Mrs Hunter, Miss Redich, Dr Hollenbecher, and her guest, Mrs Meiers, with her three children. As Mrs Sanders' guest room was more than full with the four latter, it was arranged that we should go to Mrs Hunter's and stay with her and Miss Redich for the night at least.

Dr Sanders told us there were only two trains in the week, Monday and Friday, and as the boat usually came in about the middle of the month he thought we had better take the Monday train. He was a bit doubtful if we had not better go on tomorrow with the porters – if not, advised us to send the porters on and he would motor us over on Monday in time for the Monday train, and this last is what we eventually decided upon.

Mrs Sanders' guest is a young married woman, granddaughter of Chapman the explorer, half English, half Dutch, married to a Swiss farming in the Mombola district, which she says is very rich in flowers. She told us they were expecting the Weiss family there but so far had heard nothing of them. The eldest child, a girl, is her sister's child – the sister died and this girl, Ada Chapman, married her widower.

Mrs Sanders is very interested in flowers and was anxious for me to see a lovely 'orchid' growing nearby. After supper Miss Redich took me down and, after we had searched in vain, Mrs Sanders took us to see it. It turned out to be the *kavalanganje*, and I had to tell them that, to the best of my belief, it is not an orchid. I got out my sketches and showed them a bit later on, though it is foolish to do so by artificial light.

WEEKEND AT KAMUNDONGO

Saturday October 10th—A busy morning – last night all our loads were locked up in an outside room at Dr Sanders', and this morning we went down and spent a busy two hours repacking and rearranging loads, the porters meantime clamouring to be off. They and Kaliye were sent to camp nearby last night. I don't think firewood is abundant and food certainly is scarce. Dr Sanders suggests sending them to the trader at Vila Silva Porto (the town – the station is Silva Porto, three miles further on) which is some 15 or 16 miles off. One of the letters from Administrator Madruga is addressed to the trader. Dr Sanders will send the Administrator's letters with the porters and instructions to Ferreira (the trader) to put them up and feed them till we arrive on Monday. Dr Sanders knows the trader well, deals with him and says he is a good, reliable man – he is a settler, not merely a trader, interested in the welfare of the country, with his (Portuguese) wife and children living with him.

We finally got the porters disposed of by 11.30 after they had had a few odds and ends of cloth, tins etc., and Dr Sanders had given each half a dozen oranges. With some difficulty the position was explained to them. Dr Sanders, in common with the others here talks Mbundu and only one or two of our boys understand it at all. However, with many appeals to us for confirmation, they finally got it. Kaliye did a little needful washing for us – then I presented him with my old blue tunic and knickers, and the larger of my two billy cans, while my aluminium plates went to Makai who carried my press and Shacohemba, the tent boy, who was grievously disappointed to find his load (tent and two hammocks) considerably heavier than before. However, we told the *tipoia* boys they must help with it – hope they did so! The *tipoias* went in quite easily, rather to our surprise.

That finished, we were free to enjoy a bit of a rest and see something of the working of this mission station which is run by the American Board, not the Anglo-American. Dr Sanders has been here since 1877, Mrs Sanders since eight years later! They are indeed veterans – he is a D.D., the other doctor, Dr Hollenbecher, an M.D. His spare time is given to language study and he is the 'evangelist' while Dr Hollenbecher superintends the technical drawing and boys' school work. Miss Redich is in charge of the girls' school, while Mrs Hunter sees to their housing, feeding, etc.

After lunch which we had at the Sanders' (she and Mrs Hunter exchanged guests), Mrs Sanders showed us round the garden which she supervises – peaches, bananas, persimmons, plums, figs, etc. – in addition to the 120 orange trees, naartjies, lemons and limes – roses and other flowers – all sadly in need of the delayed rains.

They had been home on furlough and only returned at the beginning of the dry season, so the garden needs a lot of attention. We were in the garden on the other side of the path when Dr Sanders joined us. He said he had just been requested to go and perform the marriage ceremony for one of the school teachers. The teacher had come to Mrs Hunter's the evening before to announce that he was going to be married today, but they leave it to the 11th hour to inform the officiating clergyman!

A MARRIAGE CEREMONY

We decided to go and see the ceremony and set off for the church. There the bride, very coy and demure, was standing waiting, dressed in a new cotton robe – sleeves and yoke with a long kind of skirt that wraps round under the arms – of pale mauve print edged with purple – round her hair a broad snood of purple with a yellow flower in her hair. An elderly maid of honour carried the inevitable umbrella. As the groom teaches the older girls, they were to precede the bride, singing. It was rather an ordeal for the maidens who don't take to singing as readily as the men, and the singing wavered considerably. However, they managed to process up the church to their places, followed by the forlorn little bride, who was finally poked into her place. She and the groom sat in two chairs set in front of the rest of the congregation. She sat with her back half turned on her man, feet pressed tight together, arms folded tight round her, as though she feared her dress would slip off, eyes downcast, lips trembling – she looked on the verge of tears! The groom was resplendent in new suit, new shoes, high collar, tie and all complete. He made his responses quite clearly – the bride's were quite inaudible, except for her first assent which wasn't 'yes' but an 'm!' which she just managed to get out. After the ceremony Miss Redich and I slipped out at the back door and went round to the front as I wanted to take a snap. The light was so poor, however, that I don't think it will come out.

Then Mrs Hunter took us to see the girls' quarters, kitchen, etc., after which I went down to the stream to collect. I'd seen tree ferns growing there and wanted to photograph them and get specimens. Besides several ferns, I found a beautiful tall red *Gladiolus* and one or two other nice specimens. Then I went back and started packing away specimens and changing papers – a work that took up the whole evening. Mrs Hunter kindly let me use the kitchen for drying the papers, and when finished, let me put the press there too. I could only pack away very few – they are so wet and are going mouldy.

Sunday October 11th—In the morning there was a rather long service in the church. They have a form of service – these people are mostly Congregationalists – which is printed in the front of the hymn book. The sermon was preached by one of their boys. He was a slave boy and does not know who his people were. He was brought north by some Portuguese when quite young and left sick at this place and has been there ever since. He, like most of the other chief boys, school teachers etc., wore full European dress, even including stiff white collar, and to my mind looked very absurd. They lose their native dignity entirely and don't obtain that of the white man in its place. However, it is a difficult question. I don't think the present day missionaries are so foolish as formerly as regards the subject of hiding the human form from view, all but head and hands, but they certainly don't discourage the adoption of European clothes!

Service was followed by Sunday school, but we did not stay for that. I went back to the house and enjoyed a good read, 'Mother Mason', a light but entertaining series of sketches of the life of a large family, and then parts of 'Mine own People' (Kipling), not all in the morning. After lunch I had a good read and laze in the sitting room. The house is delightful – so fresh and clean with its white paint and simple but comfortable furniture. I am sleeping outside in a newly built summer house of Mrs Hunter's – much nicer than being indoors. The fame of my 'hole in the ground' at Cwelei preceded us here, brought by Mr Gale. By the way, Dr Sanders tells us that Mr Gale missed both the boats he was aiming for, but a Clan Lines arrived, whose captain was born at the same place as Mr Gale, and he let him sign on as purser, share his bathroom, meals etc., and occupy one of the officers' cabins – a piece of good luck which I'm afraid isn't likely to come our way.

In the evening after supper everyone (except Mrs Sanders who had a touch of 'flu caught from Mrs Meiers' baby) gathered at Mrs Hunter's for a service in English, at which I had to play the hymns – awful tunes I'd never seen before with such catchy, queer time. Dr Hollenbecher took the service and read a sermon. Afterwards we sat chatting for a time and then went to bed.

BY CAR TO THE STATION
Monday October 12th—Up early, to finish our last bits of packing. After breakfast Mrs Hunter showed me her collection of baskets, made by the Dondi girls in most cases, a few made locally. It is beautiful work and I much regret that I have not been able to get some of it to take back with me. Mrs Hunter very kindly gave us each a little mat as a souvenir. She is collecting these to take home with her when she goes on furlough in a few months' time.

About half past seven, Dr Sanders was ready to start and we transferred ourselves bag and baggage to the Sanders' back yard, where a fine Russian (white) mulberry stands in the centre of the grass plot, while on the side away from the house is a garage at present housing two Ford cars, one belonging to one of the other missions. There we packed into the car, Mrs Hunter gave us a box with lunch and a dozen oranges or so. Dr Sanders and Dr Hollenbecher (the latter at the wheel) got on in front, we said goodbye, the Doctor pressed the starter and it refused to start. So once more man-power was applied and with several people pushing joyously behind, we wheeled out of the yard, turned to the right down the slope to the church till we had got up speed, then when well under weigh turned and went back past the house where we finally waved goodbye to our friends and turned off to the right along the Silva Porto *tapalo*.

THE TOWN OF SILVA PORTO
It was very pleasant to do the last few miles of our trek by motor and our experience of Angola certainly would not have been complete had we not done so. As usual of late, the morning was fresh and overcast and I was quite glad of my coat. The road is pretty

good, having a few bumps in critical places, e.g. over a stream where one wants to get up speed to climb the hill beyond. We soon reached the town of Silva Porto, well placed on a hill, free from swampland and with a fine view of the sweep of the valley below and the hills beyond. On the latter is the station of Silva Porto, and here too is a cluster of houses; traders and so on. We went first to the trader at Vila Silva Porto, Figuereido, I think, not Ferreira. There we found everything correct – the porters' food cost $50.00 and they were all ready to go on to the station. We got our money (£13) changed at $120.00 to the £1, of which we got enough in nickel to pay the porters. They prefer metal to paper, particularly in the wet season.

Then we went to the post office where Miss Bleek got her letters, and Dr Sanders did his business – then to the administration where again we sat in the car while the Doctor got our passports visé'd once again. While waiting there we saw our porters cross the end of the street *en route* for Silva Porto. Then back to the trader's and on to the station. We got our tickets and then sat in the car waiting for the porters to arrive. They took a short cut across country so we had not long to wait. The procession was headed by Kaliye, in a pair of blue trousers. Dr Sanders had gone into the trader's store across the road – Dr Hollenbecher was in front of us reading his mail. 'Hallo!' I exclaimed, 'Kaliye's got a pair of blue trousers. Wherever did he get those? He hadn't any money'. 'Well', quoth Miss Bleek, 'they are yours. I suppose you meant him to wear them?' Whereupon a little smile stole across Dr Hollenbecher's broad countenance and I felt exceedingly foolish!

Then, still with Dr Sanders' aid, we got our luggage labelled, weighed, booked, placed ready for the train – some for the van, some for the carriage. We paid off the porters and Kaliye, gave the latter letters for Cwelei, including a present for Muyeye and food money for his journey back to Kunjamba. The porters and Kaliye departed forthwith, and the two Doctors, after having left us in the charge of the trader opposite, Silva I think, also departed. It was very pleasant to have their aid.

THE TRAIN JOURNEY

After a rather tedious hour – there is no such thing as a waiting room, not even a bench to sit on – the train arrived. We got an empty compartment into which we were assisted by the trader and several small boys. It was quite an interesting journey. Till dark we were on high ground and all the way there were proteas – the actual limit, unfortunately, I could not see, as dark fell before we reached it.

At Huambo the train stops 45 minutes so that one can get a meal. All was darkness, and though we tried to 'follow the crowd' as instructed, the crowd went every which way and we got no nearer the dining hall. Finally, wandering down the station, I heard a welcome English voice and asked the owner thereof, a well-dressed Englishwoman, if one could get a meal. She called to her husband whose indignant reply was,

'Of course! That's what the train stops for!'

'Where?'

'At the hotel of course!'

'But she doesn't know where the hotel is!'

Whereupon, he at last grasped the position and explained. We stumbled across rails and over ditches and at last reached the hotel, where we got the usual Portuguese dinner ($30.00 for the two of us) – not bad but of course very hurried.

When we got back to the train we were horrified to find our compartment literally overflowing with men! At Miss Bleek's indignant '*E nosso trem!*' however, they meekly cleared out, after embracing one member who stayed behind with another fat man in white. However, two was not so bad and as they were quite willing to have the light out and go to sleep, and did not object to my keeping one window open, we had quite a good and comfortable night – Miss Bleek up aloft, I below. Only one snored, and he quite softly, so we fared better than we had expected from the accounts we had had. The carriages are built after the Cape fashion but without table and wash basin.

Tuesday October 13th—By morning the scenery had quite changed. We were going down and had left the plateau behind us. Instead, the country was broken up into rounded hills with here and there an abrupt kopje of granite – one was a miniature Table Mountain and Devil's Peak – and, later on, distant ranges of mountains. The proteas had disappeared and instead the conspicuous features of the vegetation were giant baobabs slung with their brown velvet fruits, the great trunks bastioned on all sides. Later still, mingled with them were pachypodiums of all sizes from a couple of feet to about 15' or so and covered with trusses of large white flowers. They looked very like young baobabs and at first I took them for that. A hillside covered with them was quite spectacular. Then there was a trailing *Acacia* with finely divided leaves and long spikes of yellow flowers, several *Convolvulus*-like flowers, and two mallows – one white and creeping, the other erect with handsome mauve flowers with a dark purple centre. Altogether I badly wanted to stop and collect. By and by, we started climbing up again, winding in and out of the rounded kopjes and leaving baobabs and pachypodiums behind. Soon, however, we were again on the downgrade and baobabs and pachypodiums reappeared.

BENGUELLA TO LOBITO BAY

Before we reached Benguella, however, all this interesting flora was a thing of the past and we were on level, barren-looking ground. At Benguella the train stopped half an hour and then went on to Lobito Bay. On arrival at the latter place, we left our booked stuff at the station, got half a dozen boys to carry the rest, and started off, as we fondly hoped, for the hotel. They landed us at a shanty where men were eating but which didn't quite seem to fit the bill. However, the proprietor produced a man who could speak English, a young Afrikaner in the employ of Pauling Bros., who are building the new harbour works. He told us that the hotel closed down on Saturday. The manager, unsatisfactory for a long time, finally did something which gave the railway the necessary opportunity to dismiss him. Whereupon they turned him out bag and baggage and closed the hotel for cleaning – unfortunately for us! This eating house was Pauling's canteen, and though we could get a meal there, we couldn't stay.

A MEAL AT PAULING'S CANTEEN

By this time we badly wanted the meal, so our five boys were instructed to dump the things on the verandah and wait. Then in company with our Afrikaner friend and a blue-eyed old German, a mason by trade I should think, we sat down to dinner. The former had the usual self-assurance of the Cape youth, and a good deal of linguistic accomplishment – English, Portuguese, High Dutch, the Taal, German and some vernacular! He was at one time in the railway employ at the Cape and is by trade, I think, a mechanical engineer. The old German was rather amusing – he complained that everything was dirty, and, suiting the action to the word, added that you looked for the cleanest spot of the tablecloth and then used it to wipe your knives, forks, plates etc. As a matter of fact you only have one knife and fork which you keep through all the courses. They told us a German cargo boat, the *Wolfram*, and a Portuguese boat were expected at the weekend, and our informer believed all the shipping companies had offices in Benguella. There was a train back at 1.30 and another at 4. Having finished and paid for our lunch ($40.00) we returned to the station, found the train (1.30) about to leave and returned to Benguella, leaving the heavier stuff at the station.

RETURN TO BENGUELLA

A slow, tedious and very dirty hour and a half brought us back to Benguella, where we decided to try the Hotel Benguella. Having collected another train of five porters, off we trudged, it seemed for miles, to a café kind of place which called itself Hotel Benguella. It did not look too promising, but they said they had a room with two beds, and very foolishly without seeing it, we said we'd take it and dismissed the porters. A long wait followed and then Miss Bleek went to inspect. She found it was a barn-like place with a large door and window opening on the street and another door into the courtyard. We did not much like it, but the people seemed pleasant and willing and we decided to stop the night at least. Later, when we discovered various other things, or rather their lack, we decided we must move first thing next morning and made enquiries at the Hotel Suisso, carefully inspecting room etc. before engaging it for the following day.

IN SEARCH OF A BOAT

Before this, however, we went first to the agents for the German line who didn't know when the *Wolfram* would be in, and expected no regular passenger boat till the 10th November. Then, shown the way by a local man, we trudged over a mile down to the seafront to the agent for the Portuguese boat. Here we learnt that a boat was expected on Sunday but couldn't learn the fare – that we must enquire for at Lobito! However, some pressure led to the information that the Benguella agent would get the necessary information by telephone.

'When?'

'Oh, the day after next'.

We said that would not do and that we must know in the morning. Finally he promised us the necessary information by 10am next day. Then, having done what we could, we returned to the hotel, with some difficulty got baths – in my rubber bath – after which one of the management came in to fix up the electric light which was on strike. Doesn't it sound civilized! The uncivilized parts cannot be particularized in a polite diary. Finally we got a good meal and retired to bed and a very welcome rest.

Wednesday October 14th—Directly after breakfast I went over to the Hotel Suisso to see if the room were ready. Needless to say, it wasn't. I waited until it was, obtained the manager's promise to send for the luggage (needless to say, he didn't) then, after waiting an hour or so, returned to see what had become of Miss Bleek, and found her impatiently waiting for the luggage to be fetched. Finally we left it, saying that we'd fetch it later and went together to the Suisso.

In search of the British Consul
In the meantime, I'd bethought me of the existence of a British Consul, and that when stranded in a foreign port it is advisable to get in touch with that personage. I also suggested that we ought to get our passports *visé*'d. Miss Bleek didn't think either necessary, but, however, we enquired the whereabouts of the British Consul. The manager said he was at the Cabo Submarino and said he'd call a boy to show me the way. This was while I was waiting for Miss Bleek, so I said I'd wait until she arrived. When we were ready to go, the manager was engaged and we didn't get the boy. However, enquiry of various people – including our travelling companion from Huambo, whom I stopped in the street to ask – took us down the same way as we had gone the previous night. Near the end, however, we went wrong, and were ushered into a small office where a youth seated at a desk seemed quite to expect us, gave us seats etc. – then we enquired if this was the Cabo Submarino and were told no. What it was they did not inform us and we left with further directions which finally landed us at a large wood and iron building, built rather like a ship and standing back in grounds planted with coconut palms, bananas and so on.

Here we marched in, knocked at the door and enquired in our best Portuguese of the tall, rather weary-looking man who came out, if this were the Cabo Submarino – 'Si', 'British Consul?', 'No' – Thank goodness! An Englishman! He proved to be Mr Ward of the Eastern Telegraph Co, invited us into his office and gave us useful information. The Consul is still Duthie – at Lobito! Cables, deferred rate $11.55 per word, must be sent through the Post Office which is just beyond the shipping office. He was just about to have morning tea and invited us to share it – really good tea, such a treat.

Much refreshed, we went on to the shipping office where we learned that the rates were £17/10/ – second class, as against £15 by the German boat. The latter, however, is a very uncertain quantity. Mr Ward tells us she is held up somewhere in the Congo, so I'm afraid we have no choice. We have decided that one of us, Miss Bleek probably, as she is better up in the language and also is doing the finances, must go down to Lobito tomorrow by the morning train, and see to things. Everything here is too vague.

MISS BLEEK'S MISSION TO LOBITO

Thursday October 15th—I saw Miss Bleek off by the 8.30 train. We saw one of the Germans whom we had seen at Pauling's canteen, a man named Kramer, and asked him if he could tell us where Mr Duthie lived. He said he could, and when they got to Lobito showed Miss Bleek the way. He turned out to be painting the hotel.

I spent yesterday afternoon going through my specimens and packing what I could in a petrol box which we bought for $3.00 in a little shop nearby. This I intended to finish this morning but instead the whole morning was taken up in writing letters to Mrs Procter, Mrs Hunter etc. Miss Bleek got back, very tired, soon after 3, having spent an exhausting but profitable morning. Fortunately, she did go to Mr Duthie and found that not only must our passports be *visé*'d both by him and by the government, but without these *visé's* we should not be able to leave! As regards the passage money, he advises us to have it paid at Cape Town. Finally he gave her note to a Mr Mears who might be able to arrange for us to camp in a room at the hotel. The latter she had to leave unsettled as he had to refer it to someone else and by the time her train left (1.30) he had not returned to his office. She had also interviewed the steamship company.

We started off at once for the post office, stopping on the way to ask Mr Ward if a deferred cable would go through tonight. Reassured on this point, we went on to the post office and sent off the following cable:

'Hepburn Bright. Mowbray. Pay Thompson Watson two second class fares per *Moçambique*. They advise Irmaôs Lobito. Urgent. Bleek'

TRAVEL ARRANGEMENTS

That done we went to have our passports *visé*'d. It proved to be the office we had gone to by mistake yesterday, and the youth in charge is staying at the Suisso! He was rather funny – elaborate calculations to find out our respective ages were followed by furtive glances to see if we really looked them! As he glanced at me he caught my eye and quickly looked back at his books. Having done everything we could, including the changing of a £1 for $110.00 at the shipping office, we went back to the Suisso very tired, particularly Miss Bleek. We tried the hotel manager for changing notes and he kindly offered $100.00 for them! Needless to say, we refused it. I think we may be able to get them changed by Sr Benoliel – Miss Redich gave us a letter to be delivered to Mrs Hastings, and a card to Sr Jose Benoliel who would take charge of it.

Friday October 16th—We went directly after breakfast to the Benguela branch of Benoliel Stores and were fortunate to find Mr Jose Benoliel himself. He speaks English – told us the boat was expected Saturday or Sunday and would change our notes for us at $120.00. We decided to change £4 at once, more later if necessary. That has settled matters quite nicely. We shall now just have to wait for the ship to arrive. I thought I should have to go down to Lobito today but we've decided to wait and go down tomorrow. I spent most of the day packing away specimens. Alas! – nearly all have gone mouldy and I've had to throw away the bulk of specimens and paper – all my work at Kamundongo worse than useless.

When we got back from seeing Mr Benoliel, the hotel porter said an English gentleman had been asking for us. We guessed Mr Ward, so walked down to the Eastern Telegraph Co. He had called to say they'd been advised that the *Moçambique* would leave Luanda at 4.30pm today, so she would probably be in tomorrow, Saturday, afternoon or early Sunday morning, and he strongly advised us to get down to Lobito in the afternoon.

Just before dinner we went down again to see if there was any more news of her. He was out, but on our way back we met him and the other English inhabitant of Benguella, who lives with him (he's in the bank) and the latter told us the bank was advised that the *Moçambique* would sail tomorrow afternoon. Now it remains to be seen if our reply cable will arrive in time.

This is a queer people. Yesterday at lunch, the piano suddenly started. I looked round and there perched at the piano halfway down the great dining and billiard hall was a small half-caste boy, bare-legged, with the inevitable bandage round one leg, banging out lesson-book tunes, not badly at all, while the manager entertained a party of friends to lunch. Beyond, in the entrance hall, a small child, about two years old, in a tiny white dress solemnly danced his own little dance round the three-way seat in the centre.

Today at breakfast our neighbour at the next table produced a packet of delicious little pears, which he made the waiter hand first to us, then to the others breakfasting. After calling at the Eastern Telegraph Co. this morning, we walked along the beach. A ship was lying off the pier, taking in or discharging cargo. Along the shore was a line of coconut palms. On the sand were the quaintest little long-legged sand-coloured spider crabs, which ran like a leaf blown by the wind, sometimes down into the holes they had dug, or into the water. When we left the beach and turned back to the town we crossed a bit of wasteland on which was a giant baobab. After lunch I went back to take a snap of this tree, the last of my Vest Pocket Kodak films.

Back to Lobito

Saturday October 17th—Up early, finished packing, then while Miss Bleek did hers, I went to the fruit market and bought four pawpaws, 16 bananas, a dozen huge guavas and a couple of lemons for $6.50. The dining hall was being washed, but we managed to get our breakfast in the entrance hall. Then, with some difficulty, we managed to get the management to produce a handcart for our luggage and departed for the station.

At Lobito we left the things at the station and went first to the post office and did our business there, then to the shipping office. No cable yet, but we left our names etc., which Miss Bleek quite forgot to do on Thursday (didn't seem to think it at all necessary!), and then walked down the length of the peninsula to Mr Duthie's. He and his wife were busy writing their mail, but made us welcome and after signing my passport, invited us in to wait and have lunch with them.

The boat was advertised to arrive at 2pm and sail six hours later or thereabouts. Mr Duthie took us upstairs and left us to rest on the beautiful wide balcony till lunch time. In the sitting room they have three delightful paintings by Hemmens – most restful and pleasing – all Cape scenes. We amused ourselves with papers and magazines for an hour or so and then had lunch. Just as we were finishing, the *Moçambique* came in sight, well up to time. Much refreshed by the rest and lunch, we wended our way back to the office. Still no cable, so having obtained luggage labels from them, we went to the station and labelled all our goods, hoping it was not labour wasted.

To our great relief, when we returned to the office, the cable had arrived and all was well. We got our tickets and then went back to the station, where, after some time, we again collected a train of boys and in two journeys conveyed our goods to the boat. She is said to be a very good boat, but the second class is right aft and we've an inside cabin – so stuffy, but fortunately a good size (four berth) and with a fan which slightly mitigates the stuffiness.

START OF THE SEA VOYAGE

We settled down and changed, and I had a bath, then I went aft and tried to do a sketch. While I was standing there, I saw a small rowing boat approach and in it Mr and Mrs Duthie, who hailed me, saying they were glad to see we'd fixed things up all right. Dinner was at 7.30 and, while we were dining, we left Lobito. It is really a wonderful harbour. There is a small pier parallel with the shore and only a short way out. The *Moçambique* lay on the outer side of this, and by and by in came the *Pedro Gomes* (the boat which was lying off Benguella yesterday), and calmly lay on the inside of it. She wasn't much more than her own breadth from the shore! I was glad to get to bed soon after dinner.

MOSSAMEDES

Sunday October 18th—We got our baths early, and then I went down to breakfast. Afterwards I was glad to 'seek the seclusion which a cabin grants'. I roused from a re-freshing doze with the feeling of absence of motion. We were gliding into position in the open roadstead of Mossamedes, an oasis at the foot of the desert – sand cliffs to the right, greenery to the left, behind the town – sand, sand, sand to a line of sandhills. The colouring was beautiful – shades of purple and mauve on the hills, yellow ochre on the sand in the foreground.

We landed in a motor boat ($10.00 *ira e retour*) and walked straight up the first street, through the town and out into the desert – real desert quite close to the town. I hoped to see *Welwitschia* but we could not get far enough, though we did get to a part where weird euphorbias dotted the sandy waste. We passed a quarry of recent sandstone – shells and pebbles in a sandy matrix, cemented, I suppose, with lime, then there was just sand. To the left was a deep *donga* along which crows were circling, beyond purple hills. After walking out about a mile and a half, we turned to the right and cut across the sand. As we trudged along, I looked up suddenly and there, away on our left, was a lake shimmering in the sun! Above it clouds of sand streamed, or so it appeared – desert, mirage and all complete on our left – on the right, a mile away, the sea! It was really rather wonderful. Besides the euphorbias there were a few scanty dry plants scattered about, and half a dozen oxen wandered about, apparently browsing, though what they could find to eat is beyond me.

Mossamedes does not offer many attractions – if it has any, it closes them on Sunday! After a long search we discovered a small kiosk where we managed to get some so-called lemonade to drink. As we have only one chair, Miss Saltmarsh's, we hoped to buy another here – the ship has none, but nary a chair did we see. However, we asked the keeper of the kiosk (who by the way had a few words of English) and behold! he had a chair! His small assistant was sent off to get it and after about half an hour appeared with quite a decent deck chair. Price? 'The price is one hundred escudos!' Pretty stiff, but beggars mustn't be choosers and we took it. We got to the pier just as the boat was about to start for the ship, and got back on board. We left between 7 and 8, and again bed was very welcome, to me at any rate.

Monday October 19th—Have felt wretched all day – I've only actually been sick once (yesterday morning), but feel I must keep flat as much as possible. However, meals are not so bad – the saloon is amidships and the boat is really very steady. I forgot – yesterday after getting aboard, I tried to do a sketch and perched myself on a bollard so that I could see over the bulwarks. In a few moments I had half the second class passengers and several stewards congregated round me in breathless interest – most trying to the nerves of a very amateur draughtsman. Poor folk, they must be hard up for entertainment! This afternoon there was an hour's music in the smoke room – an orchestra of piano, fiddle, 'cello and basso. It was quite nice to have them.

Tuesday October 20th—Still far from happy, but in the afternoon the weather, which has been dull and cloudy, changed and the sun came out bright and strong. It made us all feel much better. It got cold towards evening and we sat in the saloon writing. In the afternoon, again, there was music in the smoke room. We have put our chairs down in the waist of the ship on the hatches of the forward hold. Miss Bleek discovered it. It is warmer and there's less vibration than on the upper deck.

While I was sitting there in the afternoon, a long-haired young man was taking snaps of the passengers. By and by, I heard a voice addressing me and there was a nice-looking young woman with a baby in her arms. 'Do you speak Portuguese?' she asked, and on my replying '*Poco, poco*' she tried in English. Would I be photographed? – 'My married photos all the passengers'. I thanked her, but declined. He has a Kodak and tank and is, I suppose, trying to make an honest penny on the way.

There are two cats, a big black tom and a kitten, which are rather amusing. At Mossamedes a party, two men, a woman and a small black dog, came aboard. Miss Bleek has talked to them a good deal. They are German, were planting in Spanish Guinea for three years but found it too unhealthy, have been staying near Mossamedes for three months and are now on their way to Beira.

HEADING FOR CAPE TOWN

Wednesday October 21st—We may get in to Cape Town late tomorrow or early Friday, probably the latter. Today we saw land, quite high land too, in the neighbourhood of Luderitz. It has been bright and sunny all day but cold, and now (5pm) the seas are getting a little less kind. I'm still feeling far from happy and shall be very glad when we get to shore. I'm anxious for news – hoped to get a line from Jeannie at Lobito, but was disappointed. Perhaps I told her to address to Benguella. Like a fool, I forgot to ask at the post office there!

Now the only difficulty we may have to meet is the customs at Cape Town. I hope they don't object to my specimens! The orchestra is just going to play again – I suppose they give the second class an hour daily.

Thursday October 22nd—[No further diary entries.]

POSTCRIPT TO DIARY WRITTEN BY MAP:

Notes added June 1961 after working three mornings in Salisbury Herbarium.

Specimens etc. allowed in without question.

Remainder of 1925 and 1926, worked at specimens at Bolus Herbarium (BOL) [University of Cape Town]

Arranged sets: first, most complete, to BOL. Mounted very beautifully by Miss Arbuthnot.

Second set bought by Dr Marloth for PRE. Nat. Mus. Pret. [National Herbarium, Pretoria]

Third set bought by Dr Marloth

Fourth set (woody plants only) Dr Marloth

Work on material continued at Kew and BM [British Museum] in 1927 and thereafter at BOL again. BOL set eventually parcelled up ready to take to herbarium before leaving for overseas in 1936, stolen from car.

Hence Pretoria set is the most complete surviving one. Odd specimens finally sent to STE [Stellenbosch University Herbarium]. A few, plus drawings (coloured) given to Rhodes University Herbarium. In Pretoria, Miss Verdoorn drew my attention to one of my specimens (Gentianaceae), on which she was working for the new *Flora of Southern Africa*. (May 1961). A few records in the new *Flora Zambesiaca*, others in *Flora Angolensis*.

ENDNOTES

Introduction

1. Paynter, S. 1969. 'Digs' Algae at 80. *Seattle Post Intelligencer.* August 24.
2. Bleek, D.F. & Lloyd, L.C. 1923. *The Mantis and His Friends: Bushman Folklore.* T. Maskew Miller, Cape Town.
3. Bleek, D.F. 1956. *A Bushman Dictionary.* American Oriental Society, New Haven, Connecticut.
4. Bleek, D.F. 1928. Bushmen of Central Angola. *Bantu Studies* 3,2: 105–125.
5. Ibid.
6. Jacot Guillarmod, A. 1978. Obituary – Mary Agard Pocock (1886–1977). *Phycologia* 17,4: 440–445.
7. Figueiredo, E., Soares, M., Seibert, G., Smith, G.F. & Faden, R.B. 2009. The botany of the Cunene-Zambezi Expedition with notes on Hugo Baum (1867–1950). *Bothalia* 39,2: 185–211.
8. Huntley, B.J. & Matos, E.M. 1994. Botanical Diversity and its conservation in Angola. In: B.J. Huntley (ed.) Botanical Diversity in southern Africa. *Strelitzia* 1: 53–74.
9. Balarin, M.G., Brink, E. & Glen, H.F. 1999. Itinerary and specimen list of M.A. Pocock's botanical collecting expedition in Zambia and Angola in 1925. *Bothalia* 29,1: 169–201.

Volume One

1. Main, M. 1998. *Zambezi. Journey of a River.* Southern Book Publishers, Halfway House.

Volume Two

1. Figueiredo, E. & Smith G.F. 2008. *Plants of Angola. Plantas de Angola. Strelitzia* 22. National Botanical Institute, Pretoria.
2. Watt, J.M. & Breyer-Brandwijk, M.G. 1962. *The Medicinal and Poisonous Plants of Southern and Eastern Africa.* E. & S. Livingstone, Edinburgh & London.
3. Bossard, E. 1996. *La Medecine Traditionnelle Au Centre et a L'Ouest De L'Angola.* Instituto de Investigação Científica Tropical. Ministerio da Ciencia e da Technologia. Lisboa.
4. Gilges, W. 1964. Some African Poison Plants and Medicines of Northern Rhodesia. *Occasional papers of the Rhodes-Livingstone Museum* 1–16.
5. Falola, T. & Jennings, C. (eds.) 2003. *Sources and Methods in African History: Spoken, Written, Unearthed.* University of Rochester Press, New York.
6. Figueiredo, E., Soares, M., Seibert, G., Smith, G.F. & Faden, R.B. 2009. The botany of the Cunene-Zambezi Expedition with notes on Hugo Baum (1867–1950). *Bothalia* 39,2: 185–211.
7. Ibid.
8. Gilges 1964 *op. cit.*

Volume Three

1. Bleek, D.F. 1928. Bushmen of Central Angola. *Bantu Studies* 3,2: 105–125.
2. Ibid.
3. Ibid.
4. Cunningham, A.B. & Terry, M.E. 2006. *African Basketry. Grassroots Art from Southern Africa.* Fernwood Press, Simon's Town.

Volume Four

1. Bleek, D.F. 1928. Bushmen of Central Angola. *Bantu Studies* 3,2: 105–125.
2. Ibid.
3. Ibid.
4. Smith, A., Malherbe, C., Guenther, M. & Berens, P. 2000. *The Bushmen of Southern Africa. A foraging society in transition.* David Phillip, Cape Town.

5. Sparrman, A.1785. *A Voyage to the Cape of Good Hope, towards the Antarctic Polar Circle, and Round the World but chiefly into the country of the Hottentots and Caffres, from the year 1772 to 1776.* G.G.J. & J. Robinson, Paternoster Row, London.
6. Bleek 1928 *op. cit.*
7. Ibid.

Further reading

Anon. 1977. Botany expert dies. *Evening Post.* July 12.

Anon. 1977. Tribute to world famous scientist. *Grocott's Mail.* July 19.

Jacot Guillarmod, A. 1977. Dr Mary Agard Pocock. *Forum Botanicum* 15.1: 1–2.

Jacot Guillarmod, A. 1987. That amazing woman – Mary Agard Pocock, *The Elephant's Child* 10,3: 14.

Skotnes, P. 2007. *Claim to the Country. The Archive of Wilhelm Bleek and Lucy Lloyd.* Jacana Media, Auckland Park.

Wynne, M.J. 1994. Phycological trail-blazer; Number 4: Mary Agard Pocock. *Phycological Newsletter* 30,3: 4–5.

APPENDIX ONE

Newspaper Articles

A Trip into the Interior.

A VISIT TO THE VICTORIA FALLS

Seen at the Apex of their Glory.

The Courant, 5th June 1925 (Part I)

A Trip into the Interior.

TRAVELLING UP-TO-DATE IN
N. RHODESIA.

Railway Speeds 7 Miles in 40 Minutes.

The Courant, 8th June 1925 (Part II)

A Trip into the Interior.

DRIVING OVER A BLACK MAMBA.

Motoring Through Mud and Sticking.

In the Wastes of Barotseland.

The Courant, 10th June 1925 (Part III)

A Trip into the Interior.

REGULAR BOAT SERVICE ON
ZAMBESI.

Some of the Camping Spots Described.

Good Work and Hospitality of Missions

The Courant, 12th June 1925 (Part IV)

A Trip into the Interior.

A TREK BY MACHILA.

Travelling Up-to-date.

The Courant, 17th July 1925 (Part V)

A Trip into the Interior.

AMONG THE "PEOPLE WHO ARE
CHASED."

In Camp on the Kutri River, Central Angola.

The Courant, 17th August 1925 (Part VI)

A Trip into the Interior.

IN THE BEAUTIFUL VALLEY OF THE
QUITO.

How the Porters were Induced to Accompany
the Trekkers.

The Courant, 26th October 1925 (Part VII)

Newspaper Articles (continued)

OVERLAND TO LOBITO BAY.

In the Footsteps of a Vanishing People.

THE VENTURESOME JOURNEY OF TWO LADIES.

Cape Times, 5th December 1925 (Part I)

OVERLAND TO LOBITO BAY.

The Venturesome Journey of Two Ladies.

BY M.A.P.

In three articles, of which this is the second, there is described the venturesome journey of two ladies across Angola. The first article took the record of the tour to the Anglo-Portuguese border.

II.

Cape Times, 19th December 1925 (Part II)

OVERLAND TO LOBITO BAY.

The Venturesome Journey of Two Ladies.

END OF A LONG TREK.

BY M.A.P.

III.

Cape Times, 1st January 1926 (Part III)

May 27. Kutsi camp

The Courant, 5ᵗʰ June 1925 (Part I)

A Trip into the Interior.

A VISIT TO THE VICTORIA FALLS.

Seen at the Apex of their Glory.

[Specially written for "The Courant"]

Miss M. Pocock who was on the staff of the Girls' High School last year as Science mistress has written specially for The Courant, a series of articles of a journey into the interior. She is accompanying Miss Bleek into the little known country of Angola where the latter is to study the Bushman language and folklore. Miss Pocock will make a study and collect specimens of botany in the country. We publish the first article which describes the Victoria Falls as they have not been seen for many years.

April 6ᵗʰ at 4.15 p.m. and farewell to Cape Town for six months or so!

The four days in the train, varied by stops at stations where we alight to stretch our legs; further north these grow more interesting, for instance at Artesia the train stops to take in water not only for the engines but all along the train, and there are traders selling karosses, beauties too, of wild cat of various kinds, hyaenas, lions, besides the usual native vendors of curios – carved and burnt wood, lions, tigers, camels, elephants, and an aeroplane!

Half a day is spent at Bulawayo, and then the train passes through parkland, which at this time of the year, after the recent heavy rains, is fresh and green as an English country side.

Next morning by 7 o'clock we were up and watching eagerly for the first sign of the

GREAT "SMOKING WATERS"

The train is not very late – it is an excursion from Bulawayo and very heavy (18 coaches) so that one hour behind time is not bad.

Finally at about 8 o'clock, the banks of cloud which have been hanging low on the northern horizon became clearer, and we saw that, while some follow the line of the Zambesi gorge which now becomes visible, the main mass hangs athwart the line of the gorge and is rising out of the gorge itself.

We arrived at the Victoria Falls Hotel in time for breakfast, then donning our oldest clothes, waterproofs and such like, we set off for the rain forest. Those who have seen the falls only in June can have no conception of the stupendous volume of water which rushes into the gorge in the rainy season, and this year was exceptional – the rainfall a record and as a consequence the Zambesi higher than it has been known to be at any rate for many, many years.

On our first visit to the rain forest, being morning, the mists were at their very thickest, we were

SOAKED THROUGH AND THROUGH

several times over, and only now and again could one even catch a glimpse of the tumbling mass of waters for the clouds of mist and spray tossed up out of the gorge. Tossed up, to descend on our heads as torrential rain – it was hard to realise that a couple of hundred yards away the sun was shining and it was a fine, hot day! Nevertheless it was worth it all – when we did see that indescribable mass of roaring waters it was more than compensation for a dozen wettings. Besides, we enjoyed our shower bath and felt much refreshed thereby.

The next two days we visited the rain forest again, but in the afternoon, and each time enjoyed it more. The second day (Easter Sunday by the way) we started out about 3 p.m. and suitably clad in bathing suits and mackintosh, went first to the Devil's Cataract, which as it leapt forward into the gorge was

SPANNED BY A MIGHTY BOW.

Reluctantly tearing ourselves from this first glorious sight we passed into the rain forest and proceeded to paddle (literally) out to each successive view point, getting magnificent views, alternating with tropical down pours as we did so. Then came a part where we could walk (or wade) along the grassy edge from one view point to another, and as we did so each member of the party had her pet rainbow with her – starting at her feet on the left and ending on her right. Yet there was never more than one rainbow! Each sees but one at a time. This day the beauty of the whole was beyond description and it culminated at Danger Point. Here the mist was very thick and the rain particularly heavy and after waiting some time we turned to go, giving up hope of another view, when suddenly the

MIST IN FRONT CLEARED

and we looked across at the main central fall showing clear from summit to seething cauldron below, and the whole spanned by the arc of a rainbow – the mist descended and the picture faded, only to reappear once more with even greater clearness, the sun shining full on the tumbling waters while all round us was mist, and the colour of the bow scintillated, waxed and waned as the mist shifted to and fro. A memory to be a joy for the rest of one's life.

Rainforest for a daily joy. What in the evening? Well, the first evening there was a thunderstorm, so instead of the Falls, Bridge! What bathos! However, the next two nights were clear and moonlight – the moon, at the full, rose about 8 p.m. and we walked down to the bridge getting a gentle shower as we passed the end of the rainforest. Now, strange though it be, the bridge does not detract from the beauty of the Falls, even by moonlight, rather does it add to it – giving an enhanced beauty of well proportioned line and curve. As we crossed it, beneath our feet, an immense distance down, the water boiled and foamed in the cauldron, and there, on the swirling mist below the bridge was

OUR FIRST LUNAR RAINBOW

– the faint ghost of a rainbow, an earnest of what we were to see later.

The night was so bright and lovely that it lured us on, on, till we had reached the far end of the eastern Falls, and there in ethereal beauty, hung the lunar rainbow, one foot resting on the mist low in the gorge, the centre curving up and back over the tip of the falls where it was lost to sight as the mist ceased.

At first the bow seemed white, then as one gazed, suddenly the hues of the spectrum colours appeared, trembled and vanished and one doubted if they were actually there or existed only in imagination.

It was difficult to tear oneself away, and all the way back to the bridge, the bow appeared and disappeared, now the foot rested on the boiling waters of the gorge and the summit curved against the bridge – now finally as we entered the bridge, a perfect bow spanned it from end to end, resting on the spray on the far side from the Falls.

Yes, the Falls in April are even grander and more beautiful than in June, but infinitely wetter! (To be continued.)

_____

The Courant, 8th June 1925 (Part II)

A Trip into the Interior.

TRAVELLING UP-TO-DATE IN N. RHODESIA.

Railway Speeds 7 Miles in 40 Minutes.

[Specially written for "The Courant"]

Miss M. Pocock is writing a special series of articles for The Courant on a trip she is undertaking with Miss Bleek, through Northern Rhodesia into Angola, Portuguese West Africa. The second article of the series is published below, the first article having appeared in last issue.

From Victoria Falls to Livingstone – 7 miles: Forty minutes by train. Impossible! But true, we had to get to Livingstone on Monday and there is no passenger train on that day, but there is a goods train. What time? Oh, about 4 pm.

At 3.30 up to the station we went – train late, not due till 5. So we went back to the hotel and spent a pleasant hour on the stoep watching the colours of the spectrum spread on the mist below the Zambesi bridge and then travel slowly up the gorge on to the spray above the falls as the sun travelled westward. Five o'clock approached, so back to the station we went. Train not due till 5.15. There followed a long wait, while passengers gradually collected – most of them from Livingstone spending Easter Monday at the Falls. As we waited we caught scraps of conversation. "No carriage." "Have to travel in the guard's van." "Impossible! – all that luggage!"

Finally about 5.30 we heard a distant noise and after a few minutes,

GRUNTING, WHEEZING, CREAKING

and groaning, the goods train actually arrived and pulled up in the station.

Only too true, there was no coach, and all the 20 odd passengers, plus all their luggage, made a rush for the guard's van which looked absurdly inadequate. However, looks are sometimes deceptive, and everyone and everything finally got in and settled somehow. Then the creaking and bumping and jerking started again, and shaken like peas in a pod we gradually got under weigh and slowly progressed to the bridge where our 10-12 miles an hour dropped to under 5, and we slowly crept across, getting a last fleeting glimpse of the Falls as we did so. Then after speeding up a bit we slowed down again and finally stopped at Palm Grove siding, where another dozen picnickers waited for the train – when they saw the existing state of the van there were

SHOUTS OF LAUGHTER.

However, where there's a will there's a way, and everyone managed to squeeze in, though where the guard went is a question.

Rumble, clank and jolt, on we went and on. The sun set in a golden glory, lighting the spray above the falls like a blazing fire. The glow faded, darkness began to descend, the sky going a cold blue-grey.

Finally, as night fell, we clanked and jolted to a standstill at Livingstone.

Seven miles in 40 minutes, between 30 and 40 people in the guard's van and both doors off their hinges, at the bottom at any rate.

Travelling in April 1925!

LIVINGSTONE TO KATOMBORA.

"Livingstone to Katombora!" "But the road is impassable. Besides, it has been shut to the public while it is repaired for the Prince's visit." Cheering news when we had to get to Katombora – 36 miles from Livingstone – to start our trip up the Zambesi.

At the hotel we were met by a weary looking young man, who informed us that he was driving us out to Katombora on the next day but one, that he had just brought in a party on their way from Mongu and that leaving Katombora on Thursday, they had reached Livingstone on Saturday! That however, since there had been three hot days since, he hoped by making an early start, sending half the luggage by carriers, and taking plenty of boys to push the car out of mud holes, to get us through in a day.

We hoped so too.

Next day was spent in getting stores, – a cooking pot, paraffin for our lamp, salt to trade with and so on – and generally preparing.

We tried to start at 6 a.m. next day, but reckoned without our host – no coffee or refreshments of any kind were forthcoming till long after 6. However, we did get away at 7.

LAST GLIMPSE OF FALLS.

The first 12 miles were good going comparatively. As we left Livingstone we turned southward, and for the last time saw the smoke of the Falls. Then turning sharp to the right we went westward through open parkland, in which we occasionally saw the red and yellow spike of a very fine gladiolus.

(To be continued.)

The Courant, 10th June 1925 (Part III).

A Trip into the Interior.

DRIVING OVER A BLACK MAMBA.

Motoring Through Mud and Sticking.

In the Wastes of Barotseland.

[Specially written for "The Courant"]

The following is the third article of a series written by Miss M. Pocock specially for The Courant on a trip through Northern Rhodesia on the way to Portuguese West Africa. Below she describes a motor journey of only 36 miles, but where getting stuck again and again in the mud was all in the day's march. Next they take to a boat on the Zambesi.

Our chariot was a Ford lorry on which half our goods were piled; two of us sat beside the driver, the third was perched aloft on top of our belongings.

The first part of the road had been well repaired. It seems that the Paramount Chief of Barotseland is to have an audience of the Prince of Wales at Katombora, and so in preparation of the Prince's visit, the road is being repaired.

The Chief is to come down the Zambesi from Lialui to Katombora and there meet the Prince. Our driver had just told us that we were getting to the bad part of the road, when we saw our boys ahead. On we rattled across an apparently well-metalled piece of road, covered with stones, evidently newly laid when splosh! The lorry's left hind wheel had sunk through the "road metal"

INTO A QUAGMIRY BOG

on the surface of which the stones were laid. Fortunately the boys, twenty odd, were at hand, off came all the lorry's load including the passengers, all the boys pushed behind, Mr Ford's much-abused engine chugged in front, and with a squelch and a splutter up and out came the lorry and on to a sound piece of road a hundred yards or so further on.

Then on went the load and the passengers and we proceeded on our way. This, however, was merely a foretaste – twenty times and more we were bogged; sometimes we got out easily, the six boys we carried at the back jumping off and pushing; other times it was again

A CASE OF UNLOADING.

Then the road disappeared under water altogether and we had to take to the bush, making our way in and out of the trees, over grass, bushes and even a huge black mamba who chose to cross our track just in front of us – unfortunately we did not kill it, although the front wheels passed right over it, but it wriggled away into the bush. By noon, we were seven miles from Katombora and the worst was over, so our driver stopped at his farm where he is growing cotton and entertained us to lunch. We were warmly welcomed, almost devoured by his family of three dogs – two a bull-terrier x wire-haired cross, the third bull-terrier x great dane, both breeds are

GOOD FOR LION HUNTING.

They do not like natives and were wild at seeing their master after five days' absence. Then after the welcome rest our trek began again; again we stuck, were extricated, stuck again, then had to betake ourselves to the bush once more, this time with a boy running ahead to show the way. Suddenly, in an innocent patch of green, the car stuck again, both hind wheels in up to the axle, the front nearly so. Here after much effort we finally had to unload again, only to find that the way was blocked by a shallow and narrow ditch. So after being finally extricated the car had to make a detour of several hundred yards and return on a parallel course to circumnavigate the ditch and finally load up again, only

A DOZEN YARDS FROM THE SPOT

where she had stuck.

Well, to cut a long story short, about tea time we arrived at the trader's house, about a mile and a half from Katombora proper. There his wife gave us a welcome cup of tea, and after a short rest we started on the last lap of our journey – here again we had to leave the road which was under water, and take to the bush. However the worst was over and a little rough going brought us to a small collection of shanties – Katombora's store etc., and near these in an open glade we made our first camp.

Fortunately a good "boy" on leave from service was there waiting to go up the river. His services were promptly secured, and so almost as the tent was up we had hot water ready for baths – and very welcome was a warm bath (in a most compact little canvas bath) after our rough day's travelling.

Our next experience is to be

BOAT TRAVELLING ON THE ZAMBESI

What will that be like? Time will show.

[The intervening period spent at Katombora and the actual procedure necessary to hire a boat, etc., is not described and the writer simply continues below by stating that they were in the wastes of Barotseland in a barge.]

Well, here we are, somewhere in the wastes of Barotseland, sitting in a barge with a tarpaulin to keep us dry while round us falls the rain and as far as the eye can reach is a waste of waters – that is, nearby it looks like a waste of waters (as it is) but in the near distance it is apparently a beautiful green meadow, varied here and there by an occasional bush or tree. The water is navigable for such crafts as this and indeed, is much preferred by the boatmen to the actual Zambesi, up which we are really travelling. Owing to the phenomenal rainfall of the past few months, however, the flat country is flooded for miles on either side of the river. As I wrote that last, the rain ceased, our

PADDLERS CRAWLED OUT

from under the tarpaulin and with cheerful cry and song (the latter the reverse of euphonious) poled the boat through a few yards of grassland and behold! the river, or at any rate one branch of it, up which we are now paddling, keeping close to the side to avoid the swift current of the centre of the stream.

The middle of the Zambesi on the 5th of May! What a place to choose! No doubt people would exclaim.

THE OBJECT OF THE MISSION

Well, the objective of this little trip is a small mission station called Muié, in the centre of Angola, and its object to study the language of certain families of Bushmen forming the remnant of what was once a large tribe of those interesting and fast-disappearing people. A subordinate purpose is the collection of botanical specimens in the little-known corner of Angola to which we are going.

(To be continued.)

———————

The Courant, 12ᵗʰ June 1925 (Part IV).

A Trip into the Interior.

REGULAR BOAT SERVICE ON ZAMBESI.

Some of the Camping Spots Described.

The Good Work and Hospitality of Missions.

[Specially written for "The Courant"]

The fourth and last article to hand from the pen of Miss M. Pocock of her journey into the interior is published below. It describes their long boat journey of over 200 miles and the incidents on the long voyage which occupied 18 days. When it is recalled that the two ladies are travelling in the heart of the wilds without any European male escort, the courage necessary must be appreciated by the reader.

YOUNG UMFAAN CARRIED OFF BY LION

To reach Muié there are two possible routes, the one by sea to Mossamede and thence overland with native carriers, the other by train to Livingstone thence by barge up the Zambesi to Mongu, and thence westward with carriers either on foot or by machila (i.e. canvas hammocks carried by native porters). The first route was ruled out by the fact that the coast natives will not go into the bush, so though it is possible to come out by that route, going in is next to impossible.

The missionaries, at whose invitation the trip was planned, always make use of the Zambesi route so that it was comparatively easy to obtain information and to make arrangements with various traders for a boat, boys, etc.

I wonder how many people in Cape Province know of the existence of the Zambesi Trading Company? Yet there is such a company, carrying on a regular service of boats up and down this mighty river, between Katombora, 36 miles to the west of Livingstone, and Mongu the trading centre of Barotseland, the native capital being Lialui a little further north.

BOAT AS HOME.

One of the Zambesi Trading Company's boats, or rather barges, has been our home for the past 18 days, and in that time we have travelled in it upwards of 200 miles as the crow flies – very much more as the boat goes – in the course of which we have passed up some dozen or more rapids, up some of which the barge had to be towed, in most cases loaded, though twice we and all our goods had to go round on shore while the empty barge was towed precariously through the "hard" water, in and out of the rocks, and also past one very high fall, the Gonye Falls near Seoma, which necessitated a long portage of boat and all. Part of the time we have been in the river itself but for more of it in the inundated land, chiefly to the west and we have not seen one crocodile, and only

A SINGLE HIPPO.

The latter however, afforded us all including the paddlers considerable amusement as he came up with a great spluttering (on the far side of the river be it said!) and watched us with much surprise, waving his ears at us as he did so.

Well to return to our boat. It is a long flat-bottomed affair, rather of the nature of a barge, about 30 feet long and 5 feet wide. In front are six paddlers, at the back seven more, while in the centre is piled our luggage and over it and us is an awning made of grass mats supported on roughly cut branches curved to form an arch. In front of our luggage we sit at our ease, our seats being the cushions etc., which compose our bedding. At the back of the luggage is the induna, (or chief) who is in charge of the boat, and our personal boy, who does the cooking and rejoices in the name of Kandu – not inappropriate as it happens! He is the only one of the whole lot who speaks or understands English, except the induna who is confined to about a dozen words.

WHERE THE PARTY CAMPS

At night we camp, either on the river bank, or if we are in vlei land, on the nearest "island". As a result our camps are varied – usually those on the banks are wooded, with grassy glades between widely spaced trees. Usually their chief drawback is the grass which is most horribly burry as a rule – in fact most of the plants here seem to possess burrs or prickles of some kind.

The "island" camps are sometimes very nice, but sometimes the reverse. Our first experience was the worst – a tiny patch of dry land, some 30 by 10 feet, scarcely raised above the water, bearing 2 trees, several bushes (very full of burrs) and at the back a muddy pool with fresh hippo spoors! That was easily the worst. The best we had two nights ago – fairly extensive and although the highest part was only 5 feet or so above water, very hard and perfectly dry but we

HAD TO BE CARRIED

to it from the boat as there were several yards of water between the nearest the boat could get and the dry land. The carrying

process we never enjoy – by the way at the start, the river was so high that we could not reach the landing stage to embark until we had been carried across several yards. Other camps of special interest – each one is different and nearly all have their respective points – are those at Seseke, below the M'Tomba rapids, Sinanga and Nololo. The first and last of these are mission stations run by the French Mission.

WHERE LIVINGSTONE SAT

At Seseke, after a wretched night spent in the boat (it was our first Island Camp!) we arrived at midday, and landed just next to the neat brick church near to which is the giant fig tree under which Livingstone sat to address the people. We wandered past one attractive house, (with Pride of Barbados in full blossom in front of it) which however was shut up, on to another set in a grove of orange trees, and there we were made welcome by two ladies, Mlles Dogimont and Guiglier, who have been working at Seseke for the last 5 years. After dinner and a rest they showed us round the mission station, a most interesting place, where girls and boys are taught handicrafts in addition to other forms of instruction.

The two ladies run the girls' school, have several boarders who come in from outlying parts (one little native girl had just lost a small brother: he was out herding cattle and

A LION SEIZED HIM

and carried him off) and also have a dispensary and several hospital huts, where natives come to them to be treated for various ills.

The missionary in charge and his wife were away so we did not have the pleasure of meeting them. But the hospitality and warm welcome which we had in Mlle Dogimont's household have made Seseke a very pleasant, friendly memory.

The next two camps were particularly beautiful ones on the river bank, where our tents were pitched towards evening, and fires lighted round them to keep away possible prowling beasts.

My bed is placed outside the tent and from it I watch the pageant of the stars overhead – the Great Bear hanging in the north; to the south our Cross with the two Pointers blazing beneath it, Orion sinking to the west, and the Scorpion trailing his sinuous length up the eastern sky.

The sun sets and rises in a golden glory, while in the morning the western sky is a delight of delicate colours.

We have been fortunate in the weather – in spite of the lateness of the rains, this is only the second time we have had rain. Now we are getting to the end of the river part of our journey.

Tomorrow or the next day we start trekking from Kama (a few miles from Mongu, and almost 15 degrees south of the equator) westward to Ninda and thence Muié, our headquarters in Angola.

_____

The Courant, 17th July 1925 (Part V).

A Trip into the Interior.

A TREK BY MACHILA.

Travelling Up-to-date.

[Specially written for "The Courant"]

Miss M Pocock, whose articles of a trip into the interior and a 200 mile voyage up the Zambesi River by boat, have been read with interest, below gives her experience of travelling by Machila. Miss Pocock, who has accompanied Miss Bleek, is making a study and collecting the flora of Angola, a part of the Sub-Continent which has not been traversed by the botanist.

"You must bring machilas with you." So we were instructed four days before we started. A machila, be it said, is a kind of canvas hammock or some other contrivance slung upon a pole which is carried on the shoulders of two or more men.

A conversation over the telephone resulted in the promise of two machilas to be delivered at the station on the day of our departure, a promise which was duly fulfilled, the machilas, packed with the tent in a canvas bag were ready, and with the tent they remained throughout the river trip.

SEVENTEEN DAYS ON RIVER

At last, after seventeen days, the river part of the journey was over and at Kama the day arrived when the machilas were to be tried out. A train of carriers, a couple of dozen or so, armed with two ten-foot poles of curious shape, descended upon us, and certain of them proceeded to rig up the machilas while the headman or induna, allotted the loads to the remainder. This was not so easy as all the camp outfit and personal goods had to be sorted into groups of about 50 lb each, that being the usual load per man in Northern Rhodesia. However, at last it was done and the machilas declared ready. Then came a shock – only the hammock part! – a good strong canvas affair, had been provided by our firm of sail makers, and there was nothing to keep off the rays of the sun, no joke, fifteen degrees south of the equator, even in winter. For the time that difficulty was got over by draping a blanket across the pole.

CRUCIAL MOMENT ARRIVES

At last the crucial moment arrived and the entry to the machilas was essayed. "Get into a hammock! Nothing in that!" No, but when the hammock is suspended not many inches below a long pole, supported upon the shoulders of two sturdy Barotses, it is not so easy as it sounds. Besides, what goes in first, does one sit, or put one's foot in first? Or how does one enter a machila? Somehow, anyhow, we were in and off we went!

Jog! Jerk! What an experience – full length, heels up, head, if anything, down, – and if it came up, bump against the pole – a stiflingly hot blanket falling on both sides, – no, the first experience of machila travelling is not pleasant. At first what it felt most like was riding a horse, lying full length, head to his tail, feet on his head, while he proceeds at that most uncomfortable of gaits, a jog, which shakes all one's bones loose.

LIKE LEARNING TO RIDE.

By and bye however, one grew a little more accustomed to the curious motion and ventured to move a bit, the blanket was pushed aside, the position altered a bit to lower the heels slightly and raise the head end – one's hat went anyhow, and it was possible to begin to enjoy the scenery.

At first the path wound in and out through widely spaced trees, through which the light flickered; herds of cattle wandered about with their black attendants, grazing on the sparse growth beneath the trees. Then suddenly the scene changed, the wood ended and was replaced by a wide grassy plain over which the path passed.

TRAVELLING IN THE WATER.

Soon the dry sandy soil of the wood became black and peaty, then our bearers were wading; anon the water was up to their ankles, then halfway to their knees and the centre of the machila began to feel decidedly cold and one wondered how soon the whole contraption would go plump! Splash! Into the water. Suddenly the bearers stopped, a swing, a jerk, and the pole no longer rested on their shoulders but on the crown of their heads, and on we went, the water still getting deeper.

Splash, splash, feeling their way, it was slow work; but at last the worst was over, the water grew shallower, beyond the grassy plain (all this swamp, or vlei land although covered with from one to five or six feet of water, is a thick growth of various kinds of grass) appeared another belt of trees on higher land, and strung out in a line, just above the limit of the plain, we could see

A NUMBER OF VILLAGES

– these are Lukona and to one of them we were bound. As the waters ended the grass disappeared and the path was bare and sandy, winding in and out of the villages, round this man's courtyard, past that man's front door, round somebody else's mealie patch and then avoiding a clump of paw paws, – bush above on the left, grassy plain away to the right.

Then came a hedge of a compact Euphorbia – mark of a white man's work; a few steps more and there was the store on one side of the compound, and there awaited us a warm welcome and the end of our first experience in machila travelling.

AN INTERESTING POINT

An interesting point about our machilas is their poles, these are fetched from a place miles away on the far side of the Zambesi; they are the central stalk of the leaf of a certain species of Palm from which the lateral fronds are stripped, and cost the large sum of half-a-crown each. They are almost three inches in diameter, flattened along one side and are very tough and for their size remarkably light.

It is hardly credible, but one gets quite to enjoy machila travelling. It has to be learnt like riding. After the hammocks had been properly adjusted and canopies rigged up to give some shade, they were quite comfortable, and one got to know the various carriers, to machilas of this type, (there are of course others) there are six boys who take turns, usually about half an hour at a time, though a specially good boy did a spell of one and a half hours early in the morning and took turns again later in the day.

TRAVEL AT GOOD SPEED

They go at a good pace and often sing or call out words or phrases to one another – for instance our second try was an hour's journey to a neighbouring mission station and as we returned a magnificent thunderstorm burst upon us. The boys hurried as much as they could, shouting out, and chanting a weird refrain, which being translated ran something like this. "The rain comes, hulu! Make haste, make haste, wah, wah! We must not let our people get wet. Hurry! Hurry! Hoorooh!"

They did hurry, and we did not get wet, or hardly at all, and it was really quite thrilling, both to see, hear, and feel.

Yes, even machila travelling has its points.

———

The Courant, 17th August 1925 (Part VI).

A Trip into the Interior.

AMONG THE "PEOPLE WHO ARE CHASED".

In camp on the Kutsi River, Central Angola.

[Specially written for "The Courant"]

[In a further letter from Angola Miss M. Pocock, whose previous letters have been read with interest, gives some few details of the Bushmen, which her fellow companion, Miss Bleek, is studying, some pen pictures of the surroundings in which they are camped, and the description of some of the flowers. Of particular interest not only to botanists but also to the general reader are the descriptions of the insect catching plants, which catch and eat their catches! The sundews, she mentions, catch quite big butterflies. Then there are the bladderworts and other insectivorous plants.]

Hearing that there were Bushmen in the neighbourhood of this part of the Kutsi, we decided to camp here for some weeks. We were disappointed to find that the Bushmen – their native name is Vasekele, or the people who are chased – were not very near, but the promise of some salt and food and the hope of other benefits induced two families to come and settle near us for a few weeks. Salt, by the way, in this part of the world, almost replaces money, for small purchases at any rate.

Our Bushmen include 2 men, 2 women and a small boy, they are Kavikisa, with a tuft of black wool on his head, and a small woolly beard, his wife, !Ko (English spelling does not convey it – It starts with a dental click and ends with a nasal sound) who is by way of a magic-maker, Golli-ba (father of Golli) a hunter, collector of wax with a good admixture of Bantu blood in his veins, his wife Baita and eldest son Golli.

Both these men, contrary to usual Bushman habits, have a second wife, they however have not been staying here though Golliba's second wife, !Nishe with her two little boys, has come over twice to see us. All except Baita, talk Bushman; she however, has lived with Mbunda (the local Bantu tribe) since her childhood and consequently knows only a few words of the Bushman language.

A COMFORTABLE CAMP

Our camp is comfortable though simple. The day after we arrived we selected a site at the edge of the woodland, showed our Mbunda boys what we wanted, set them gathering poles and grass, and in two or three hours they had built us a large shelter or nsinge, 15 ft. by 9 ft., the long wall facing north; open to the south. On the south it is about 10 feet high, the roof sloping down to about 6 feet on the north side.

Next morning a similar but smaller shelter was built over the tent – really an extension to the first, the western wall of the two being continuous. Then with split sticks and bark (Bark is the "nails" of this country, the whole framework, not only of our nsinge but of large buildings such as the mission churches, schools, etc is tied together with strips of flexible green bark of a certain leguminous tree known as Mukovi) a shelf was made along the long wall; with the same materials plus crooked sticks, a table in the centre and a seat against the west wall followed.

A HOUSE FOR 6 SHILLINGS AND 8 PENCE.

Two mornings' work saw the whole completed for the stupendous sum of 30 escudos – six and eight pence at the present rate of exchange!

This has been our home for the last month. Behind us rises the wooded slopes of the divide between our river, the Kutsi, and the next large river, the Kwandu, and round us are outposts of the woodland proper, of which our part is merely the outskirts. In front, for nearly a mile, is spread out the width of the river plain, soft shades of tawny brown grasslands, with a streak of vivid green on the far side where the river actually runs. The woods behind are dark green with autumnal shades of brown and gold, and now and again a splash of red, mingled with the green of those trees which do not drop their leaves.

In early morning as the sun rises above the height on the far side of the river, its rays tinge with pink the upper layers of the clouds of mist which lie over the valley at night; then as its heat reaches the soft white clouds, they dwindle and soon disappear and by nine o'clock nothing of the mist remains – the sky is cloudless, deep turquoise blue, the sun pleasantly warm, the breeze cool and refreshing.

WHEN THE SUN IS HOT

By noon, the sun is hot, and one seeks the welcome shade of the nsinge or a nearby tree; the sky is no longer cloudless – small white clouds are gathering all round and later fleck the whole sky. This happens regularly every day. Sometimes the wind is cold and remains so even through the heat of the day, but more often noontide is quite hot.

The bird life around is interesting – one morning early I heard a "tap, tap! tap- tap-tap-tap!" in the tree behind the nsinge and went out to see the small hammerer – too late, however, as he had heard me move and taken fright. A couple of days later, however, towards evening, he came again and I had an excellent view of Mr Woodpecker at work; upside down; he is a pretty creature – back dark olive-green, tawny

breast, head with a bent bar of black above each eye, and between the bars a scarlet pate, and a velvety black throat. His beak is long and the whole head and throat massive to deliver his powerful hammer blows.

BIRD OF STRIKING PLUMAGE.

Another bird of striking plumage is about the size of a spreeuw, with white breast and black back and wings – when he is at rest. As soon as he spreads his wings to fly, however, the back and upper side of the wings flash out in scarlet so that as he darts through the woods he is a vivid streak of colour.

Our Bushman friends take turns to come morning and afternoon to do their work; to us it sounds easy – only answering questions and talking in their language with its curious clicks, but to them it is hard work and the food and salt they are given in return thoroughly well earned!

FLORA OF THE PLACE

As for flowers, as this is the dry season, they are not numerous; still there are a fair number to be found if one looks – two tall red ones, rather like thistles, but scarlet and not prickly, several different kinds of pea-like blossoms, a tall handsome orchid, yellow with red or brown inside the two back petals are found in the woods behind us, as well as several sweet-scented white blossoms. But the most interesting part to a Botanist is the swampy river plain in front, where there are a number of insect-catching plants – or rather plants which catch and eat small animals, or even plant organisms. The sundews here catch quite large butterflies

– the bodies are digested and absorbed by the leaves, on which the wings and hard parts of the body remain for quite a long time. More interesting still are the bladderworts, small plants with minute,

BEAUTIFULLY CONSTRUCTED TRAPS,

the size of a pin's head, or even smaller, which catch and digest various microscopic plants and animal organisms. Some of these float freely in the pools of the swamp, but most of those found here grow in wet mud in which fine ramifying branches bearing the bladders spread in all directions, sending small green leaves above ground here and there. Of these, one has a delphinium blue flower about the size of a large violet, the stalk of which twines round grass blades carrying the flower a foot or more above the ground – several hundreds of these waving in the wind with their grass supports make a sight not to be forgotten. Another climbing one has small deep blue flowers with a greenish-blue centre. Two others stand a foot high unsupported – one has dark violet, the other pale mauve linaria-like flowers.

SOME RARE FLOWERS

Then there are several with yellow flowers, some free-floating, others growing in the mud, and finally one with tiny white flowers one or two inches high, and most beautiful little round green leaves just above ground, each leaf with a hemisphere of a clear, transparent jelly-like substance to protect it.

Another type of insectivorous plant in the swamp has wonderful spirally twisted traps, arranged in pairs, each pair leading to a single hollow chamber. In the river are numbers of fish ranging in size from minute creatures to large full-grown fish a couple of feet or more in length. Some are excellent eating – are caught by the natives in traps made of reeds and bush, and brought to us for sale. They make a welcome change from the daily dish – chicken.

What with Bushmen, collecting, bread making (the bread sometimes rises but more often doesn't!) the days slip by and before we realise it our month has passed and it is time to move on to a fresh Bushman district – this time, we hear, one where lions are also abundant!

The Courant, 26th October 1925 (Part VII).

A Trip into the Interior.

IN THE BEAUTIFUL VALLEY OF THE QUITO.

How the Porters were Induced to Accompany the Trekkers.

[Specially written for "The Courant"]

In another interesting instalment of her series of articles Miss M Pocock who together with her friend, Miss Bleek, is now in the heart of the little-known country of Angola, relates in entertaining style their further experiences in the hunt for the elusive Bushmen. Reading between the lines of playful humour, one gains an insight into the real difficulties of transport which these stout-hearted tourists have encountered. The letter, which is printed below, also contains a graphic description of the magnificent scenery in the valley of the Quito.

Everyone will remember how Alice, when she visited Wonderland, played a strange and wonderful game of croquet where, when the flamingo which served as a mallet was ready for use, the hedgehog which formed the ball, had strolled away, and when the ball was nicely curled up, the mallet was looking round. Well, I always sympathised with Alice, but never so much as now, when we have tried to get porters to transport us and our goods from the Kunzumbia to the Quito.

To begin with, we visited all the five small villages which are strung out on this side of the river, and enlisted porters. Then when we were ready to start, we sent messengers giving two days notice. On Monday, when we had planned to start, our tipoia boys turned up, but no porters. Next day, we sent boys out again, and they brought in the porters, too late however to start that day. On Wednesday, the porters were still there, but the tipoia (i.e. machila) boys had disappeared! Finally by putting a load into one of the machilas, and binding several into two, the half-loads being carried by boys, we did manage to get under weigh, and after spending nearly a month on the Kunzumbia, left it and trekked across to the Quito, which we crossed in dugouts just below its large tributary, the Cuanavale.

A WONDERFUL PANORAMA

There we camped in the wood, just above a great bend in the river where it has cut a high bank in the hard yellow sand. Above us, perched on the heights is the government post of Cuito-Cuanavale, and from it one sees a wonderful panorama of the lovely plain through which the Quito winds its serpentine way. To the north are the fertile valleys of the Quito and the Cuanavale: to the east is a beautiful little river, the Tumpu, and on the west, south of the post, is yet another tributary, the Lengu, from which the post gets its native name.

We had plenty of time to enjoy the beauties of the Quito! We paid our porters, and like the hedgehog and flamingo in the story, they disappeared – even more entirely than the Cheshire cat – not even the grin remained! We asked the authorities of the post for porters, and were promised them, but it took five days to get them, and there we had to stay till they appeared. There, however, our troubles ended – for the time at least – and we continued our trek westward.

A PICTURESQUE ROUTE.

For a whole day the route lay up the valley of the Quito, which would delight the heart of any artist. The valley bends in and out, the river turns and twists through it, and the path now runs through bush, now follows the grassy edge of the plain, then cuts across a bend and runs along the bank of the river, translucent, green and cool. Here a patch of purple iris greets the eye, there a shrub with dark vivid green and tiny orange pea flowers. Or a tall yellow orchid waves gracefully as we pass, while long slender stems bear scarlet heads, thistle shaped. The woodlands show dark green, delicate spring green, and vivid emerald green, mingled with autumn tints of gold, red and brown. Every turn of the path brings a different scene, each beautiful and each crying out: "Paint me! – if you can!" I can't unfortunately, and moreover there is not time, but many times I longed for the skill of an artist.

THE OLD TRICK

The second day out our porters tried the old trick of stopping in the middle of

the day's trek, near a village. Fortunately we were well posted as to the route and were able to take up a firm stand. We announced that we were going on, and went on. Then followed an awful half-hour when we did not know whether they would come or not. Imagine a wide rolling landscape, cut right across by a twenty foot wide, grassy road or "strada" and on this, toiling up a mile-long hill, two lonely figures – not a living thing besides in sight! There you have our predicament – village and porters being hidden in the bush. However, after some time, first one and then another figure emerges from the bush and starts up the hill. Joy! the incipient strike is ended, the porters are coming to heel!

After that, never a murmur or grumble – to us at any rate, and many mystified comments on our strange knowledge of the route. To these the boys who have been with us longer, answer with a great air of superior knowledge that we know, we have written it all down. An awestruck audience gathered round us to watch us write. "Dona sonneke!" (The lady writes!) they tell one another.

A TREK ON FOOT.

The trek from Quito Cuanavale takes seven days, or rather and six and a half, all but the last meaning a full six or seven hours' trek, not counting rests and halts for food each day, no light matter, when it all has to be done on foot – the porters carry loads of 50-60 lbs., and machila carrying is quite heavy work. Our machila or tipoia boys, however, had an easy time – mine especially, as I walked quite three-quarters of the way.

On the last day the character

of the country began to change, sand gave place to rock – and what a joy it was to see rock once more after three months or more of sand, sand, nothing but sand!

Near Menongue there were tall proteas, small trees, in flower – large heads of pink and silver, most beautiful, the first I had found in bloom.

Menongue itself is a small town – really hardly that – on the river Cweve which rushes down in a series of beautiful rapids, just below where the post is placed. On its banks we made our camp, while we were most hospitably entertained by Senora Madruga, the wife of the Administrator, during our short stay.

THE FINAL STAGE.

Fortunately for us one of the men from the Cwelei mission station, happened to be at Menongue and he acted as our interpreter, talked gently, but firmly, to our porters who didn't at all want to go the further day's trek to Cwelei with us. However, when he had told them they must go, and we had given them a present of meal, they quite meekly went – then we got permission from the Government to take them, – the Government saying certainly, they could go, if they liked! By that time, of course, they did like. They had to, like Old Man Kangaroo.

Next morning we started on the final stage of our journey – through beautiful country, up one open valley, with great granite boulders scattered about at intervals, and down another; after four hours, a party from Cwelei, bound for Menongue, met us and we all made camp together, high above the river – a pleasant meeting in the wilderness. Then a short three hour trek brought us at last to Cwelei, where the mission station barely a year old, stands above the rushing torrent of the Cwelei, which is here broken up into several streams by a series of islands; on these the missionaries have made their gardens, among palms and orchids and strange tropical plants, watered by streams led in from the Cwelei. A leopard lurks in the bushes of the further island and crocodiles in the river above, but these only add to the attractions.

Cape Times, 5ᵗʰ December 1925

OVERLAND TO LOBITO BAY.

In the Footsteps of a Vanishing People.

THE VENTURESOME JOURNEY OF TWO LADIES.

A trek across Angola! It sounds a venturesome undertaking for two unprotected females, yet it is surprising how easily it can be accomplished and that without a single event worthy the name of adventure – new experiences we had in abundance, but of adventures never one. The trip was undertaken for the purpose of studying some scattered remnants of one of South Africa's most interesting native races, those strange little yellow-brown people, relics of an extremely ancient and widespread race, the Bushmen. Few people in this country perhaps realise what a treasure from a scientific point of view these few surviving groups of people represent, and that not merely to South Africans but to all in every part of the world who are interested in the history and development of mankind. Further, it is a treasure which is rapidly dwindling and in a few decades more will probably have entirely vanished; hence it follows that all investigations into the life and language of these people, if they are to be made at all must be made without delay.

The Bushmen live in small groups scattered among the black races and more or less subservient to them. A people, still to all intents and purposes, living in the Stone Age, the Bushmen of South Africa retain many of their ancient habits and customs as well as their own distinctive language; but, whereas in Angola their contact with the dominant black race is close, many of their ways are being abandoned for those of their neighbours, their folklore forgotten and their beliefs changed.

The organiser of this expedition has devoted herself to the study of the Bushman languages and has spent much time among the Bushmen of South-West Africa, but had not previously visited those further north, so when certain missionaries working in Angola reported the presence of Bushmen in the neighbourhood of some of the mission stations and invited her to go up and investigate them, she decided to avail herself of their invitation at the earliest opportunity. Thus originated our trip, which resulted in the collection of much valuable material and incidentally involved a most interesting and varied journey through very beautiful country.

The First Stage.

The first stage of our journey was by train to Livingstone, with a visit to the Falls. Then followed an interlude of buying stores and other final preparations. On the following day came a trip by motor lorry, which though only 36 miles in distance, took nearly nine hours in time, thanks to the action of long-continued rain on a road never at the best of times very good. That it did not take more than one day was only made possible by supplementing the normal horse-power of the Ford lorry by five man-power, to which end we carried at the back of the lorry five strong native "boys", whose business it was to get off and push when a specially bad piece of road was reached. If as happened several times the back wheels sank through the surface of the road up to their axles in mud, the boys had to off-load, carry the goods several yards up the road, and then come back and push till the lorry was free of mud – for the time! Even so, we had had to send half the equipment (a very modest one, too) on ahead by carriers the previous day.

Finally, however, after running over a 6-foot long black mamba, Katombora, where the boat journey was to start, was reached, and stage number two of the journey successfully accomplished. Here we made our first camp while we waited for the carriers with the rest of our goods to arrive.

Up the Zambesi.

Now came an entirely new experience. For seventeen days we were to travel up the Zambesi, in a 30-foot flat-bottomed boat propelled by thirteen paddlers, under the direction of an Induna. These paddlers, tall, well-built, broad in the shoulders and narrow in the hips, might have stepped out of an ancient Egyptian painting. They have long tough poles rudely cut at one end to form a paddle and as such they are used in deep water, while where the water is shallow they function as punting poles. The seven stalwarts in the front of the boat, led by a particularly active boy in the bows, were a never-failing source of interest and amusement as we sat on our opened bedding rolls, leaning comfortably against our luggage piled up amidships behind us, and protected from the heat of the sun by an awning of grass mats. The paddlers, who stand to their work, would work steadily and quietly, keeping time with their leader; then suddenly the latter would quicken his stroke, and with calls or rude chants urge on his fellows till all were working at the top of their speed, joining in a rhythmic chorus to which they timed the swift upward and long slow downward strokes of their paddles. Then after a spell of this swift progress, all relaxed and slow easy strokes would again propel the boat at the usual steady pace. They were a contented, jovial crowd, ready to find entertainment in the merest trifle – if a big black and gold hornet disturbed by the passing boat buzzed angrily near the head of one of the boys, making him duck down in terror, the rest shouted with delight – till their own turn came! And when as happened now and again the wet paddle slipped from the hands of one of them, his clumsiness was greeted with roars of joyous derision, which culminated when the leader himself actually lost his pole. To our amusement, next

day, No. 1 went down a place, and No. 2 was allowed to lead till he too dropped his paddle. The Induna, Sanpiere, sat perched on the luggage behind the awning, where he and the cook boy, Kandu, passed much of their time playing on a little calabash piano. His work was to direct operations, portion out the men's daily ration of meal, organise the nightly camp, and generally supervise things; naturally, on him depended much of the comfort of the trip, and the fact that we never had a moment's trouble with the boys speaks volumes for his management.

Flowers, Birds and Insects.

Above Katombora the country, low-lying on both sides of the river, but more particularly so to the south and west, is flooded for miles during the wet season. Where possible the inundated country is always chosen in preference to the river itself by upward-bound boats, so for days at a time we hardly saw the actual river, but made our way through a wonderful growth of tall grasses standing several feet deep in water. As the boat was poled through them the grasses parted on both sides, and spiders, beetles, grasshoppers, insects of all kinds, were swept into the boat where most of them immediately resumed the interrupted business of life.

Then an open stretch of deeper water would suddenly appear, starred with fragrant water lilies, white, pink and blue, or festooned with floating trails of convolvulus with deep purple-red trumpet-shaped flowers, and many other flowers, whilst delicately branched red-tinged floating water ferns added colour and variety to the

scene. Everywhere there were birds. What a country for the naturalist, the artist, or even such otherwise undistinguished folk who merely love Nature.

The bird life was wonderful; besides water fowl, such as wild duck, geese, coot, herons, white and purple, there were darters, divers, egrets, the sacred ibis, with its melancholy call of Wah! Wah!, a pair of Egyptian geese flying overhead, kingfishers, swallows, and many other species unknown to us, besides many small perching birds among the trees and bushes along the banks and in the tall grasses.

Then, in certain of the grasslands, there lives the most fascinating little frog; less than an inch long, very slender and delicate in build, translucent jade green in colour, with golden-brown eyes and a stripe of gold along each side, he makes his home on the fresh green leaves of the grass on which he rests with feet, the toes of which are ball-tipped, tucked closely at his sides. In one of these grasslands every leaf seemed to have its jade frog inhabitant, and as the light shone through the young leaves the shape of each little frog appeared as a darker spot on the blade, giving a most curious effect.

At Sesheke.

At Sesheke a most interesting day was spent at the old-established station of the French Protestant Mission. There we saw the mighty fig tree under which Livingstone sat to instruct the natives, and next to which the church now stands; there, too, the banks where in years gone by the Barotse King Sepopo, used to amuse himself by having children thrown into the river to be devoured by crocodiles – the crocodiles seem to have handed the tradition down to their children for we were told that they still lurk near the banks in wait for the women or children who, going down to fetch water, venture too far in. Unfortunately, we saw never a sign of a crocodile in all our travels, though we should have liked to do so – at a safe distance, of course!

Above Sesheke the character of the river changes – here it flows between fairly high banks, and as the level of the land falls rapidly, there is a long series of rapids, culminating in the beautiful Gonye Falls, some 80 or 90 miles to the north-west; beyond the Gonye Falls, the land flattens out again, and the inundated grass-grown flats reappear. In the region of the rapids the exceptionally high water stood us in good stead, as many of the rapids were almost drowned by the flood, and consequently we were able to work our way up comparatively easily over many a place which in the dry season can only be surmounted after a long and arduous portage.

Among the Rapids.

Now, however, all that was changed; here the river would run deep and swift between high wooded banks, then spreading out over a belt of hard resistant rock it would form a shallow foaming rapid, the roar of which announced its proximity long before it came in sight. Sometimes so full was the river that the only visible sign of a rapid was the "hard water" as our Induna called it, when our paddlers, putting forth all their strength and urged on to renewed effort by the cries of their leader, had to fight their

way inch by inch against the stream.

More often ridges of rock or a line of bushes and trees showed across the river which made its way through the obstruction in a series of narrow rushing torrents. Here, creeping along under the bank, hard paddling might just suffice to keep the boat abreast of the stream till a slight change in the direction of the current gave the paddlers their chance, and with triumphant shouts the boat moved forward.

A Long Portage.

As we made our difficult way up one of the larger rapids a downward-bound boat was sighted sweeping along in mid-stream. As she shot swiftly over the rapids and disappeared in the distance the contrast with our slow progress was very marked.

To pass the Gonye Falls a long portage was necessary; in a quiet little bay to the right the boat was unloaded, the loads carried across rising ground for nearly a mile and then a team of oxen dragged the boat up and round into a primitive kind of canal up which, after reloading, the boat was laboriously poled back into the Zambesi above the falls. Leaving the boys to make camp at Seoma we walked across to the head of the falls to see what could be seen of them – not very much as the bush was thick. To see them properly one has to go in the dry season when they can be approached by canoe from below.

The Only Hippo.

The Gonye Falls once passed, we were beyond the region of rapids and once more moved through quiet backwaters or across flooded grasslands with now and again an interval of

open river and the floodland. It was in one of these quiet backwaters above Seoma that we saw our one and only hippo. We were gliding leisurely along towards evening when we heard one of the boys exclaim "Kooboo!" pointing ahead and to the right; this being the one word of the local tongue which we had then managed to learn we were on the alert and were just in time to see the hippo feeding in the distance. As the boat approached, however, he submerged and when we passed only the bubbles rising to the surface marked his lurking place. Very quietly we edged past, keeping as near to the far bank as we conveniently could, then, when we were well past up came Mr Kooboo with a mighty puffing and blowing, shaking his red-lined ears to get rid of the water and gazing after us apparently much surprised at the derisive epithets hurled at him by the boys – very bold now that he was well in the rear.

The river journey with all its varied interest at last came to an end: our time, 17 days from the time we left Katombora to Kama where we finally left the Zambesi, was unusually short for the upstream voyage. Kama lies on the right bank a day's journey south of Mongu; there we left the river and started the westward trek, and from the time we left our boat until we reached Silva Porta in Bihe five months later, every step of the way, except an occasional river crossing in dug-outs, had to be done by foot – whether our own feet or someone else's! A train of porters carried our belongings in loads of 40 or 50 pounds, while a dozen more carried the machilas, six to each machila, working in relays of two at a time. The machila is a canvas hammock slung on a long pole which is carried by a couple of sturdy boys, either on the shoulder or the head.

Machila Travelling.

The first experience of machila travelling is somewhat nerve-racking, especially when as in our case, on the hour's trek from Kama to Lukona it includes the passage of a long stretch of floodland where the porters have to wade knee-deep in muddy water. Soon, however, one grows used to the motion and learns to enjoy this mode of progression – lying at one's ease there is leisure for much meditation, and to enjoy the constantly changing scene with the never-ending play of colour in grassland and bush. Indeed, it soon becomes so enjoyable that one is tempted to ride all the time, a most demoralising state of affairs! Later on in the trek we used to walk most of the way, only using the machilas for short spells of rest, and in the middle of the day when it was too hot to be pleasant for walking.

For the first five days after leaving Lukona we were in British territory, and the way led through large and prosperous villages built at the edge of the bush, where it met the riverland. Both here and in Angola, this was the usual position for the villages – the former provides the necessary building material and fuel, while in the latter, even if the river itself is some distance off, water holes sunk a few feet down soon reach water level, and provide a sufficient supply for the needs of the village, even in the dry season. At the beginning of our journey, however, scarcity of water was not a difficulty we had to contend with – on the contrary!

Our First Trek.

A friendly Magistrate, whom we met on the second day going the round of the villages under his jurisdiction, taking pity on our ignorance of this type of trekking, lent us a messenger to escort us as far as the border; his presence certainly kept our porters up to the mark, and helped us considerably. On this first trek, too, we had a cook boy who understood and spoke English, a luxury which we never enjoyed again! He acted as interpreter, and had it been possible we should have liked to keep him, but we could not do so. Two days before we reached the border we had our first experience of taking our train across a river in dugouts – it is rather an ordeal to have to watch all one's goods carried across a wide, deep river in one of these unsafe-looking little boats, still more so when one's own turn comes! Kneeling in the middle with a paddler before and behind, if the dugout is a fair-sized one, or if it is small, with one behind only, the unhappy passenger holds his breath least an unduly deep breath should overbalance the frail craft! Yet in the end even the dugout becomes attractive, and the short journey has its own special delights – the clear translucent green water so close to one's eyes, through which every plant, every ripple of the sandy bottom, every tiny darting fish is visible, has a beauty which would be missed in a larger and more substantial vessel.

The Border Reached.

About midday on the fifth day, the border was reached, and there the messenger bade us farewell and returned. A small grass hut was all that marked the spot where British rule ended and Portuguese began; there are no villages anywhere near the border in this part of the country, and after crossing it there is a marked decrease in the number and size of the villages. Angola is a sparsely populated country, and there is a constant stream of emigration from it into the neighbouring British territories in spite of all that can be done to stop it. Yet it is a fertile and well-watered country. As in Northern Rhodesia, west of the Zambesi the country is chiefly wooded, the bush land being cut across at intervals by belts of grassland; the latter is the mark of a river valley or plain, and the grass belts vary in width from a few hundred yards to a couple of miles. Through the grassland meanders the river, and often springs are thrown out near its edge. Much of such grassy stretches consists of swamp, in which grow many interesting flowers – Sundews, which catch and devour slow-flying butterflies as well as lesser insects; Bladderworts with sky-blue flowers twining themselves round the stems of grasses in their efforts to obtain light and room for their blossoms to expand; others with mauve or yellow flowers, and some tiny ones growing low on the ground with white, yellow and purple flowers, scarcely larger than a pin's head.

Cape Times. 19ᵗʰ December 1925

OVERLAND TO LOBITO BAY.

The Venturesome Journey of Two Ladies.

BY M.A.P.

In three articles, of which this is the second, there is described the venturesome journey of two ladies across Angola. The first article took the record of the tour to the Anglo-Portuguese border.

II

The messenger and his attendant satellite having left us at the Anglo-Portuguese border, we proceeded on our way westward into Angola, now following a winding path through bush country where the machilas were carried with difficulty through thick bush or between closely growing trees, now across swampy river plains covered with tall golden-brown grasses. Two days later, topping a slight rise, we suddenly came in sight of a strange looking, 30-foot wide line of cleared country which puzzled us at first till we found it was the beginning of one of the unnecessarily broad "roads" which the Portuguese delight to cut through the bush. These are simply cleared of bush and all plant growth, leaving the bare loose sand; consequently they are usually far from pleasant to walk on, and unless constantly cleared soon get overgrown once more; the narrow footpath often follows the road, meandering from side to side of it much as the rivers in this part wander about in their own plains. This first little stretch led us up a short hill on the summit of which was a group of grass-thatched buildings, outside which a small group of men, including two Portuguese, stood to watch us pass; another half mile and we

had passed through the village and had reached the mission station of Ninda, perched high on the hillside overlooking the beautiful plain of the little river from which it takes its name.

Here we spent four days at the mission guest house, the first of the Angolan Mission stations with which we were to become acquainted. All the stations we visited belong to the South African General Mission, with the exception of the last, Kamundongo in Bihe, which is under the American Board. Here, too, we first came in contact with the Portuguese rulers of the country and had our passports examined, which ritual, together with the endorsement of the same was to be repeated many times before we left the country – in fact the endorsements were so many and so voluminous that on one of the passports there isn't room left for a single word more!

The First Bushmen.

From Ninda our train of porters and servants, all British subjects, had to return to Lukona, and from now on our porters were drawn from tribes under Portuguese rule – chiefly Luchasi and Wambunda, usually speaking their own language only, with perhaps a word or

two of Portuguese. As a rule, however, their knowledge of the latter tongue did not go beyond "Si Senhor" which though pleasant hardly forms a basis of conversation! Fortunately we were able to borrow a vocabulary of Mbunda and Luchasi words and phrases from our hosts at Ninda, and it proved of great value to us for the rest of our journey.

As soon as we reached Ninda we began to make enquiries about Bushmen, and on Sunday one of the village men brought the news that there were Bushmen at a village a couple of miles off, on the far side of the river, so next morning we hired tipoia boys and started off to try and find them. As we reached the gate of the mission compound, however, word came that the Bushmen had gone away, and we were forced to turn back disappointed. Our porters for the next stage of our journey were already hired for the following day, so that we feared that our only chance of meeting these Bushmen had gone, but the next morning, as we were in the midst of giving out the loads to the porters, an interested crowd appeared before the guest house, following three genuine Bushmen – two men and a small boy. They had heard of our promise of presents of salt (in these parts far more precious than gold!) and beads and had come to see what was toward.

This our first meeting with the folk we had come so far to study was, of course, most exciting, and the business of packing etc, was put on one side while the Bushmen were questioned and notes made of their speech and what information could be obtained of their mode of life. They all speak the language of the neighbouring Bantu tribes, in addition to their own, so, with the missionary to interpret and a large and interested crowd looking on, the work proceeded apace. The two men were small, spare, with well-shaped, though small, thin limbs and tiny hands and feet. In colour dark yellow-brown, both cast of feature and colour of skin, in addition to build, distinguished them at once from any of the Bantu races among whom they dwell.

Mild Interest.

They answered questions readily, seemed quite unafraid, but mildly interested. After a very short time, however, their attention began to wander, and it was evident that not much could be gained by detaining them longer; the Bushman is like a child, his attention is easily caught, but soon lost, unless something very interesting is toward, and answering questions such as: What is your name? Who was your father? And your mother? What do you call this? And that? though at first intriguing, on account of its oddity, very soon becomes boring! However, a small gift, particularly a handful of the delicious much-desired salt, amply rewarded them for their little morning stroll of four or five miles. They were photographed, and then sat about talking with the local natives for a time before they departed, walking, as they had come, in strict order of seniority.

Four and a Half Days' Trek to Muié, the Chief Mission Station.

Much pleased with this first encounter with our little people, we set off with our new train of porters for the four and a half days' trek to the chief station of the central Angolan mission field at Muié. Here there is a river system, consisting of a number of nearly parallel streams, draining in a general direction to the west of south and further still to the west this becomes even more marked. Between the narrow winding river plains are belts of higher ground covered with thick growth of "bush", i.e. tree land composed of comparatively small trees with fairly dense undergrowth. Everywhere is sand, both on hill and in valley – from the Zambesi to beyond the Kwandu we never saw even a stone except at Muié, where pebbles from the river had been used to make a road. It was quite a treat when we eventually reached a land where there were occasional rocks to be seen!

The Nightly Camps.

The nightly camps were always interesting. Once, the night before we reached Ninda, camping in the bush rather high above the river we noticed our Induna (a wild and woolly individual!) directing the porters to make their fires in a circle round the tent. "Lions!" we thought with a thrill of excitement. This was in the days when we had an English-speaking cook so we asked the reason – was it lions? No, not lions, only hyenas! It was rather a come-down, but at Ninda they gave the hyenas such a bad name – one of these usually cowardly beasts had actually chased a man down the village street in broad daylight – that we felt less "had". Between Ninda and Muié we continued to make the porters camp more or less in a circle round the tent, but it was usually evident that there was not much to fear from wild beasts in these parts; when later on we did come to lion country, the gathering of logs for fires and the arrangement of the latter showed a very different attitude on the part of the boys – no spreading out of the camp there!

When possible our porters made camp near a village where they could buy their meal and a little "belela", or relish. The latter might be beans, or dried fish, or a piece of evil-looking, strong-smelling meat, while the meal was either millet or manioc; further west maize replaced the local corn. As soon as camp was made visitors from the village would arrive – a woman with a conical basket of meal or sweet potatoes on her head, followed by a small boy carrying a skinny and protesting fowl, or an old man with half a dozen eggs.

Presents of Salt.

Occasionally the head man of the village would send some of his henchmen down with a "present" of meal for us, receiving in return a "present" of salt! It is quite possible to "live on the country" as one passes through, but not advisable to trust entirely to doing so – if one has ample, provisions are sure to flow in, but as soon as one runs short there is almost certain to be little or nothing to buy, not even a fowl!

Sometimes when the natives refused to sell for either money or salt, a piece of calico will purchase all that is needed. At the present time stuff, even the commonest of print, is so expensive in this part of Angola – a yard costing as much as 20 days' wages – that it is practically never seen except at the mission stations or trading posts. The natives dress either in skins or less often in bark cloth; the adults usually wear two skins, either of buck or goat, belted by the hind legs round the waist and hanging down before and behind, leaving the limbs free. The dandies wear large skins so arranged that the forelegs (the head is usually cut off) barely clear the ground as they walk, but when trekking the skins are hitched well up or even folded neatly under the belt. Several of our porters had small boys with them to carry their possessions – food and a bark blanket – and these boys all wore a single small skin; one little fellow had a very nice soft wild cat skin, of which the front legs were tucked into his belt in front while the hind legs were similarly disposed at his back, the bushy tail swinging from side to side as he walked along with his load on his head.

Occasionally we passed a post where a few houses advertised the presence of a "Chefe" or local official in charge of a district, and a trader. Several of these posts were deserted and already falling to pieces; the houses were all built of large sun-dried bricks and thatched with grass from the river, so that if not cared for, a couple of rainy seasons will go far to demolish them.

On this trek we first noticed a phenomenon with which we were to become very familiar – the intense blueness of the distant river valleys. Very soon, if as the machila made its way in and out among the trees, a line of blue gleamed through them on the skyline, one at once called out to the boys to ask the name of the donga (river), and when, later, rivers were several miles apart one grew to watch for the glimpse of deeper blue in the distance and to welcome it when it appeared – it always meant water and a welcome rest, whether that of the evening camp or the midday halt.

A Month's Work among the Bushmen.

At Muié we spent a week-end and then moved on half a day's journey to the river Kutsi, of which the Muié is a tributary. Even up here the Kutsi is fairly large, and we had to cross it by dugout; it is a nerve-racking experience to watch all one's most precious possessions wobbling across a swift deep stream in a dugout with a large piece out of one side.

Next day we selected a suitable spot near the edge of the riverland, and proceeded to make camp for a prolonged stay. The first thing was to build a shelter, not so much in case of rain, but to protect us from the heat of the midday sun. Men were at hand only too glad to be hired and, with the help of three boys we had brought from Muié, we soon had what we wanted. This was a roomy shelter, some 9 by 20 feet, with the long wall to the north and open to the south, while an extension to the south at one end served as a roof to keep the sun off the little tent. The materials used were ready to hand for the workers – poles cut from the bush formed the framework, grass from the river and the leafy branches trimmed from the trees cut down for the poles filled in the walls and thatched the roof. The addition of a long shelf, a table, and a bench, all made of round twigs tied with strips of bark to cross poles laid on four crotched sticks planted firmly in the ground, and a thick carpet of grass gave the finishing touches to our abode. It had taken two mornings to build and had cost us the large sum of six and eight pence!

Here we settled down for a month's work. At first we were disappointed to find that there were no Bushmen actually living on this part of the Kutsi, but the local headman soon got in touch with a party on the next river – hardly more than a series of muddy water holes at this time of year – and after visiting us once or twice, two families moved over, put up a couple of their little leafy huts near our camp and settled down quite happily. We soon fell into a regular routine – the Bushmen came by turns, or all together if they were allowed, for a couple of hours in the morning and again in the afternoon to do their daily task (a very hard one for them) of answering questions, telling the names of their relatives, familiar objects of their daily life and surroundings or describing their hunts, travels, and in fact talking about anything they could be got to take an interest in; then they were given a portion of meal and few spoonsful of salt and departed to their huts.

Meanwhile the river plain and the woody hill behind us provided a rich field for the collection of botanical specimens. Thus in Bushman work, collecting, writing and painting, the time passed quickly and profitably. Our camp was between two villages not half a mile apart so, in addition to the Bushmen, we had frequent visits from the village folk either to see pictures in the few books we had, to sell us fowls or to barter eggs for safety pins, needles, thread or salt.

Typical Bushmen.

The Bushmen deserve more particular description. Chief of them and perhaps the most typically "Bushman" was Kavikisa, a little spare hardy fellow of coppery red-brown in colour, with a tuft of black hair under his chin and another standing up on the top of his head, his only clothing a dilapidated old skin, his weapons a bow and arrows. His name among the black people was Kasindzela, and he was apparently widely known. A small sketch of him proved amazingly popular, not only with himself (he capered with delight when he first saw it), but with Bushman and Bantu alike; henceforth all visitors who wanted to see pictures always started by asking to see "Kasindzela". Incidentally we were a perfect God-send to the villagers near which we camped, a rare show of the first order! His wife, little !Kõ, was very different in appearance – a small neatly built little woman, light yellow brown in colour, her face was broad across the cheekbones, which as in all Bushmen were high, and rather short with a rounded chin, whereas his face was of a longer narrower type. She was by way of being a "medicine" woman or rainmaker among her own folk, and was in many ways the best of the little group for our purpose; she was by far the most regular in attendance, coming practically every day and giving much information not only from a linguistic point of view but also as to the habits, and to some extent the folk lore of her people. The other couple, Golli-ba (i.e. father of Golli, with his eldest son Golli), and his wife Baita-de, were less interesting – the latter indeed having been brought up among the black people and being in consequence ignorant of the Bushman language, was not of much use to us.

The Bushmen are as a rule monogamists, but Golli-ba had another wife (!Kõ's younger sister I believe), who with her two small children came over from their camp on the next river to visit us one day. Other members of this little family group also came from time to time, spent a few hours at our camp and then returned – chief among them an elderly woman named Shové, unusually tall for her race and with a remarkably fine old face. In colour she was similar to Kavikisa, but lighter and more of a yellow tone.

The Bushmen faces are strangely attractive and full of character, with bright alert eyes; they have a strong sense of humour, but in repose there is nearly always a tragic look in them, particularly in those of the older folk. The name by which they are known among the black people, the Vasekele, or people who are chased, may go far to explain the reason for this.

On Trek Again.

When the time came to leave the Kutsi and move on, we felt quite sad at leaving these little Bushmen friends of ours – they, I must say, took it quite philosophically, and as our departure was signalised by the presentation of gifts of beautiful new red print, were probably rather glad than otherwise! We had planned to return via the Zambesi as we had come, but hearing of the presence of Bushmen still further west, eventually decided to go on in that direction and return by sea down the West Coast, a project which in the end we successfully accomplished. From Kutsi our next move was to a river some days to the south-west, the Kunzumbia, a tributary of the Lomba. We tried to recruit our porters from the neighbouring villages, but failing to get sufficient, had to apply to the Chefe at Muié for the remainder. We never had much trouble with our porters, but this special lot was certainly the most troublesome, but even they at their worst were merely like a lot of unruly schoolboys.

Our way lay first southward along the Kutsi, where another group of Bushmen was found, then westward to the mighty Kwandu, which though here comparatively small was still considerably larger than anything we had seen since leaving the Zambesi. We crossed just below its junction with its almost equally large tributary, the Kembu; the crossing of the main stream by dugout took only a few minutes, but was followed by over an hour's trek through the swampy ground beyond. Large hairy caterpillars were swept off the grasses into the swift little stream, but seemed able to skim the surface till they came in contact with a floating twig or leaf, from which they could crawl back on to the grass. Small birds flitted from reed to reed stopping to watch the curious procession advancing, then darting further off to alight once more for a snatch of song or to seize a succulent morsel from the herbage. The last half mile or so before we finally left the river the way led along the west bank of the Kembu, here dry and firm though scarcely raised above the level of the water. The river, full to the brim, was lovely – a swift deep stream, clear translucent green above its sandy bottom, it swept by to join its brother stream a little lower down.

These Angolan rivers are very beautiful, and later on in the same day we reached one which was in some ways perhaps the most lovely of all, the little Kutiti, a tributary of the Kembu.

All Kinds of Game.

From now on we saw traces of all kinds of game; along the path was spoor of buck both large and small, birds of many kinds including guinea fowl and some large black birds the size of a turkey, and occasionally lion or leopard. All alike seemed to prefer to walk on the path whenever they can, and the

footprints stand out clearly on the smooth windswept white sand. Occasionally, though rarely, the sound of breaking underbrush startled us, and we would glimpse the fleeing form of a duiker. On the whole, however, we actually saw very little game – in parts of the country it is certainly abundant, but in others very scarce.

A brief stay at Kunjamba, a small station in a poor country, brought us our nearest approach to an adventure: we were about to leave when a great uproar arose in the valley below and boys armed with spears dashed out down the hill. Later we learnt that a lion, probably young and inexperienced, had sprung at a cow, missed its hold, and made off down river. As it happened, the lion did us good service – all that afternoon and the next morning it acted as a spur to our carriers, who had been giving us some trouble – there was no lagging, and in camp no difficulty in obtaining plenty of wood for our fires!

Next day we had the longest trek between waters which we had yet experienced. From 10 a.m. till nearly 5 p.m. when we reached the Kunzumbia we trekked with only an hour's halt at midday without meeting a stream or even a water hole. Needless to say, the river was very welcome when we did reach it, and the boys were not sorry to get the unavoidable wetting awaiting them – there is no bridge across the river, and at the ford, such as it is, the water was nearly up to their waists; consequently there were elements of excitement about the crossing. We made camp in the bush near the river, and as we were told there were plenty of Bushmen about, we decided

to camp for some time. After a day's rest the porters returned to Muié, taking our mail with them, while we proceeded to select a suitable camping spot halfway between the village, Kamundonga, and the river. As at Kutsi we had a shelter built, but here as we had been warned that it was lion country, in addition we erected a "cepatonga" or kraal round our camp – really an extensive fence made of poles and leafy branches. It turned out to have been an unnecessary precaution, but gave one a certain amount of privacy which had been lacking at the Kutsi camp. Here again we were destined to spend nearly a month.

———

Cape Times. 1ˢᵗ January 1926

Overland to Lobito Bay.

The Venturesome Journey of Two Ladies.

End of a Long Trek.

By M.A.P.

III

This is the final of three articles, of which the first two appeared on December 5 and 19, describing the venturesome journey of two ladies across Angola. The writer describes how a Bushman settlement has taken to ironworking and the making of knives and spears. This is believed to be the only record of Bushmen who have learned the smithy work.

THE LAST SECTION TO BENGUELLA.

Kunzumbia was a beautiful spot. The plain here was two or three miles wide, and the river wound down the centre; further south a large bend brought it close up to the tree land fringing the plain to the west, out of which it has carved a slice so that a high, steep, tree-clad bank overshadows it. Under it form still dark pools shaded by the trees above, and decked with lily pads and blooms.

This, however, was an hour's walk from our camp, which we made about a furlong from the edge of the plain on a level stretch of land where the trees had grown tall and stately. Their tall trunks, white-flecked like poplars, and large fluttering green leaves, made a screen on all sides. Now and again a red-crested woodpecker visited one just next to our hut, or two long-tailed blue-black birds perched on a bough just above it, and bowing and swaying towards one another whistled their dainty song. At night the night-jar's curious creaking cry sounded, now close at hand, then far off among the trees, or a small owl raised his melancholy voice.

Past the nsinge (shelter) led the path from Kamundonga's village to the edge of the riverland, where were several large water holes. Soon half-a-dozen paths from different parts of the village – rather a scattered one – converged on the nsinge, made by the women bringing meal, men with a haunch of venison, knives, or other wares to sell. More often still, our visitors were brought solely by the keen interest they took in us and our doings. A couple of young men would come down in the afternoon, and, standing nearby, would watch with absorbed interest while we wrote or made sketches of the wild flowers we had collected in the morning. These sketches delighted them almost more than the pictures in the book or the painting of Kasindzela. The joy of recognising them and being able to put a name to them was great. They had different names for nearly all, but occasionally there would be a discussion as to the correct one, or, as happened in a few cases, they would decide merely that they were "wisone" (flowers).

Elusive Buck.

In the stretch of grassland between our camp and the river fed a large herd of lechwe, known locally as vansonge, large buck, with powerful hind quarters, adapted for leaping through the tall grasses of the swamp land, in which they take refuge should any danger threaten. They could always be seen feeding morning and evening, but in the heat of the day they entirely disappeared, hidden by the friendly grass. Often we got within a few yards of one or two feeding near the edge of the woodland, and on several occasions, accompanied by Telosi, the cook boy, an early morning stalk provided plenty of excitement and interest, if nothing else. We had no firearms, except a small pistol, whose "safety catch" was difficult, but our boys were very anxious that we should try to get a buck – a hopeless task, of course, but to satisfy them we went out armed with the said pistol. Then followed a most exciting chase. Telosi, by no means an expert hunter, tried to head the buck toward the mighty "gun", the vansonge, apparently quite indifferent to his manoeuvres, went on quietly feeding until he was fairly near, then made off with leaps and bounds into the swamp. We followed, crashing through the reeds, waving well above our heads, splashing knee-deep in red muddy pools, drawn on and on by the buck, of which one, reluctant to go far from his feeding ground, would lurk near the edge of the swamp till we were within ten yards of him. Then with a crash start up and go further on, completely screened from sight by the surrounding grass.

Stalking buck in lion country, armed with a pocket pistol! It sounds foolhardy to put it mildly, but it was certainly exciting, and later we learnt that the lions were all on the other side of the river, and there so far as we were concerned, they appear to have stayed.

River Birds.

Other game was fairly abundant, though less often seen; the birds were particularly attractive. A tall, long-legged black and white fellow, somewhat of the crane type, with an enormous red bill, was often seen stalking out of the wood to his feeding ground in the river, or flying with wide sweeps of his mighty wings;

purple herons, long-tailed grey cranes, and several more of that ilk fed in the grasslands, while in the woods were many beautiful little feathered creatures – some songsters, others flashing jewels of turquoise blue or emerald or polished blue-black.

Taken all in all, the Kunzumbia camp was a most attractive one, save for one melancholy fact: we could get no Bushmen! It was tantalising in the extreme – we knew they were there, plenty of them; we found their camps, the ashes of the fires still warm. We walked up the river to the next village – they were gone. Down to the three villages in that direction – still without success. Then, one day they came, and we thought our difficulties were at an end. Not a bit of it – day after day passed and never a glimpse of a Bushman! The annoying part was that so far as we could make out, they themselves were ready enough to come, but were prevented from doing so by the people of Kamundonga (Wankangalla).

Short of Supplies.

Finally, as stores were getting low, we decided to give it up and move on. But first we had to collect porters and wait for the return of messengers we had sent to Muié for fresh supplies. Day after day went by and they did not come. Finally, our flour, eked out though it had been with local meal, was reduced to one more baking; sugar was finished, salt nearly so, and the various little extras, such as tinned milk, dried fruit, jam, all showed signs of a speedy end. The nearest post was three days' off so we decided it was time for us to go, whether our messengers returned or not. With extreme difficulty we scraped together enough porters

to take us to Cuito-Cuanavale, our next stage. When our procession actually started it was a queer one – one tipoia loaded with two packs, one old man, a few young ones and a crowd of boys of thirteen or fourteen carrying half loads!

Of course, when we were all ready to go, we got some Bushmen! One of our porters brought them in the day before we left, two youths and a mother and daughter; we were able to photograph them and get some information from them, but it was very tantalizing to have this happen just as we were departing.

A Further Delay.

It would take too long to go into the detail of our westward trek – our objective was the recently opened mission station on the Cwelei, a tributary of the Kuchi, which eventually joins the Okavango. The first stage – three days, was to Cuito, where there is an important post, there we were held up for five days waiting for porters. It is lovely country – the junction of the Cuito and its almost equally large tributary is a mile or so above the post which stands on a hill high above the river and overlooking the beautiful valleys of the Cuito and its tributaries. We camped in the wood high above a large bend of the river, and enjoyed our five days' enforced stay. It is a perfect paradise for artists, and when, having eventually got our porters, we continued on our way, up the winding valley of the Cuito, the beauty of the views that met our eyes at every turn of the path beggars all description.

Lazy Porters.

On this section of our way we had some trouble with the

porters – their idea of a day's trek differed radically from ours, and things came to a head when they announced their intention of camping at a large village after about three hours' march. We announced ours of going on immediately, and proceeded to do so on the spot – outwardly firm, but inwardly quaking. The relief was great when, after trudging in lonely and dignified state up a long straight stretch of tapalo for nearly a mile, a backward glance revealed our porters coming out one by one from the bush where we had made our midday halt. That was the end to all incipient mutiny – thereafter the "power of the pen" made itself felt. They discovered to their utter astonishment that these two strange females, whom they had thought entirely ignorant of the way, actually knew the names, not only of the rivers, but of many of the villages! How in the world was this, they enquired with bated breath of our personal boys, who proudly returned that we had "written" it! Henceforth, if ever notebook and pencil appeared, admiring glances greeted them, while the act of writing was watched with awe and wonder.

From Water to Water.

After leaving the Cuito and its tributaries our way led across a system of rivers some distance apart from one another, and all flowing south. This determined the length of the treks – sometimes two a day, at others, one long trek of six or even seven hours (actual trekking without counting halts for rest) from water to water. Finally, on the eighth day after leaving Cuito-Cuanavale, we reached Menongue or Serpa Pinto as it has recently been renamed,

where we received much help from one of the Cwelei missionaries on his way to the yearly conference at Muié, and were hospitably entertained by the Administrator of the district and his charming wife. Just before reaching Serpa Pinto we suddenly came upon a grove of proteas – small bushy trees some 12 to 15 feet high – in full bloom. The heads of the flowers are large and most beautiful – pale pink with a veil of silver. Among them a lovely gladiolus, pink with deep red flecks all over it, was growing, while here and there in the tapalo a tiny scarlet hibiscus formed patches of brilliant colour against the white sand.

Changing the "Guard".

From Menongue a day's march along beautiful river valleys, where were actually great granite boulders, brought us to the beautiful station of Cwelei. Here our porters turned back and after a pleasant weekend with our friends at the mission, a fresh lot of porters having arrived from Serpa Pinto, we pushed on, down the Cwelei – for once our way lay south and east – to a large village named Kaiongo, where we hoped to find Bushmen.

On the way we were able to get two of these much-desired folk – a girl of about 16 and a small boy of eight or nine, slaves at one of the villages we passed on our way. Poor children! The little boy was evidently very frightened. He was brought to us with tears still wet upon his cheeks, having washed small patches where the true light yellow-brown colour of his skin showed through. He had forgotten nearly all his mother-tongue; the girl remembered

rather more, and could tell vaguely where she had come from and what her parents' names were, but even she had nearly lost her speech – leaving out a click here and there and filling up many a gap with the Bantu speech.

Bushmen Found at Last.

At Kaiongo we found ourselves in luck's way. The chief promised to bring us Bushmen at 9 o'clock next morning and he proved as good as his word. Soon after we had begun to expect him a wave of excitement passed through our camp, and there coming from the village came first our friend Kaiongo, dressed complete in overcoat and hat, in addition to his striped vest and cotton petticoat; behind him came a boy carrying his chair. Then most of the men from the village, followed by a crowd of Bushmen, and finally the women and children from the village came for a morning's entertainment.

This was something like – over a dozen adult Bushmen, old and young, and half a dozen more young boys. We set to work at once, getting their names and relationships, measuring and photographing. They were a well-fed set, and the average height was much greater than usual among Bushmen – evidently there was much Bantu blood in them, more even than in those we had met further east, though they still keep their Bushman speech and to a certain extent, manner of life.

BUSHMEN WHO MADE KNIVES.

The "ten" days we spent at Kaiongo were in many ways the most interesting, and certainly the busiest of the whole trip – every day either several Bushmen – both men and women – came to our camp, or we walked over to "Bushman Town" where a couple of miles from Kaiongo the Bushmen had their camp. Apparently they have given up their nomadic life and live more or less permanently in this one spot. Their little huts, made solely of leafy branches in winter, while in summer a thatching of grass is added to keep out the rain, were built in a large circle round an open space, where to our immense surprise we actually found two Bushmen doing ironwork like their neighbours, who are great ironworkers. They had built a little shelter for their forge, and while one old Bushman worked the bellows (they had only one pair instead of two, presumably their resources only ran to one) the smith, Kahorta, was busy forging the blade of a knife. When one realises that this, so far as we know, is the only record of metal work by Bushmen, one can understand how interesting a sight it was. The Bushman in his natural state is still in the Stone Age, so that it came as a great surprise to find him working iron. The knives they forge, though of rough metal work, have a good cutting edge and are fitted with very well-worked wooden sheaths.

It would take too long to tell all we saw and heard here – the visits to the camp were even more interesting than their visits to us. Before we left we presented them with a goat – such a tough old billy goat he must have been too – in return for which they danced for us; the whole community, over thirty adults plus a number of children and small babies, the latter on their mothers' backs, had a day of merriment and feasting, no doubt a very great and unique event in their lives.

Tree-covered Kopjes.

The country round Kaiongo was very beautiful, the stream from which the village drew its water, was a tiny one and round it are rocky kopjes covered with trees; signs of spring were all about, blue moraeas along the stream, tiny blue and yellow flowers among the grass above it, gardenias, orchids and many other flowers in the woods. But the chief glory lay in the colours of the trees themselves, standing on the kopje behind our camp. Our little valley ran south-east and merged in the blue of the distant Cweve beyond which Serpa Pinto lies, and all around was the woodland, a mosaic of colour, golden yellow and browns left from the autumn, bright emerald and palest tender green of young foliage and blazing out the brilliant scarlet of the spring shoots of many of the trees. These reverse the colour changes of colder climes – the leaves start deep red, then turn from bright scarlet to bronze and finally grow green.

We always enjoyed the walk to the Bushman camp, but arranged to take it either early or late in the day – midday now was a time to stay quietly in the shade of our nsinge.

After the Dance.

The "dance" marked the end of our stay. Next day we had planned to leave early; very soon after sunrise we heard voices, and there was the greater part of the Bushman population come to bid us farewell – no doubt drawn partly by the hope of presents, but not entirely so.

Porters were scarce, so we pressed four of the younger Bushmen into our service, promising them various desirable articles – blankets, old cotton frocks – if they would go with us to Cwelei. They carried well, and were most entertaining on the two days' trek, again a very beautiful one. We hoped they would consent to stay at Cwelei for a day or two, but they were determined to return at once; after they had been given their blankets and frocks, and had sold all that they had to sell or that we would buy, they set off delighted with their possessions, and that was the last we saw of them.

Only once more did we see any Bushmen – two men passed through Cwelei on their way to hunt "bambi" (duikers) and we had a few minutes conversation with them as they passed.

Again on Trek.

We had planned to spend a few days with our Cwelei friends, while we sent a request to Menongue for porters; the few days lengthened out into weeks, and still the porters did not come. There was plenty to do and see at Cwelei, with its glorious rushing river, dividing into several streams and cascades just above the station, but time was getting on, and eventually we decided to send to the next post to the west to try for porters there. No sooner had we done this, however, than the long-awaited train came in from Menongue, taking us fairly by surprise.

A hectic morning's packing of specimens and goods; an afternoon, long-planned, spent wading about the river, exploring the streams and islands above the station, then farewell to Cwelei and all the kind friends there, and the beginning of our last trek.

This last trek was almost due north, lying between the Cwelei and the Kakuchi (i.e. "Little Kuchi"); the way was quite unknown to our friends. The usual route from Menongue to Silva Porta in Bihe, lies to the east of the Cwelei, but we had heard from some Portuguese traders of this shorter route to the west and decided to try it. It proved most enjoyable, well-watered, through lovely country, and though it took us longer than we had been led to expect – our boys took us round a long way at the end instead of directly across – we were quite pleased that we managed it in ten days, of which two had been really only half days, whereas before it had always taken either eleven or twelve days from Cwelei to Kamundongo, the mission station just south of Silva Porta.

Back to Civilisation.

One might well have lingered by the way; many a lovely spot, flower-clad, was merely glimpsed and then passed by. Most beautiful of all was the Kukema, here a rushing torrent, foaming through narrow rock passages from one deep pool to another, its rocky brim fringed with strange little plants which grow and flower only in the rushing water of torrents or falls. By this time we had left the Okavango system, and the Kukema, like the smaller streams we had lately been meeting, flows northward to join the Kwanza.

Kamundongo is on the fringe of civilisation. After a restful week-end there, we were motored into Silva Porta, where we got the train to the coast, Benguella and Lobito Bay. Now for the first time we saw mountains in Angola – that was when we began to go down! Cwelei is nearly 5,000 feet in altitude, and we continued at much the same height in Bihe. Then the railway begins to descend to the edge of the plateau, winding in and out among rocky mountains till it reaches sea level, where beaches fringed with coconut palms made one at last realise that this was well within the tropics.

There after various adventures – life in so-called civilisation is far more difficult than in the wilds, and the "hotels" of Benguella made us long regretfully for our pleasant camps of the past six months – we finally got on board the fine ship "Moçambique", and five days, broken by half a day at the desert port of Mossamedes, and finished by two days' pitching and tossing, brought us safely back to Cape Town after an absence of six and a half months, full of interest and crowded with new experiences.

———

APPENDIX TWO
Career of Mary Agard Pocock – Synopsis

YEAR	ACTIVITY
1886	Born 31st December in Rondebosch
1899	Sent to Bedford High School for Girls, UK; achieved a 1st Class pass in Oxford Junior Exam.
?	To Cheltenham Ladies' College. 1st Class in London University Matriculation.
1906	Certificate of Honour from the Royal Drawing School.
1908	Working from Cheltenham College, gained BSc in Botany, Geology and Mathematics from London University.
1911	Teachers' Diploma, London University, by private study.
?	Became an Associate of Cheltenham College.
1909–1912	Taught science and mathematics at Pates Grammar School for Girls in Cheltenham.
1913	Returned to South Africa.
1913–1917	Taught science and mathematics at Wynberg Girls' High School.
1915	Requested the University of Cape Town to grant her a BSc degree *ad eundem gradum*.
1916–1918	Examiner in Botany for the Cape Matriculation Examinations Board.
1917	The University of Cape Town granted Mary Pocock an *ad eundem gradum* degree.
1919–1921	Went to Cambridge, UK. Did advanced work in the Botany School as a student of Newnham College and for Natural Science Tripos Part 2. Wrote London University exam to gain BSc Honours in Botany.
1923 (in part)	Temporary lecturer in Dept. of Botany at the University of the Witwatersrand for seven months.
1924	Temporary lecturer in Dept. of Botany at Rhodes University College.
1925	Angola expedition with Dorothea Bleek.
1926	Examiner in Botany, Cape Matric Board.
1927	Temporary lecturer, University of Cape Town.
1927	Worked at Kew and British Museum, identifying her Angolan specimens. Studied South West African algal material under Prof F.E. Fritsch of Queen Mary College, London University.
1928	Began full time studies of Volvocales and other freshwater algae.
1929	Temporary lecturer, Rhodes University College, Grahamstown.
1930	Expedition to the Linyanti, Botswana.
1931	Taught at Wynberg Girls' High School, Cape Town.
1932	Gained PhD on topic of genus Volvox from the University of Cape Town.
1934	Temporary lecturer at the University of Cape Town.

YEAR	ACTIVITY
1934–1935	Temporary lecturer at the Huguenot College, Wellington.
1936–1937	British Association of University Women, Crosby Hall Fellowship. Worked on algae at Queen Mary College and Birkbeck College of London University. Two month tour of Europe – Belgium, Germany, Czechoslovakia, Poland and Russia.
1937	Temporary lecturer at the University of Cape Town.
1938	Temporary lecturer, Rhodes University College, Grahamstown.
1939–1945	Mary and her sisters did war work in Cape Town as members of the South African Women's Auxiliary Services.
1942	Acting Head of Department, Botany Department, Rhodes University College. Established Rhodes University College Herbarium (RUH) with Miss E. Archibald (later Dr Eily Gledhill).
1948	South African Council for Scientific and Industrial Research (CSIR) travel and study grant for work in Australia and America.
1954	Delegate to 8th International Botanical Congress in Paris.
1955	First president of Grahamstown branch of the South African Association of University Women.
1957	Crisp Medal and Award by Linnaean Society of London for outstanding work on algae.
1964	Attended International Botanical Congress in Edinburgh.
1967	DSc *honoris causa* from Rhodes University.
1969	Elected Vice-President of 11th International Botanical Congress held in Seattle.
1977	Died 10th July in Grahamstown.

APPENDIX THREE
Award of Honorary Doctorate

RHODES UNIVERSITY
GRADUATION CEREMONY
1967

AKKERSDYK STUDIOS   CAPE TOWN

Mary Pocock graduating D.Sc. *Honoris Causa*, 1967

Public Orator's Address in Proposing Dr M.A. Pocock for a D.Sc. *Honoris Causa*, of Rhodes University, 1967.

Mr Chancellor, I have the honour to present to you Dr. Mary Agard Pocock, Research Associate in the Department of Botany in this university.

The Romantic poets claimed to find moral wisdom and social ethics in "some impulse from a vernal wood." It seems that the 20th century might well consider as a point of more prosaic contemplation those humbler organisms, the Algae. They are twice blessed. Scooped from pool to pot, in the form of plankton soup they literally nourish and provide food for the body. In their natural state, they provide food for thought. We should probably enlarge the meaning of the verb "to ponder", since ponds and their Algae are full of social paradox and drama. For Algae form "the most intricate networks" – yet have neither radio nor electricity: they "make colonies", but claim no empire. And the behaviour of the daughters of the Algae suggests moreover, that teenage problems are older than time. I quote from the first of Dr. Pocock's papers: "when the daughter is ready to escape ... the daughter shows a kind of boring movement ... escaped daughters of all ages are seen moving among mature colonies". One need only add, "*Et in Arcadia, Algae*".

It is not, of course, at the level of the layman, but of the academic scientist that Dr. Pocock has conducted the researches which have made her world famous. And this is no matter of botany à la Squeers, "When he has learned that bottiney means a knowledge of plants, he goes and knows them." To the contrary. High intelligence, physical stamina and grit; unflagging industry and the detective eye; meticulous precision and infinite patience in the exercise of it; great skill with microscope and camera – these are the qualities which have gone into the making of the truly remarkable achievements of my colleague, Mary Pocock.

When in 1931 Mary Agard Pocock first published her identification of *Volvox capensis* as a separate species, the study of Volvocales, a group of fresh water algae, was in its infancy on the continent of Africa. From that point onwards, with herself as pioneer, researches have continued, and the pattern of advanced studies in Africa has changed beyond recognition.

Three years later, Dr. Pocock held a fellowship at Crosby Hall, London. But her pursuit of her new prey, *Volvox tertius*, took her across the Channel to Belgium, to Germany, Poland and Russia. In 1948,

a grant from the Council for Scientific and Industrial Research made possible further researches both in Australasia and in America. At Berkeley, California, she worked with Dr. Cave, counting the chromosomes of microscopic uni-cellular algae. In the very year in which she finally retired, she was the Rhodes delegate at the 8th International Botanical Conference, in Paris, where in 1954 she presented a paper on Phycology. She had in fact been three times around the world and dabbled in most of the ponds and puddles of it: we who know her, have no doubt she will board a rocket for the moon at the first rumour of a puddle there. One hesitates to point to a zenith in a career so eminent as Dr. Pocock's still is: at a venture it would seem to be in 1957 when her work on the minute parasitic red algae won the Crisp Medal and award of the Linnaean Society, London, for the best paper involving microscopical work.

Sometimes the work of the scientist involves tenacious planning. Sometimes there are lucky breaks, as when in 1936 she gazed with awe at that great stone circle of Stonehenge. She might perhaps have become an archaeologist. But in the side of a great fallen monolith, there was a little puddle, just big enough to scoop with a tablespoon. Out came her apparatus and when she returned to base she identified *Haematococcus draebakensis*, never before discovered outside Norway. Cultures were made, and the descendants of that historic venture into archaeology cum ecology still thrive, at Göttingen, at Cambridge, and in Indiana.

Mary Agard Pocock has devoted her life to the study of Plankton: in particular but not solely, the Volvocales order of

Algae: their morphology, their culture, taxonomy and ecology. She has completed no less than twenty-six scientific papers; and the presses of four continents have been proud to publish for her. There is hardly a laboratory of renown but has placed its facilities at her disposal.

But my presentation would not be complete, Mr Chancellor, did it not include our own -ology: a kind of doxology. Since 1929, when Mary Pocock first lectured at Rhodes, through the war when she acted as head of department, and as recently as 1954, she has taught and wrought with us, in the unarmed conflict zone of laboratory,

lecture theatre and administration. She has made her home among us, and we count her friend. Particularly in 1967, when plans are afoot for the enlargement of the Schönland laboratories at Rhodes, and the fuller development of microbiology, we are proud and glad to have her with us. We wish her long enjoyment of "the peace that is in libraries and laboratories."

Mr Chancellor, I have the honour to ask you to confer on Mary Agard Pocock, the degree of Doctor of Science, *honoris causa.*

Professor Winifred Maxwell

APPENDIX FOUR
Obituary[1]

Mary Agard Pocock (1886-1977)

Algology has lost, in the death of Mary Agard Pocock (10 July 1977, Grahamstown) one of its great characters and a worker known throughout the world. Born on 31 December, 1886, at Rondebosch, South Africa, she came of English stock but her parents were born in South Africa. Her forebears were 'landowners and people of substance' (Ashworth, 1974) from Berkshire and later London, and the Pocock family motto is *'Nil desperandum'*, a fitting one for Mamie Pocock. Her paternal grandfather, Lewis Greville Pocock (1823-1888) emigrated to the Cape Colony in 1842 and Ann Elizabeth Agard, her paternal grandmother, came as governess to the Rev. Croumbie Brown's children. They were married in Cape Town in 1847 by the Rev. Croumbie Brown (later

first official botanist to the Cape Colony), and had eight children, the sixth, William Frederick Henry, being Mary Pocock's father. Lewis was a licensed chemist and druggist, prospected for gold, copper, etc., ran, with his wife's expert aid, a school and also tried several other occupations. William (1857-1922) married Elizabeth Lydia Dacomb of Natal, a particularly clever pupil of Fairleigh School, Durban, and later herself a teacher, in 1885. He trained as a chemist, first with a German pharmacist and later with his uncle, John Thomas Pocock of Cape Town, elder brother of Lewis. John Thomas married, after his first wife died, Parthenia Mary Martin, who was to become a valued and influential member of Mary Pocock's home life. Parthenia Pocock inherited the

[1] Reproduced with permission from *Phycologia* (1978) Volume 17 (4), 440–445.

druggist's business on her husband's death, and William, her right-hand man, was sent to qualify as a chemist in London. This he did with distinction, gaining the bronze medal for chemistry and he eventually inherited the business. Such is part of Mary Pocock's background, including the explanation of her second name, so close to that of another renowned algologist, J. G. Agardh.

Sent in 1899 to Bedford High School for Girls (England), she obtained a first-class pass in the Oxford Junior Examination and went to Cheltenham Ladies' College, where she again got a first-class in the London University Matriculation. In 1908, working from Cheltenham College, she gained her BSc degree in Botany, geology and mathematics from London University. She later (in 1921) took the honours degree of this university in botany. In January 1911 by private study she gained the Teachers' Diploma, also of London University. She became an Associate of Cheltenham College and, with a degree behind her, taught science and mathematics at Pates' Grammar School for Girls, Cheltenham, 1909-1912.

Returning to South Africa (1913), she taught mathematics and science at Wynberg (Cape) Girls' High School for almost four years. In testimonials from this and Pates' School, she was described as an enthusiastic teacher who put herself wholeheartedly into work, unsparing of time and effort, a good disciplinarian and keenly interested in outside matters – school gardening, nature rambles, historical trips, etc. She had prepared herself for this generous giving of her talents, while at Cheltenham and later,

at Cambridge University. She did not confine herself to academic courses only, but was a keen photographer, developing and printing her own studies in the early days, though later the modern 35 mm camera was a mystery to her, others having to load the film or untangle the peculiar stoppages she could cause – but she could press the release button herself to good effect! She became a competent water colourist and studied sculpture, gaining diplomas for these and a Certificate of Honour from the Royal Drawing School in 1906. Some of her well-executed brass rubbings are today in the Albany Museum, Grahamstown, while an album of tree studies made while at Cheltenham shows her photographic skill (this is in Rhodes University Herbarium archives, Grahamstown).

In 1915, obviously planning the way ahead, she requested the University of Cape Town to grant her a BSc. Degree *ad eundem gradum* which was acceded to in 1917. However, in 1919 she went to Cambridge, England, where, in the Botany School she did advanced work under Professor A. C. Seward and others for two years, as a student of Newnham College and for the Natural Science Tripos Part II. She wrote the London University examination to gain her BSc. (Hons.). The then Principal of Newnham College (Miss B. A. Clough) wrote that she was 'much liked by her fellow students and by members of the staff. She has interests outside her special work and has given some time to antiquarian studies, as a result of which she was awarded the Creighton memorial Prize for an essay on Brasses.'

She acted, during part of 1923, as a

lecturer in the botany department at the University of the Witwatersrand, for seven months, replacing a staff member absent on leave. This set a pattern she was to follow for many years, during which she was a temporary lecturer or in one case, acting head of department, at Rhodes University College (1924, 1929, 1938, 1942 and in the 1950s when the College became a full university, the University of Cape Town (1926, 1934, 1937) and the Huguenot University College, Wellington, Cape Province (1934-35). She also acted as examiner in botany for the Cape Matriculation Board (1916-1918), and again (1926) and was a moderator for the same subject in 1928. A further spell of high school teaching, again at Wynberg, occurred in 1931. During these periods of teaching she was always keen to organize or join expeditions and increase students' awareness of the subject in the field. She led one trip from Wellington to the Wilderness Lakes and was also instrumental, with Miss E. Archibald (now Dr E. Gledhill), in setting up as a student teaching device, the Rhodes University herbarium (1942). This now numbers some 28,000 specimens, many of them algae and other plants of Mary Pocock's collecting.

Her inquiring mind and adventurous spirit led her to join Miss Dorothea Bleek, in 1925, in a six months' expedition across Northern Rhodesia (now Zambia) and Angola to Luanda. The trip was made on foot, other means being unavailable, but with porters to carry equipment and food. The menu was meagre and monotonous, so much so that one mere male who accompanied them for a short while asked plaintively if there were no other food

than lumpy, burned maize (corn) meal porridge. The reply, reminiscent of Dr Pocock at her forthright best, was 'We are far too busy to cook and that is all the porters know how to make.' Miss Bleek studied the peoples they encountered, especially the Bushmen, while Mary Pocock collected plants, painted delicate and accurate water colours of these and the scenery, took photographs and made rough notes on her collection, and kept a journal. Such a record should be almost invaluable today.

1927 saw Mamie Pocock back in Britain working at Kew and the British Museum, identifying as far as possible the material collected on the Angolan trip. Some of this proved to be new species, others extended locality records; (a portion of the collection is housed in Rhodes University Herbarium as are many of her water colours of plants). Later that year she studied South West African algal material under Professor F. E. Fritsch of Queen Mary College, London University. Also during this year in Britain, she cheerfully set off with young Inez Verdoorn (now Dr Verdoorn) of the National Herbarium, Pretoria, seconded to work at Kew, to watch an eclipse of the sun from Yorkshire, where it would be at its greatest. They drove up, found accommodation fully booked so slept out on the Yorkshire moors (fortunately it was summer!) and rose with dawn to see the eclipse.

In 1928 Mamie Pocock (Mamie to relations and many friends always), began her full-time studies of the Volvocales and other freshwater algae. The genus *Volvox* provided her with the topic for her Ph.D. thesis and also with lecturing material that has been ingrained in generations of South African first year students until Dr Pocock became known as 'the *Volvox* queen'. This same genus brought her the friendship of Dr Florence Rich and of Dr G. F. Papenfuss of Berkeley, California. It was he who first interested her in marine algae, while he was on the staff of the University of Cape Town.

Her Ph.D. gained in 1932, Mary Pocock soon started travelling again, this time with a Crosby Hall Residential Fellowship (1936-37) granted her by the British Association of University Women. During her year in Britain she worked on algae partly at Queen Mary College and partly at Birkbeck College, both of London University. She carried out field work in Britain and made a two-month tour of Europe, visiting Belgium, Germany, Czechoslovakia, Poland and Russia, working especially at the universities of Marburg, Freiburg, Prague, Moscow and Kharkov. In Russia, when all other means failed, she had recourse to Latin during discussions with some colleagues! She had a fair knowledge of French and German also, and her friendliness would always overcome many obstacles.

Apart from teaching and research, Mary Pocock did much collecting of higher plants for the Botanical Survey of South Africa, concentrating mainly at first on the mountainous parts of the Cape Province, especially above 2,000 m. and including the Swartberg range near Oudtshoorn. The description of this work, together with other manuscripts she had among her papers, may provide useful information today. Many herbaria are indebted to her for material donated and one algologist, Dr H. W. Johansen,

said as recently as 1977 (Johansen 1977), 'Particular mention should be made of the great reliance I placed on the excellent specimens collected and prepared by Dr Mary A. Pocock.' Some years before she died, she had housed her algal collection in the Albany Museum Herbarium, Grahamstown, where it was specially curated for four years by Miss S. C. Troughton, who, under Dr Pocock's guidance, compiled various lists and indexes. It is thus easy to locate the East African material or that from America, etc. while identification was made either by Dr Pocock herself or various experts in certain taxa, e.g. Dr R. Norris for the Kallymeniaceae, the genus *Plocanium* by Mr R. Simons etc. Lists are also kept of her collecting localities and of the reprints available. There are 13 holotypes, 12 isotypes and 2 lectotypes in this collection, all, bar two, of South African material. The total number of plant specimens which Mary Pocock collected during her ninety years must be high but she had a separate numbering system for algae, pteridophytes and angiosperms and the final count is difficult to assess; if must be over 15,000, however.

For a while, the Second World War changed life for Mamie Pocock. She and her sisters, members of the SAWAS (South African Women's Auxiliary Services) did what they could to relieve the tedium and boredom of the long journey troops had in confined quarters on board ship round the Cape of Good Hope to war centres. Good, home-cooked meals and drives to show the scenic beauties of the Cape were welcomed, while Mamie added another feature. She took photographs and sent these happy reminders, together with a cheerful letter giving news of their menfolk, to the families back home. She received many very grateful letters of thanks in return.

When she felt inclined, Dr Pocock was a very good cook and any outing with her was sure to be well-provisioned. Almost inevitably too, she was accompanied by one of a series of long-haired dogs, all tending to be spoiled. One was renowned for snatching up dropped cover-slips in the laboratory, chewing them as eagerly as if they were the barley sugar on which his mistress regaled him. Local outings with her were, though vastly instructive, latterly somewhat of a trial. With one functional eye and a conviction that she was as right, in driving as in algology, she would speed along the road on the centre line, muttering, "What does that fool think he is doing, coming straight for us?" Meanwhile, the on-coming driver, scraping the bank, cowered, and passengers in Dr Pocock's car, braking madly, would push holes in the floor while waiting for the crash. Arrived fortunately without incident at the destination, one would offer help to unload the impedimenta. "No, no, I can manage, thank you." Out would come several plastic buckets, seawater containers, old and bent silver spoons, empty but bulging haversacks and bamboo sticks and nets. Dr Pocock would then load herself up with all this, still refusing aid and tramp off down to the sea. Behind would creep a furiously blushing strapping healthy young person, only too aware of the thoughts and comments of other beach users. However, matters would improve in the water – I have been soaked dozens of times through encouraging shouts to go "farther, farther out, you'll get the best

material out there" while fingers and toes were scraped to bleeding – but at least the blushes were cooled!

Entertained at home by Mary Pocock, one prepared for a long, long session. First, the meal, good always, would be picnic-fashion in the sitting room, the dining room table being invariably deep in chipped flasks of algal cultures, slides in various stages of preparation, a dish of dog food of some antiquity, dust the maid would not dare to remove, a microscope fit for a museum but through which Dr Pocock could see more accurately than we with modern equipment. Afterwards, out would come the colour slides – trays one, two, three, four, five, six. They were excellent and most interesting but by 11 pm, with tomorrow's heavy programme of lectures and practical classes ahead, one gently suggested it was time to go, even though it seemed we were only one third

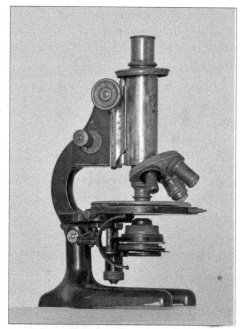

Mary's Carl Zeiss microscope, 'fit for a museum but through which Dr Pocock could see more accurately than we with modern equipment'.

round the world on this particular tour. 'Nonsense, the slides to come are the most interesting,' we would be told, and poking vigorously at the projector (not switched off!) with a metal nail file because the tray had stuck, Dr Pocock overruled our arguments. On went the show, while those in front stifled yawns, those behind gently snored till pinched awake. To Dr Pocock, used to studying the reproductive processes of *Volvox* through the night, bedtime was any time and she was enthusiastic about her slides!

The honour of which Mary Pocock was most proud was the Crisp Medal and Award given her in 1957 by the Linnean Society of London for outstanding work on algae. She was a Fellow of this society, as she was also first a member and then Fellow, of the Royal Society of South Africa. A delegate to the Eighth International Botanic Congress in Paris (1954), she also attended the Congress in Edinburgh in 1964. In 1969, she was elected a Vice-President of the Eleventh International Botanical Congress held in Seattle, U.S.A. In 1948 she received a travel and study grant from the Council for Scientific and Industrial Research (South Africa) for work in Australia and America. This brought her into contact with Dr M. S. Cave of Berkeley, California, and they published together on chromosomes of the Volvocales. Rhodes University, Grahamstown, honoured her with a D.Sc. (*Honoris Causa*) in 1967, a well-deserved acknowledgement of her scientific work and recognition of what she had done to bring renown to Grahamstown, her final and long-time home. She supported the work of the South African Association of University Women always, and was

unanimously elected first president of the Grahamstown Branch in 1955, later becoming an honorary member of the Association, 'for services rendered'.

The arrival of quick air transport in the post-war period gave Dr Pocock a chance she eagerly took, that of travel round the world in easy fashion, giving her plenty of time to visit friends and relations, consult and work with colleagues and collect in most unusual places the algae that still held her attention. It was the depression in a monolith at Stonehenge that caught her eye, and from the dust scraped up, she cultured *Haematococcus*. In this new travel era, she visited many countries, including Britain again, New Zealand (where her brother L. G. Pocock, was Professor of Classics at Christchurch University, thus providing her with a chance of family reunion and opportunity for collecting algae in new localities), Australia, United States, Brazil, Canada, Sweden, France, Greece, Italy, Malawi, Mozambique, Rhodesia and places in the Far East. Her colleagues in Grahamstown were only faintly surprised, when she was almost eighty, to receive a postcard from her from Greece, describing her tour of the night clubs of Athens. Those who knew her well will smile at the latest passport description – eyes grey, hair white and height 5 feet 3 inches. In previous passports, heights had been 5 feet 1½ or 2 inches, but Mary Pocock grew in stature as one got to know her better! She was no small person in character and her last days in hospital as she gently faded away, did not lessen her in any manner – she retained her old fashioned courtesy and at ninety still remembered to offer a visitor a chair or a share of the slab of chocolate she so dearly loved.

Those who knew her well will smile at the latest passport description – eyes grey, hair white and height 5 feet 3 inches.

Enthusiastic, forthright, compassionate, helpful, not bearing fools gladly but enjoying a keen discussion, well-read and artistic, a gardener who appreciated, long before they became fashionable, South African plants, completely indifferent to clothes except as a covering and to time except as a nuisance, Mary Pocock's ninety years were very well spent. The world is the poorer for her going but the richer for her work and herself.

A. Jacot Guillarmod, Institute for Fresh Water Studies, Rhodes University, Grahamstown, South Africa.

References

Ashworth, M.G. (1974) *The life and fortunes of John Pocock of Cape Town*, 1814–1876. Compiled from his journal and letters. College Tutorial Press (Pty) Ltd., Cape Town.

Jacot Guillarmod, A. Personal information collected over 20 years.

Johansen, H.W. (1977) The articulated Corallinaceae (Rhodophyta) of South Africa: 1. *Cheilosporum* (Decaisne) Zanardini. *J. S. Afr. Bot.* **43** (3), 163–185.

Levyns, M.R. (1977) (ed. J.E.P. Levyns) *Insnar'd*

with flowers. Bot. Soc. S. Afr., Cape Town. Rhodes University Herbarium Archives.

Troughton, S.C. (1969) Four years with the Pocock Collection. Typescript report in Albany Museum herbarium, Grahamstown (Pocock Collection).

Personal papers belonging to Dr M.A. Pocock, use of which was graciously allowed by Mrs A. Evans (niece) and Dr J.V.L. Rennie (executor of Dr Pocock's will).

Note

To honour the memory of Dr Mary Agard Pocock, the Grahamstown Branch of the South African Association of University Women has established a capital fund to enable a bursary to be given to a woman student of any race, studying botany at any university in South Africa, including such territories as Transkei. Those wishing to contribute to the memory of this outstanding woman algologist, may send donations to:

The Mary Pocock Memorial Fund,
c/o Dr A. Jacot Guillarmod, SAAUW Convenor,
P.O. Box 342,
6140 Grahamstown,
South Africa.

Publications

Thoday, D. & Pocock, M.A. (1928) On a *Myosurus* from South Africa, with some notes on *Marsilia macrocarpa. Trans. Roy. Soc. S. Afr.* **16** (1), 23–30.

Rich, F. & Pocock M.A. (1933) Observations on the genus *Volvox* in Africa. *Ann. S. Afr. Mus.,* **16,** 427–471.

Pocock, M.A. (1933) *Volvox* and associated algae from Kimberley. *Ann. S. Afr. Mus.,* **16,** 473–521.

Pocock, M.A. (1933) *Volvox* in South Africa. *Ann. S. Afr. Mus.,* **16, 523–646.**

Pocock M.A. (1937) Studies in South African Volvocales. *Pro. Linn. Soc. London,* **149,** 55–88.

Pocock M.A. (1937) *Hydrodictyon* in South Africa. With notes on the known species of *Hydrodictyon. Trans. Roy. Soc. S. Afr.* **24,** 263–280.

Pocock M.A. (1938) *Volvox tertius* Meyer. With notes on the two other British species of *Volvox*. *J. Quekett microsc. Club Ser.* **4** (1), 1-26.

Pocock M.A. (1938) Volvox-hunting in Gloucestershire ponds. *Proc. Cotteswold Nat. F.C.* **26**, 318-323.

Pocock M.A. (1939) A phenomenal drift of seaweed in False Bay. *J. S. Afr. Bot.*, **5**, 75-79.

Pocock M.A. (1947) *Volvox* in culture at the Cape. With special reference to *Volvox tertius* Meyer. *J. Indian Bot. Soc.*, (Prof. M.O.P. Iyengar Commem. Vol.) 1946: 151-165.

Cave, M.S. & Pocock, M.A. (1951) Karyological studies in the Volvocales. *Am. J. Bot.*, **38**, 800-811.

Pocock M.A. (1951) A rare alga from Stonehenge. *Nature*, **168**, 524.

Pocock M.A. (1952) Observations of the occurrence of certain South African seaweeds. *S. Afr. J. Sci.*, **49**, 189.

Pocock M.A. (1953) South African parasitic Florideae and their hosts: 1. Four members of the Rhodomelaceae which act as hosts for parasitic Florideae. *J. Linn. Soc. Lond.* (Bot.), **55**, (356), 34-47.

Pocock M.A. & Martin, M.T. (1953) South African parasitic Florideae and their hosts: 2. Some South African parasitic Florideae. *J. Linn. Soc. Lond.* (Bot.), **55**, (356), 48-64.

Pocock M.A. (1953) Two multicellular motile green algae, *Volvulina* Playfair and *Astrophomene*, a new genus. *Trans. Roy. Soc. S. Afr.*, **34** (1), 103-127.

Pocock M.A. (1955) Seaweeds of the Swartkops Estuary. *S. Afr. J. Sci.*, **52**, 73-75.

Pocock M.A. (1955) Studies in North American Volvocales. 1. The genus *Gonium*, *Madroño*, **13** (2), 49-64.

Pocock M.A. (1956) South African parasitic Florideae and their hosts: 3. Four minute parasitic Florideae. *Proc. Linn. Soc. Lond.* **167**, 11-40.

Pocock M.A. (1958) Preliminary list of the marine algae collected at Inhaca and on the neighbouring mainland. In: *A natural history of Inhaca Island, Mozambique*, pp. 23-27. ed. by W. Macnae and M. Kalk, Johannesburg, Witwatersrand Univ. Press.

Pocock M.A. (1959) *Letterstedtia insignis* Areschoug. *Hydrobiologia*, **14** (1), 1-17.

Pocock M.A. (1960) *Haematococcus* in Southern Africa. *Trans. Roy. Soc. S.Afr.*, **36** (1), 5-55.

Pocock M.A. (1960) *Hydrodictyon*: a comparative biological study. *J. S. Afr. Bot.*, **26**, 167-319.

Pocock M.A. (1962) Algae from De Klip soil cultures. *Archiv für Mikrobiologie.* **42**, 56-63.

Pocock M.A. (1968) Harvey's *Nereis australis*. *Taxon*, **17** (6), 725.

Additions to the list of Dr M.A. Pocock's publications, kindly supplied by Prof-Dr P. Bourrelly, Museum National d'Histoire Naturelle, Paris:

1937. Studies in the South African Volvocales in *Proc. Linn. Soc. London* 149, 55-58: 1. A new *Sphaerella* (Hematococcus). 2. Fertilization of *Eudorina elegans*. 3. A new species of *Volvulina*.

1951. Notes on the occurrence in New Zealand of *Volvulina steinii* Playf. and species of *Volvox* 1. *Rec. Cant. Mus. Christchurch*, 6, 1-8.

1951. Cave, M. S. & Pocock, M.A. The aceto-carmine technic applied to the colonial Volvocales. *Stain Technol.* 26, 173-174.

1959. Two little-known species of *Volvox*: *V. powersi* and *V. obversus*. *IX Intern. Bot. Congress, Proceed. II, IIA* (Canada), 302.

Dr M. A. Pocock is commemorated in the following plant names: the genus *Pocockiella*, *Oxalis pocockiae* L.Bolus, *Amphithalea pocockiae* Bolus, and *Lampranthus pocockiae* (L.Bolus) N.E.Br.

APPENDIX FIVE
Vernacular Plant Names and Ethnobotany

Mary recorded vernacular names and local uses of many of the plants that she collected in her collector's notebooks and occasionally also in her diaries or as captions to her paintings. These notes were meticulously copied to her five herbarium specimen registers while she was identifying specimens at Kew and the British Museum in 1927. We have compiled this information as an important record for this still poorly-known region. Vernacular names are given verbatim and language is given only where indicated by Mary.

Botanical name	Vernacular name	Plant use
Acokanthera sp.	*Minyeli* (Mbundu)	
Aframomum sp. indet.	*Litundu*	Fruit eaten (see text box May 23rd)
Aloe sp.	*Welwe* [spelling?]	
Alvesia rosmarinifolia	*Shivé-shivé* (Bushman)	Leaves used to clean hands; Bushmen smoke the flowers
Ancylanthos fulgidus	*Sendanda* (Chokwe)	
Ansellia africana	*M'posa* (Chokwe)	
Anthochortus insignis	*Mukuku*	
Asplenium adiantum	*Tshila wuli*	
Asteraceae sp. indet.	*//Gōngo* (Bushman)	
Baikiaea plurijuga	*Mukusi*	
Baissea wulfhorstii	*Muliansefu*	
Baphia massaiensis subsp. *obovata*	*Mwana* (Mbundu), *Tshonde* (Chokwe)	
Brachystegia gossweileri	*Mushovia*	
Brachystegia obliqua	*Mutiti*	
Burkea africana	*Musesa* (Mbundu), *Musesi* (Chokwe)	
Burkea 'paniculata' [species name not traced]	*Munyumba*	Used to make bark cloth
Chlorophytum stoltzii	*Wusundu, Mutundu*	Bulb pounded and used for baiting fish traps
Chamaeclitandra henriquesiana	*Kambunga* (Mbundu), *Keenia* (Mbundu)	Rubber
Colophospermum mopane	*Muha* (Mbundu), *Thivi* (Bushman)	Fruits eaten

Botanical name	Vernacular name	Plant use
Combretum c.f. molle	Mumbumboshe (Chokwe)	
Combretum paniculatum	Mulanga 'ndombe (Kangalla, Nkangala in modern texts)	
Combretum platypetalum subsp. virgatum	Mulangandombe	
Combretum sp.	Muyoweh (Mbundu)	
Combretum zeyheri	Munkenge (Mbundu)	Roots used in basket making
Combretum collinum subsp. suluense	Muku	
Copaifera baumiana	Muha, Thivi (Bushman)	
Crotalaria amoena	Musokola	
Crotalaria griseofusca	Mungandu	
Crotalaria sericifolia	Muendakaka, Khina khina (Bushman)	
Cryptosepalum exfoliatum subsp. pseudotaxus	Mukuvi, Mukuve (Mbundu), Koviku (Bushman)	Bark used for tying
Dioscorea sp.	Vanyemba	Tuber used for applying to abscess of the breast [medicinally]
Diospyros batocana	Mundzongola	Fruit used to make meal
Diospyros virgata	Mukulikuli, Mutonte	
Diplorhynchus condylocarpon	Mayonga (fruit), Muwulia (plant)	
Eugenia malangensis	Musamba kambanda (Chokwe)	
Eulophia saxicola	Kasiza	
Eulophia speciosa	Wulua	
Ficus sp. indet.	Linongo	
Ficus sur	Mukuyo	
Gardenia imperialis	Kavalanganje ('cup', see also the cup-like flower of Siphonochilus puncticulatus)	
Gardenia brachythamnus	Muyeké	
Gisekia africana	K'aba (Bushman)	
Gnidia kraussiana	Kangangive	
Gnidia newtonii	Munyenge (Mbundu)	
Grewia flavescens	Akakina (Mbundu)	
Guibourtia coleosperma	Musivi (Mbundu), !Kwi (Bushman)	Fruits eaten
Gymnosporia senegalensis	Musongasonga	
Haemanthus multiflorus	Sakamkanjango (Chokwe)	

Botanical name	Vernacular name	Plant use
Heinsia gossweileri	*Ciliatu* (Mbundu), *'Tsa 'tsi* (Bushman)	
Kotschya strobilantha	*Kukuwikuwi*	
Landolphia camptoloba	*Kambungo* (Luchazi), *Bungwé* (Bushman)	Used to make rubber (see text box 5th June)
Lannea edulis	*Lonsi*	Bark used to make rope (*mukole*)
Leonotis nepetifolia	*Chituvapuku*	
Lepidagathis macrochila	*Mukombokombo*	
Leptactina benguellensis	*Mutanga* (Mbundu)	Leaves chewed in mouth and pulp applied to cuts and sores; Leaves smoked by Bushmen
Manihot esculenta	*Mwanza, Tsompo* (young leaves)	Roots eaten raw, young leaves eaten as vegetable
Monotes dasyanthus	*Muwalangonga*	
Napoleonaea gossweileri	*Muhole*	
Nephrolepis cordifolia	*Tewa tewa*	
Nymphaea sulphurea	*Linkande*	
Ochna pulchra	*Mufukofuko* (Luchasi)	
Ochna roseiflora	*Mufukulilah, Mufukulita*	
Ochna gamostigmata	*Musoko*	
Ocimum americanum	*Liniké-niké*	
Oxygonum fruticosum	*Mukokotiva*	
Ozoroa paniculosa	*Musungo*	
Parinari capensis	*Muchansi, Mefu* (fruits)	Fruit edible
Parinarium capense [species name not traced]	*Muteka* (plant) *Vinteka* (fruit) (Mbundu)	Rhizomes used in basket making
Plectranthus pseudomarrubioides	*N'tombokonde*	
Plectranthus tenuis	*Wusungu wa kalumba* (= 'rabbit's honey')	Leaves stick to anything that touches them
Pleiotaxis linearifolia	*Mukilawankima*	
Psorospermum febrifugum	*Muhorta*	
Pteridium aquilinum	*Chiselu* (= 'medicine') (Mbundu)	
Pterocarpus angolensis	*Mukolo*	
Rhus polyneura	*Lisa* (Mbundu)	
Rhynchosia gossweileri	*Mukundekunde*	
Rothmannia engleriana	*Muwanguwangu*	Sap from fruit used for staining cuts, tattoo marks
Sansevieria hyacinthoides	*!Khwi* (Bushman)	Grown around fetish sticks
Sida cordifolia	*Muhoha*	Fibre used to make rope

Botanical name	Vernacular name	Plant use
Scadoxus multiflorus	Tsani	
Siphonochilus puncticulatus	Kavalanganje (= 'cup', see also the cup-like flower of Gardenia imperialis)	
Sonchus sp.	Muliakolo	
Strychnos pungens	Mukola	The pulp of ripe fruit is rich in citric acid and is edible (see 3rd October)
Syzygium guineense	Mukunde	
Tabernanthe iboga		Black seeds edible
Terminalia confusa [species name not traced]	Muwewa (Mbundu)	
Tricalysia delagoensis	Musuli	
Tricalysia griseiflora var. benguellensis	Pumpalyamuto (Mbundu)	
Triumfetta dekindtiana	Kasakola (Mbundu)	
Tryphostemma viride	Bungeyimo (Luchasi)	
Utricularia sp. indet.	Katalandondo	
Waltheria indica	Kasakola	
Xylopia odoratissima	Mundzumba dzemba	
Xyris sp. indet.	Kamashi, Musoni	

APPENDIX SIX
Glossary

No attempt has been made to correct Pocock's phonetic spelling as it would require fluency in the many African and Bushman languages of south-eastern Angola.

!Kũ – Bushman ethnic group

//gu – water, rain

bambi – duiker

belela – relish, dried fish

blue chalk – butterfly (probably referring to the British Chalkhill blue butterfly)

bokmakierie – species of bush shrike

camba – fowls' cage (see **hok**)

cepatonga – enclosure of branches and poles (see **kraal**)

Chara – stonewort (a genus of green algae in the family Characeae)

chefe – chief

chikovelo – dress

cihembu (sing.) (**vihembu** pl.) – medicine

Cindele (sing.) (**Vindele** pl.) – great devil (applied to Portuguese trader/s)

cini na muisi – mortar & pestle

cituamo – wooden stool

Courant – Oudtshoorn Newspaper (still in circulation)

dagga – marijuana, *Cannabis sativa*

dinoke – snake

doña – lady

donga – river

doodgooi – heavy bread that has failed to rise

dzo – honey

E nosso trem! – 'this is our train [compartment]!'

foo-foo – matches

fueta – pay/wages

Gengi – Mbundu name for Barotse people

guia – travel permit

guya – sewing needle

helu – roof

helu na mbalaka – roof for the tent

hok – fowls' cage (see **camba**)

ichneumon brush – mongoose hair paint brush

imena – tomorrow

induna – designated person with authority, chief, headman (see **mueni**)

jalla – hunger

jigger – jiggers, or chigoe fleas, are sand fleas found in Sub-Saharan climates that burrow into the skin and lay eggs

kalabas – gourd

kaross – skin cloak

Kasekele (sing.) **Vasekele** (pl.) – Mbundu name for Bushman; meaning 'the chased people'

kasila – bird (noun)

klappertjie lewerik – species of lark

kraal – 1) traditional African village; 2) cattle kraal, byre

kulia – eat/food

kusweha – to hide or put away (see **sweha**)

laager – fortifying ring of ox-wagons

lelo – today

lesu – handkerchief

Lifoola (sing.) Mafoola (pl.) – foam or scum (applied to a European person)

likande (sing.) makande (pl.) – crane or stork?

liku – nightjar?

machila – litter (see **tipoia**)

Mafoola – see **Lifoola**

Maizena – corn starch (brand name)

makande – see **likande**

makoro – dugout canoe

makovi – local vegetable. Possibly *couve portuguesa*, a cabbage cultivar

manioc – cassava, *Manihot esculenta*

Meneris tulbaghia butterfly – Table Mountain beauty or mountain pride, now *Aeropetes tulbaghia*

missâo – mission

miti – plants

mongwe – salt

morena – missionary

mpembe – goat

mpulu (sing.) **vimpulu** (pl.) – wildebeest

mu musengi – in the woods (forest?)

mueni – designated person with authority, chief, headman (see **induna**)

muito caro – very expensive

mukuluntu – senior relative

munakasi – woman

mungombe – ground hornbill

municipale – government official

musambi – sweet potatoes

muswe – grey cat-like animal

mwana donga – 'child river' – tributary

Najas – water nymph, aquatic plant in the plant family Hydrocharitaceae

Nâla – crab (referring to a European person)

ndinde ndinde – 'small small' or 'little little'

ndomba – lion

ngulu – domestic pig

njamba – go

nji xaka – 'I want to buy'

nsinge – hut

nsonge (sing.) vansonge (pl.) – lechwe

ntento – species of francolin

outspan – unharness an animl from a wagon for resting or grazing

Ovimbundu – Bantu speaking ethnic group

padkos – food for a journey

Philippia – genus in the plant family Ericaceae

piet-my-vrou – red-chested cuckoo

Punch – British magazine (1841–2002) famous for its cartoons

quatta mpembe – 'catch the goat'

ratel – honey-badger

S.W.A. – South West Africa/Namibia

scrophs – snapdragon family (informal abbreviation for Scrophulariaceae)

sepoi – government official (derived from **municipale**)

sjambok – heavy whip originally made of hippopotamus hide

songe – beads

sonneke – to write or draw (writing/picture)

South West Africa – Namibia

spreeuw – starling

stoep – porch

strada – road

stuff – cloth

sweha – hide/put away (see **kusweha**)

taal – Afrikaans language

tanga – cloth

tapalo – road

Thymelaceae, Thymelaceous, Thymelaea – of the fibre-bark plant family (Thymelaeaceae)

tipoia – litter (see **machila**)

trek – long arduous journey

Vachivokwe – Bantu speaking ethnic group

vangamba – porters

vangonde – ancestral spirits

Vankalla – Bushmen

Vankangola (pl.) – Bantu speaking ethnic group

vansonge see **nsonge**

Vasekele see **Kasekele**

Vihembu see **cihembu**

vikuni – wood/fuelwood

vimpulu see **mpulu**

Vindele see **Cindele**

vlei – seasonal low-lying marshy ground

wahi – not here/gone

wanda – hand spun cotton thread

wata – boat or boat-shaped container

wata leuwanika – The royal barge of Lozi King Leuwanika of Barotseland

wisone – flowers

wunga – local meal, probably millet